Frederick James

CERN, Switzerland

Statistical Methods in Experimental Physics

2nd Edition

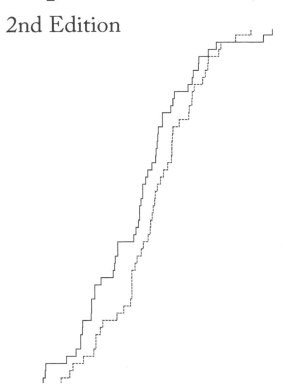

World Scientific

NEW JERSEY · LONDON · SINGAPORE · BEIJING · SHANGHAI · HONG KONG · TAIPEI · CHENNAI

Published by

World Scientific Publishing Co. Pte. Ltd.

5 Toh Tuck Link, Singapore 596224

USA office: 27 Warren Street, Suite 401-402, Hackensack, NJ 07601

UK office: 57 Shelton Street, Covent Garden, London WC2H 9HE

Library of Congress Cataloging-in-Publication Data
Statistical methods in experimental physics, 2nd ed. / [edited by] Frederick James.
    p. cm.
    Includes bibliographical references and index.
    ISBN-13 978-981-256-795-6
    ISBN-10 981-256-795-X
    ISBN-13 978-981-270-527-3 (pbk)
    ISBN-10 981-270-527-9 (pbk)
    1. Physical measurements. 2. Mathematical statistics. 3. Mathematical optimization.
I. Title. II. James, F. (Frederick), 1939–
    QC39 .S74 2006
    530.072/7--dc22

                                 2007282449

British Library Cataloguing-in-Publication Data
A catalogue record for this book is available from the British Library.

The first edition:

Statistical Methods in Experimental Physics,
by W. T. Eadie, D. Drijard, F. E. James, M. Roos and B. Sadoulet

was published in 1971 by North-Holland Publishing Co.,
Amsterdam, New York and Oxford,
and reprinted in 1977, 1982 and 1988

First published 2006
Reprinted 2008, 2010, 2012

Printed in Singapore.

PREFACE TO THE SECOND EDITION

Thirty years after the first publication of the first edition, it was decided that the continued demand justified the production of a new edition. The five original authors agreed that the new edition should reproduce as much as possible the complete text of the first edition, of course corrected for mistakes and modified to take into account recent developments. The original publisher ceded his rights to World Scientific Publishing Co. who kindly agreed to set the entire first edition in LaTeX to initiate the preparation of the new edition. Unfortunately, only one of the five original authors was ready to do this preparation, but the four others agreed to yield their rights in the interest of allowing the renewed availability of the book.

The new edition has required a considerable amount of work. For example, there are over 1000 formulae in the book, about half of which required modification, mostly for improved and consistent notation, but also to correct all the mistakes that have been reported over the years. In addition, the author of the second edition has had thirty years' additional experience in statistical data analysis, which necessarily translates into a better understanding of some problems and requires more than cosmetic changes in a few chapters. The overall result is that most of the text comes from the first edition, but the modifications are sufficiently important that the author of the second edition must take all the responsibility for the final text.

For the reader, the most striking difference between the editions will certainly be the improved typesetting. All the other benefits of computer preparation should make this edition much easier to read and more reliable than its predecessor.

F. E. James

July 2006, Geneva

PREFACE TO THE FIRST EDITION

This course in statistics, written by one statistician (W.T.E.) and four high-energy physicists, addresses itself to physicists (and experimenters in related sciences) in their task of extracting information from experimental data. Physicists often lack elementary knowledge of statistics, yet find themselves with problems requiring advanced methods, if adequate methods at all exist. To meet their needs, a sufficient course would have to be very long. Such courses do indeed exist [e.g. Kendall], only the physicists usually do not take the time to read them.

We attempt to give a course which is reasonably short, and yet sufficient for experimental physics. This obviously requires a compromise between theoretical rigour and amount of useful methods described.

Thus we are obliged to state many results without any rigorous proof (or with no proof at all); still we have the ambition to present more than just a cook-book of prescriptions and formulae. We omit the mention of many techniques which, in our judgement, seem to be of lesser importance to experimental physics.

On the other hand, we do introduce many theoretical concepts which may not seem immediately useful to the experimenter. This we think is necessary for two reasons. Firstly, the experimenter may need to know some theory or some "generalized methods" in order to design his own methods, experimental physics posing always novel questions. This is a justification for the stress on Information theory (Chapter 5), and for the attempt in Chapter 7 to define a "general" method of estimation. We hope that although the method the reader will arrive at may not be optimal, still it will be useful.

Secondly, the experimenter should be aware of the assumptions underlying a method, whether it be a standard method or his own. It is for this reason that we insist so much on the Central Limit Theorem, which is at the foundation of all "asymptotic" statistics (Chapters 3 and 7).

Quoting theorems, we also try to state their range of application, to avoid too careless use of some methods.

Among the underlying assumptions, especially important are the ones about the parent distributions of the data, since they will condition the results. In Chapter 4 we give a catalogue of useful ideal distributions; in real life they may have to be truncated (Sec. 4.3), experimental resolution may have to be folded in (Sec. 4.3), detection efficiency may have to be taken into account (Sec. 8.5). Moreover, the true distribution may not be known, in which case one is led to empirical distributions (Sec. 4.3), robust estimation (Sec. 8.7), and distribution-free tests (Chapter 11).

A very common tacit assumption in the everyday use of statistics is that the set of data is large enough for asymptotic conditions to apply. We try to distinguish clearly between asymptotic properties (usually simple whenever they are known) and finite sample properties (which are usually unknown). We also often give asymptotic expansions, in order to indicate how rapidly the asymptotic properties become true.

In general, we stress the various concepts of optimality. The justification for this is not only that this is the only way for a classical statistician to choose between different procedures, but also that experimental physicists handle ever increasing amounts of data, and therefore need increasingly optimal methods. However, there is an "optimal optimality", because the last bit of optimality can often be achieved only at great cost. This introduces the aspect of economy, which we try to stress on many occasions.

Facing the controversy between Bayesians and Anti-Bayesians ("classical" statisticians), we tend to favour the classical approach (because of professional bias), however keeping the reader partly informed about the Bayesian approach throughout. This attitude we justify as follows. In Chapter 6 we show how taking a decision from a limited amount of information leads to a fundamental indeterminacy: any decision depends on *a priori* assumptions. These assumptions being largely subjective by nature, we think that it is not the role of an experimenter to take decisions. His aim should be to summarize the results of his experiment for the rest of the physics community in such a way as to convey a maximum of information about the unknowns measured. In a certain sense this leaves to the general consensus the task to take decisions.

This is our motivation to the Information theory approach to Estimation. Logically, Test theory should then be Bayesian (since testing really is a decision). Our excuse for not being Bayesian in Test theory (Chapters 10, 11) is that physicists, as a matter of general practice, consider a confidence level as an objective measure of the "distance" of the experiment from the hypothesis tested.

A minor consequence of our professional bias is that in contrast to most (if not all) books on probability and statistics, we avoid using examples from gambling. Physicists often find it frustrating trying to convert such examples into physics; therefore, our examples are taken from physics (mainly high-energy physics). The theory is, of course, the same and gamblers should not be discouraged from converting our examples back into card games, dice, etc!

Let us finally point out that we do not discuss numerical optimization techniques, very important e.g. in the methods of maximum likelihood and least squares. The reasons are that there exist in our opinion excellent treatises of optimization, should the experimenter want to know the details of optimum-searching algorithms, and most physicists do have powerful optimization programs at their disposal (e.g. in the CERN Computer Program Library), which save them one more worry.

W. T. Eadie
D. Drijard
F. E. James
M. Roos
B. Sadoulet
December 1970, Geneva

CONTENTS

Chapter 1

INTRODUCTION

1.1. Outline

The subject of the following ten chapters can be divided into two main parts:

- Theory of Probability: Chapters 2–4
- Statistics: Chapters 5–11

The theory of probability is needed only to provide the necessary tools for statistics, which forms the main body of the course.

Chapters 5 and 6 define two general approaches to the choice of estimators: the *information* approach and the *decision theory* approach. The former consists essentially in maximizing the amount of information in the estimate, whereas the latter is based on minimizing the loss involved in making the wrong decision about the parameter value. In the limit of large data samples, the two approaches are equivalent, but where they differ we will try to point out the distinction.

Estimation of parameters is divided into three chapters, 7 and 8 dealing with point estimation, theory and practice, and 9 dealing with interval estimation. Tests of hypotheses are divided into general testing, Chapter 10, and goodness-of-fit tests, Chapter 11.

Our reference policy is as follows. We quote literature when we have omitted the proof of an important result, or when we want to give hints for further reading. We do not usually attempt to give credit to original results. In the text the references take the form

[First author, Volume*, Chapter*, page*]
* if necessary.

At the end of the course before the subject index, the literature references are ordered alphabetically after the first author.

1.2. Language

Statistics, like any other branch of learning, has its own terminology which one has to become accustomed to. Certain confusion may, however, arise when the same term has different meaning in statistics and in physics, or when the same concept has different names. In the former case we usually imply the statistical meaning (obliging the physicist to recognize and learn the difference); in the second case we often choose the physical term.

An example of the first kind is the following:

Physicists say	Statisticians say
Determine	Estimate
Estimate	Guess

Thus the word estimate has different meaning in physics and in statistics. We use it as statisticians do. (We use three chapters to explain what statisticians mean thereby).

An example of the second kind is "the demographic approach" to experimental physics. Much of statistics has been developed in connection with population studies (sociology, medicine, agriculture) and at the production line (industrial quality control). Then one is not able to study the whole population, so one "draws a sample". And the population exists in a real sense.

In experimental physics, the set of all measurements under study corresponds to the "sample". Increasing the number of measurements, the physicist increases the "size of the sample", but he never attains the "population". Thus the "population" is an underlying abstraction which does not exist in any real sense. These "demographic" terms are therefore to some extent inappropriate and unnecessary, and we try to avoid some of them:

For the "demographic" term	we use the physics term
Sample	Data (set)
Draw a sample	Observe, measure
Sample of size N	N observations
Population	Observable space

Still, one has to be able to distinguish between, say, the mean of the data at hand, and the mean if the data set were infinite. When this distinction is necessary, we use sample mean, sample variance, etc. as contrasted to parent mean, parent variance, etc., or mean and variance of the underlying distribution. Thus

Parent mean = Mean of the underlying distribution = Population mean.

We avoid the physical term "error", which is misleading, and use instead "variance of estimate", "confidence interval" or "interval estimate". We also try to avoid the words "precision" and "accuracy", because they are not well defined.

In many books on statistics one finds whole chapters dealing with the "propagation of errors". Such a term, in our minds, is confusing. The corresponding notion here is "change of variables".

Other topics which may seem to have got lost, may also sometimes be refound under other names. For instance, the term "regression analysis" is never used, but the techniques are treated under least-squares fits of linear models.

1.3. Two Philosophies

Unfortunately, statisticians do not agree on basic principles. They can crudely be divided into two schools: Bayesian and frequentist (or classical). The name Bayesian derives from the extended use of Bayes theorem in the former group. We try to present the main results from both approaches.

The Bayesian approach is closer to everyday reasoning, where probability is interpreted as a *degree of belief* that something will happen, or that a parameter will have a given value.

The frequentist approach is closer to scientific reasoning, where probability means the relative frequency of something happening. This makes it more objective, since it can be determined independently of the observer, but restricts its application to repeatable phenomena. In particular, one can define the frequentist probability for observing data (which are random), but not for the true value of a parameter (which is fixed, even if unknown).

In the areas of parameter estimation and hypothesis testing, numerical results tend to be the same for the two approaches in the asymptotic regime, that is, when there are a lot of data, and statistical uncertainties are small compared with the distance to the nearest physical boundary. However, exact results require for each approach information which is not allowed by the other:

- Exact frequentist results require as input the probabilities of observing all data, including both the data actually observed and that which could have been observed (the Monte Carlo). This violates an important principle in Bayesian theory, and is not allowed in the Bayesian method.

- Exact Bayesian results require as input the prior beliefs of the physicist doing the analysis. This is necessarily subjective, and is not allowed in the frequentist method.

In the area of goodness-of-fit testing (testing of a single hypothesis, where no alternative hypothesis is specified) it is essentially impossible to obtain any results in the Bayesian approach, so that is the traditional bastion of classical statistics.

On the other hand, decision theory, because of its fundamentally subjective nature, is the domain of Bayesian methodology.

In this book we largely follow a classical (frequentist) approach because we feel this is more appropriate for the reporting of experimental results. We do however develop also the main ideas of Bayesian statistics and try to point out the differences wherever possible. It can be a great help in understanding each approach to see how the difficult problems are handled in the other approach.

1.4. Notation

Roman letters

$b(\hat{\theta})$	bias of estimate $\hat{\theta}$
corr $(X, Y), \rho$	correlation between random variables X and Y
cov (X, Y)	covariance of random variables X and Y
$D_N, D_{NM}, D_N^{\pm}$	Kolmogorov statistic
$e(X, Y')$	detection efficiency
$E(X)$	expectation
$f(X)$	probability density function (p.d.f.)
$F(X)$	cumulative distribution of $f(X)$

$g(X)$	probability density function
$G(X)$	generating function
H, H_i	hypothesis
H_0	hypothesis under test, "null hypothesis"
$\mathcal{L}_X(\boldsymbol{\theta}), \mathcal{L}_N(\boldsymbol{\theta})$	information matrix
K	non-centrality parameter
$K(t)$	cumulant generating function
K_r	cumulant
ℓ	likelihood ratio
L_p	norm of power p
$L(\mathbf{X}|\theta) = L(\theta)$	likelihood function
$L(\theta, d)$	loss function
N	number of random variables (events, experiments)
$N(\mu, \sigma^2)$	Normal distribution of mean μ and variance σ^2
$O(N^{-1})$	term of order N^{-1} or less
$p(\theta), 1 - \beta$	power of test
$P(A)$	probability that A is true
$P(A|B)$	conditional probability that A is true, given B
p.d.f.	probability density function
Q^2	quadratic form, covariance form
$r(X, X')$	resolution function
$s^2 = \dfrac{1}{N-1} \sum\limits_{i=1}^{N} (X_i - \bar{X})^2$	unbiased estimate of variance
$S^2 = \dfrac{1}{N} \sum\limits_{i=1}^{N} (X_i - \bar{X})^2$	sample variance
S_N	sum of N random variables, distribution function for order statistics
t, T	statistic
tr	trace (of matrix)
$V(X), \sigma_X^2$	variance
$\underset{\sim}{V}(\mathbf{X})$	covariance matrix
$\underset{\sim}{V}_{(rs)}$	submatrix of $\underset{\sim}{V}$ having r rows and s columns
w_α	critical test region of significance α
w_i	weight
W	space of test statistic
W^2	test statistic

X, X_i	random variables
$X^2_{(N)} = \sum_{i=1}^{N} X_i^2$	sum of squares of random variables
Y, Z, Y_i, Z_i	random variables
$X_\alpha, Y_\alpha, Z_\alpha$	α-point of $f(X)$, etc. [defined by $F(X_\alpha) = \alpha$]

Greek letters

α	confidence level, significance level, loss
β	contamination, confidence level
γ_1	skewness
γ_2	kurtosis
$\delta(X)$	δ "function" of Dirac
θ, θ_i	theoretical parameter
θ_0	true value of θ, null hypothesis value of θ
λ	maximum likelihood ratio
λ_α	α-point of Normal distribution [defined by $\Phi(\lambda_\alpha) = \alpha$]
μ, μ'	mean value
μ_n, μ'_n	moment of order n
ν, ν'	degree of freedom
ν_n, ν'_n	moment of order n
$\pi(\theta)$	prior density
$\pi(\theta)|\mathbf{X})$	posterior density
ρ	correlation coefficient
σ^2	variance
$\hat{\sigma}^2 = \dfrac{1}{N} \sum_{i=1}^{N} (X_i - \mu)^2$	unbiased estimate of variance, given the mean μ
$\chi^2(N)$	chi-square distribution of N degrees of freedom
$\phi(t)$	characteristic function
$\Phi(X)$	Normal probability integral
Ω	space of random variable

Symbols

bar (e.g. $\bar{X} = \dfrac{1}{N} \sum_{i=1}^{N} X_i$)	average (e.g. of random variables $X_1, \dots, X_N$)
hat (e.g. $\hat{\theta}$)	estimate (e.g. of parameter θ)

bold (e.g. $\mathbf{X}$)	vector (e.g. with components $X_1, \ldots, X_N$)
tilde (e.g. $\underset{\sim}{V}$)	matrix (e.g. covariance matrix)
upper U (e.g. θ^U)	upper confidence bound (e.g. of parameter θ)
lower L (e.g. θ_L)	lower confidence bound (e.g. of parameter θ)
T (e.g. $\underset{\sim}{A}^{\mathrm{T}}$)	transpose of matrix ($\underset{\sim}{A}$)
$[\theta_\mathrm{a}, \theta_\mathrm{b}]$	interval $\theta_\mathrm{a} \le \theta \le \theta_\mathrm{b}$
$\binom{p}{q} = \dfrac{p!}{q!(p-q)!}$	binomial coefficient

Chapter 2

BASIC CONCEPTS IN PROBABILITY

In this chapter we introduce the concept of *probability*, and define the three kinds of probability that will be used in this book. We then generalize to random variables and continuous probability distributions. The important properties of distributions are presented: expectation, mean, variance, correlation, covariance and moments. Finally, we introduce tools for handling distributions: characteristic function, cumulant generating function and probability generating function.

2.1. Definitions of Probability

Although the concept of *probability* has been known since antiquity, a proper mathematical theory was developed only in 1933 when Kolmogorov published his *Foundations of the Theory of Probability*. In this theory, probability is a basic concept, and is therefore undefined except for the fact that it must satisfy the *Kolmogorov axioms*. The theory then holds for any quantity that satisfies the axioms. We will refer to this abstract probability as *mathematical probability*.

For statistics, we require an operational definition which allows us to measure probabilities. Such a definition will of course be more restrictive than the abstract definition, but as long as the more restrictive probability satisfies the Kolmogorov axioms, the theory of Kolmogorov applies to it. We will consider two such definitions: *frequentist probability* and *Bayesian probability*.

2.1.1. *Mathematical probability*

We define Ω to be the set of all possible *elementary events* X_i which are *exclusive*; that is, the occurrence of one of them implies that none of the others occurs. Then we define the probability of the occurrence of X_i, $P(X_i)$, to obey the Kolmogorov axioms:

$$\text{(a) } P(X_i) \geq 0 \quad \text{for all } i$$
$$\text{(b) } P(X_i \text{ or } X_j) = P(X_i) + P(X_j) \tag{2.1}$$
$$\text{(c) } \sum_{\Omega} P(X_i) = 1 \,.$$

From these properties, more complex probability expressions can be deduced for non-elementary events, which are sets of elementary events, and for non-exclusive events, which are overlapping sets of elementary events. The most important results will be given below in Section 2.2.

2.1.2. *Frequentist probability*

Consider an experiment in which a series of events is observed, and suppose that some of these events are of a type X. Suppose the total number of events is N, and that the number of events of type X is n. Then the *frequentist probability* that any single event will be of type X can be defined as the empirical limit of the frequency ratio

$$P(X) = \lim_{N \to \infty} \frac{n}{N} \,.$$

We note the following properties of *frequentist probability*:

- Although the definition as a limit has been criticized because it appears to require an infinite number of experiments, this is in fact not the case. As long as it is in principle possible always to perform one more experiment, any specified accuracy can be attained, and that is sufficient to define the concept. In practice, probabilities will often be known exactly without doing any experiments.

  Many important quantities in both mathematics and physics are defined as limits. For example, the electric field is defined as the ratio of the force on a test charge to the magnitude of the charge, and since the test charge disturbs the field, this ratio has to be taken in the limit as the magnitude of the charge goes to zero. Moreover, this limit is physically impossible since charge is quantized, but that does not invalidate the concept of electric field or make it unmeasurable in practice.

- The definition does however imply an important restriction: Frequentist probability can only be applied to repeatable experiments. This means, for example, that one cannot define the frequentist probability that it will rain tomorrow, since tomorrow will only happen once and other days are not identical to tomorrow.

- It may appear that there is a problem because no experiment can be repeated in *exactly* the same conditions as before. But it is of course the job of the physicist to make sure that all *relevant* conditions are the same, and if even that is impossible, one makes corrections for the unavoidable changes. Good science should produce reproducible results.

2.1.3. *Bayesian probability*

In order to define a probability that can be applied to non-repeatable experiments, we must abandon the concept of frequency and replace it by something else. Among the various possibilities, the most important is certainly the *degree of belief* which is the basis of *Bayesian probability*. The operational definition of *belief* is due to de Finetti and is based on the *coherent bet* [Finetti]. The idea is to determine how strongly a person believes that X will occur by determining how much he would be willing to bet on it, assuming that he wins a fixed amount if X does later occur and nothing if it fails to occur. Then $P(X)$ is defined as the largest amount he would be willing to bet, divided by the amount he stands to win. Thus it is always between zero and one: zero if the person is sure that X will not occur and is therefore not willing to bet anything; and one if he is sure that it will occur and is therefore willing to bet as much as he stands to win. Defined in this way, Bayesian probability obeys the *Kolmogorov axioms* and is a probability in the mathematical sense.

We note the following properties of *Bayesian probability*:

- It is as much a property of the observer as it is of the system being observed.

- It depends on the state of the observer's knowledge, and will in general change as the observer obtains more knowledge.

- There are many other ways to define Bayesian probability; the definition we give here is the most operational (most amenable to measurement) that we know, but measurement is not without problems. In particular, it has been pointed out that this definition depends on the utility of money being a linear function of the amount, and Bayesian theorists have suggested that modifications may be needed to correct for actual non-linearity.

The probability defined above is sometimes known as *subjective Bayesian* probability. The currently dominant school of Bayesian theory among professional statisticians is certainly the subjective one, but among physicists there is another important school associated primarily with physicists Harold Jeffreys and E. T. Jaynes, which we may call the *objective Bayesian* school. This is clearly an attractive alternative for physicists, but unfortunately no one has yet been able to develop a logically consistent objective Bayesian theory that professional statisticians find satisfactory. We therefore adopt the attitude that Bayesian methods are subjective. This attitude is further justified in Sections 7.5.1 and 7.5.2.

2.2. Properties of Probability

In this section we present the most important results from the mathematical theory of probability. These results apply to any *probability* that satisfies the *Kolmogorov axioms*.

A set A of elementary events X_i, can again be treated as an event, although non-elementary. The occurrence of A is defined by the occurrence of an event X_i in the set A. We then denote by $P(A)$ the probability that an X_i in the set A occurs. For such sets, there also holds an *addition law* as well as a *multiplication law* if the sets are *independent*. We then generalize to *conditional probability* for which the important *Bayes theorem* holds. Finally, we take the step from events to random variables.

2.2.1. *Addition law for sets of elementary events*

Consider two non-exclusive *sets* A and B, of elementary events X_i; that is, some X_i may belong to both A and B. In this case, following from the definitions in Eq. (2.1), the probability of an event occurring which is either in A, or in B, or in both, is given by

$$P(A \text{ or } B) = P(A) + P(B) - P(A \text{ and } B), \qquad (2.2)$$

where "A or B" denotes the set of events X_i, which belong to either set A, or set B, or both, and "A and B" denotes the set of events X_i belonging to both A and B.

The relationship (2.2) can be generalized to the case of several sets $A_1, \ldots,$ A_N [Feller, p. 99]. Let us take

$$P_i = P(A_i),$$

$$P_{ij} = P(A_i \text{ and } A_j), \quad i < j$$

$$P_{ijk} = P(A_i \text{ and } A_j \text{ and } A_k), \quad i < j < k$$

etc., and let S_r denote the sums

$$S_1 = \sum_i p_i,$$

$$S_2 = \sum_{i<j} p_{ij},$$

$$S_3 = \sum_{i<j<k} p_{ijk}, \text{ etc.}$$

Then the probability of the occurrence of an event belonging to at least one of the sets A_i is given by

$$P(A_1 \text{ or } A_2 \text{ or } \cdots \text{ or } A_N) = S_1 - S_2 + S_3 - \cdots - (-1)^N S_N.$$

2.2.2. *Conditional probability and independence*

It is now possible to introduce *conditional probability*, the probability of A given B, written $P(A|B)$. This is the probability that an elementary event, known to belong to the set B is also a member of the set A. It is defined through the relationship

$$P(A \text{ and } B) = P(A|B)P(B) = P(B|A)P(A). \tag{2.3}$$

Sets A and B are said to be *independent* if

$$P(A|B) = P(A), \tag{2.4}$$

which means that the (previous) occurrence of B is irrelevant to the occurrence of A. From the definition (2.3) of conditional probability, we have, if A is independent of B:

$$P(A \text{ and } B) = P(A) \cdot P(B). \tag{2.5}$$

Equation (2.5) is a necessary and sufficient condition for A and B to be independent, which happens to be quite useful. We note in passing that the correlation coefficient, to be defined later, vanishes if A and B are independent, but lack of correlation does *not* imply independence.

2.2.3. *Example of the addition law: scanning efficiency*

Suppose that data has been recorded on a photographic medium which must be scanned manually to find the events of interest. The film is scanned twice for the same kind of event. A number $(C + D_1)$ of events is found in the first scan, and $(C + D_2)$ events in the second scan. C denotes the events found in both scans, whereas D_1 and D_2 are found only in the first or second scan respectively. We wish to estimate the true number of events in the film, and the over-all scanning efficiency, under the assumptions that in each scan all events are equally likely to be found, and that the two scans are independent.

The first assumption permits us to deal with the sets 1 and 2 of events found in the first and second scans, instead of with each event separately. Thus every event on the film has probability $P(1)$ of being found in the first scan and $P(2)$ of being found in the second scan.

The second assumption implies [from Eqs. (2.3) to (2.5)], that

$$P(1|2) = P(1)$$

or

$$P(2|1) = P(2)$$

or

$$P(1 \text{ and } 2) = P(1) \cdot P(2).$$

From the numbers of events observed, we can estimate the scanning efficiencies in the two scans by

$$\hat{P}(1|2) = \frac{C}{C - D_2} = \hat{P}(1)$$

and

$$\hat{P}(2|1) = \frac{C}{C + D_1} = \hat{P}(2).$$

(A hat, "^", over a quantity is used to denote an estimate).

To estimate the over-all scanning efficiency, we use Eq. (2.2), to give us

$$\hat{P}(1 \text{ or } 2) = \hat{P}(1) + \hat{P}(2) - \hat{P}(1 \text{ and } 2)$$

$$= \hat{P}(1) + \hat{P}(2) - \hat{P}(1) \cdot \hat{P}(2)$$

$$= \frac{C}{C + D_2} + \frac{C}{C + D_1} - \frac{C^2}{(C + D_1)(C + D_2)}$$

$$= \frac{C(C + D_1 + D_2)}{(C + D_1)(C + D_2)}.$$

To estimate the total number, N, of events on the film, we may use

$$\hat{P}(1) = \frac{C + D_1}{\hat{N}},$$

to give

$$\hat{N} = \frac{(C + D_1)(C + D_2)}{C}.$$

A more detailed discussion of this problem has been given in the literature [Evans], [Knop].

2.2.4. *Bayes theorem for discrete events*

The theorem which links $P(A|B)$ to $P(B|A)$ is *Bayes theorem* [Bayes]. For sets A and B (of events X_i) it states that

$$P(A|B) = P(B|A) \cdot P(A)/P(B). \tag{2.6}$$

Evidently this follows from the definition of conditional probability, Eq. (2.3). More generally, if $A_1, \ldots, A_N$ are exclusive and exhaustive sets (i.e. an observed elementary event must belong to one and only one of the sets A_i), and if B is any event, then Bayes theorem can be written as

$$P(A_i|B) = \frac{P(B|A_i)P(A_i)}{\sum\limits_{i} P(B|A_i)P(A_i)}. \tag{2.7}$$

Example

An experiment is planned to study leptonic decays of the K^0 meson, and a Čerenkov counter is to be used to detect the leptonic decays. It is desired to know if one counter is sufficient for detecting the leptonic decays against a small background from other reactions which may also trigger the counter. The relevant probabilities are:

$P(B) \equiv$ the probability of any event giving a Čerenkov count.

$P(A) \equiv$ the probability of occurrence of a true leptonic decay event.

$P(B|A) \equiv$ the probability of a true leptonic decay giving a Čerenkov count.

$P(A|B) \equiv$ the probability of an event being a true leptonic decay, if a
     Čerenkov count is received.

$P(B)$ can be measured by turning on the beam and recording the number of Čerenkov counts. $P(A)$ is assumed known from other experiments. $P(B|A)$ can be calculated from the measured efficiency of the counter and its geometrical cross-section. Then, the interesting quantity $P(A|B)$ can be deduced immediately from Bayes theorem, Eq. (2.6).

2.2.5. *Bayesian use of Bayes theorem*

When A or B is not a set of *events* but a set of *hypotheses*, the meaning of *Bayes theorem* is less evident. This is where the two schools of statistics part. In the Bayesian paradigm, $P(\theta_i)$ is taken to mean the *degree of belief* in hypothesis θ_i. In the frequentist framework however, θ_i is not a random variable, even if it is unknown, so frequentist probabilities cannot be assigned and Bayes Theorem is not applicable. Therefore, Bayes theorem involving hypotheses can only be applied in the Bayesian framework.

Let us inspect the different factors in the Bayesian case of Eq. (2.7) or

$$P(\theta_i|\mathbf{X}^0) = P(\mathbf{X}^0|\theta_i) \cdot P(\theta_i)/P(\mathbf{X}^0)\,, \tag{2.8}$$

where the θ_i are different hypotheses and $\mathbf{X}^0$ represents the data observed:

- $P(\theta_i|\mathbf{X}^0)$ is called the *posterior probability* for hypothesis θ_i, given that data $\mathbf{X}^0$ have been observed;

- $P(\mathbf{X}^0|\theta_i)$ is the probability of obtaining the observed measurements $\mathbf{X}^0$, given the hypothesis θ_i, which must be known, since it is essentially a description of the behaviour of the experimental apparatus;

- $P(\theta_i)$ is called the *prior probability* and represents the knowledge or degree of belief in different hypotheses before the experiment was performed.

- $P(\mathbf{X}^0)$ can be considered a normalization constant since the sum of the left-hand-side over all values of i must be unity if the different hypotheses form a complete and exclusive set. The more general case is discussed below.

Equation (2.8) shows that the degree of belief in a hypothesis depends not only on the results of the experiment, but also on the degree of belief before the experiment. This has been the cause of an enormous amount of study and controversy and given rise to different schools of Bayesian statistics depending largely on how they determine the prior. In particular, the obvious case of interest to physicists is when the experimenter does not have, or does not want to assume *any* prior knowledge. Both physicists and statisticians (and even philosophers) have devoted great effort over the years to finding the right way to express prior knowledge, and especially the *uninformative prior* which would express no previous knowledge or belief at all. This fascinating topic has probably been the subject of more research than all the rest of Bayesian statistics together. We feel that the search for a satisfactory uninformative prior has not been successful, and therefore Bayesian statistics is appropriate only when it is desired (or unavoidable) to put the physicist's prior beliefs explicitly into the statistical analysis (see Section 7.5.1).

$P(\mathbf{X}^0)$, the probability to get the result $\mathbf{X}^0$ given any hypothesis, may not be known. If the hypotheses θ_i are exclusive and exhaustive (they exhaust the full set θ), which certainly is a very special case, then we can write:

$$P(\mathbf{X}^0) = \sum_i P(\mathbf{X}^0|\theta_i) \cdot P(\theta_i).$$

In the general case, $P(X)$ is not known, and Eq. (2.8) takes the weaker form

$$P(\theta_i|X) \propto P(X|\theta_i) \cdot P(\theta_i). \qquad (2.9)$$

Although now we cannot compute the *posterior probability*, we still can calculate *odds*, which may also be useful. We define the odds on θ_i against θ_j by the ratio

$$P(\theta_i|X)/P(\theta_j|X). \qquad (2.10)$$

2.2.6. *Random variable*

A *random event* is an event which has more than one possible outcome. A probability may be associated with each outcome. The outcome of a random event is not predictable, only the probabilities of the possible outcomes are known. In contrast, an event with only one possible outcome is certain, the probability of the outcome being unity.

With a random event A may be associated a *random variable* X, which takes different possible numerical values $X_1, X_2, \ldots$ corresponding to the different possible outcomes. The corresponding probabilities $P(X_1)$, $P(X_2), \ldots$ form a *probability distribution*. Usually, we shall denote a random variable by a capital letter, X.

In the case of more than one random variable, or of a sequence of observations of the same random variable, the question of the *independence* of different observations must be considered. If they are independent, the distribution of each random variable is unaffected by knowledge of any other observation. On the other hand, dependence means that the distribution of one variable changes when the value of another observation is known. Dependent variables can still be random, the result of an observation being predictable only in terms of the probabilities of possible values, as described by the distribution. Only in the degenerate case of complete dependence, when knowledge of one observation exactly determines the value of a second variable, does the second variable become certain.

When an experiment consists of N repeated observations of the same random variable X, this can be considered as the single observation of a random vector $\mathbf{X}$, with components $X_1, \ldots, X_N$.

In experimental science it is common to speak of some variables as being "more random" than others. In fact this usually means either that the variables are more uniformly distributed, or that their distribution is wider, or that successive observations are less correlated. It seems confusing to use the word random to refer to what are really properties of distributions of variables, so we reserve the word random to be taken only in the qualitative sense of the above paragraph. If a variable is referred to as random, this implies only that it has a distribution of possible values, and nothing about the form of the distribution, or about any dependence which may exist.

2.3. Continuous Random Variables

We can now generalize probabilities of events to probability distributions of random variables. For discrete random variables, this generalization is obvious. For continuous random variables (whose possible values cover continuous intervals) we need the tools of the *probability density function* and its integral, *the cumulative distribution function*. We also define *marginal distribution* and *conditional distribution*, and show their application in Bayes theorem for continuous variables.

2.3.1. *Probability density function*

Consider for illustration a counter experiment in which the direction of a particle beam is determined by two consecutive arrays of counters, X and Y. Each accepted event is characterized by two discrete random variables; their values are i and j when a particle has traversed counter X_i in array X and counter Y_j in array Y. There exists a corresponding two-dimensional discrete probability distribution $P(X \text{ and } Y)$.

The physicist may frequently think that nature actually can be described by a continuous probability distribution $f(X,Y)$, and that the only reason for getting results in terms of discrete variables is that the counters have a finite size ΔX or ΔY.

The relation between $P(X \text{ and } Y)$ and $f(X,Y)$ can be written

$$
\begin{aligned}
f(X,Y) &= \lim_{\substack{\Delta X \to 0 \\ \Delta Y \to 0}} \frac{P(X \text{ and } Y)}{\Delta X \Delta Y} \\
&= \lim_{\substack{\Delta X \to 0 \\ \Delta Y \to 0}} \frac{P[(X - \frac{1}{2}\Delta X \leq X < X + \frac{1}{2}\Delta X) \text{ and } (Y - \frac{1}{2}\Delta Y \leq Y < Y + \frac{1}{2}\Delta Y)]}{\Delta X \Delta Y}.
\end{aligned}
\tag{2.11}
$$

It is clear from dimensional arguments that $f(X,Y)$ represents a probability density per unit array length X and unit array length Y. Therefore $f(X,Y)$ is called a *probability density function*, abbreviated p.d.f.. A *joint p.d.f.* applies to density functions of more than one variable.

A p.d.f. is normalized in a way analogous to Eq. (2.1). In particular, the joint p.d.f. in Eq. (2.11) must obey

$$
\int\!\!\int_{\Omega} f(X,Y)dXdY = 1 \,,
\tag{2.12}
$$

where Ω is the space of all possible values of X and Y (the total length of both counter arrays).

Probability density functions may, of course, be functions of any number of continuous random variables. They may also be mixed functions of continuous and discrete random variables. For example, a function of the form

$$
h(X) = p_0\delta(X) + p_1\delta(X-1) + (1 - p_0 - p_1)\frac{1}{\sqrt{2\pi}}e^{-X^2/2}
$$

is the density function of a random variable taking the values

$$X = 0 \quad \text{with probability } p_0 \,,$$

$$X = 1 \quad \text{with probability } p_1 \,,$$

and, with probability $(1-p_0-p_1)$, being Normally [a] distributed with mean zero. In such case the normalization condition (2.12) takes the form of integration over all continuous variables and summation over all discrete variables.

2.3.2. *Change of variable*

Suppose that $f(X)$ is known, and one would like to know the density $g(Y)$ that results from a variable transformation

$$Y = h(X) \,, \tag{2.13}$$

which maps the interval $(X, X + dX)$ onto $(Y, Y + dY)$. If the transformation (2.13) is one-to-one, one has

$$g(Y)dY = f(X)dX \tag{2.14}$$

and

$$g(Y) = \frac{f(X)}{|h'(X)|} \,,$$

where $|h'(X)|$ is the absolute value of the derivative of the transformation. In the multidimensional case, where X and Y are vectors, $|h'|$ is the Jacobian of the transformation.

If the transformation (2.13) is not one-to-one, there are many segments $(X, X + dX)$ that map onto $(Y, Y + dY)$, one then has to sum over all such segments, thus

$$g(Y) = \sum \frac{f(X)}{|h'(X)|} \,. \tag{2.15}$$

For example, if the transformation (2.13) is $Y = X^2$, the density $g(X^2)$ is obtained by adding the two branches $f(-X)$ and $f(X)$, thus

$$g(X^2) = \frac{f(X) + f(-X)}{2|X|} \,.$$

An important case is the change of variable required to obtain the distribution of the ratio of two random variables. This is treated in Section 2.4.4.

---

[a] The properties of the Normal (Gaussian) distribution will be treated in Section 4.2.1.

2.3.3. *Cumulative, marginal and conditional distributions*

The physicist usually calls a probability density function a *distribution* (e.g., a mass distribution). The statistician reserves the name distribution for the integrated probability density function. We shall call this a *cumulative distribution*. The *random variable* X is characterized either by its *probability density function* $f(X)$ or by its *cumulative distribution* $F(X)$;

$$F(X) = \int_{X_{\min}}^{X} f(X')dX'. \tag{2.16}$$

By construction, the cumulative distribution has the properties

$$F(X_{\min}) = 0, \quad F(X_{\max}) = 1$$

if the range of possible values is $X_{\min} \leq X \leq X_{\max}$. $F(X)$ is a monotonic function of X such that

$$F(X_1) \geq F(X) \text{ for all } X_1 > X.$$

Obviously $F(X_1)$ is the probability of X being smaller than X_1, thus

$$F(X_1) = P(X \leq X_1).$$

In general, the probability of X and Y having values in some region $R(X, Y)$ is

$$P(X \text{ and } Y \text{ in } R) = \int\int_{R(X,Y)} dF(X, Y).$$

The projection of a multidimensional density is called a *marginal density*. For instance, the projection of the two-dimensional density (surface) $f(X, Y)$ onto the X-axis is the marginal density function of X,

$$g(X) = \int_{Y_{\min}}^{Y_{\max}} f(X, Y)dY. \tag{2.17}$$

Example

In a reaction yielding a three-body final state the p.d.f. is a function of the two variables

$$X = m_{12}^2, \quad Y = m_{23}^2,$$

where m_{ij}^2 is the invariant squared mass of particles i and j. The joint density $f(X, Y)$ is the point density in the Dalitz plot, normalized by Eq. (2.12). The marginal distribution of X is the projection of $f(X, Y)$ on the X-axis, that is, the "distribution" of m_{12}^2.

A section through a distribution is called a *conditional distribution*. Thus the (normalized) section through the density function $f(X, Y)$ at $X = X_0$ gives the conditional density function of Y, given that $X = X_0$, which is denoted by $f(Y|X_0)$. In analogy with the discrete case, we write

$$f(Y|X_0) = \frac{f(X_0, Y)}{\int f(X_0, Y)dY} = \frac{f(X_0, Y)}{g(X_0)}, \tag{2.18}$$

where $g(X)$ is the marginal distribution of X.

More generally, the conditional density of Y, given $X = h(Y)$ is given by

$$q[Y|X = h(Y)] = \frac{f[h(Y), Y]}{\int f[h(Y), Y]dY}. \tag{2.19}$$

2.3.4. *Bayes theorem for continuous variables*

Consider the joint probability density function $f(X, Y)$ for two variables X and Y, with marginal density functions $g(X)$ and $h(Y)$, and conditional density functions $p(X|Y)$ and $q(Y|X)$. Then, from the definitions of Section 2.3.3, we have

$$f(X, Y) = p(X|Y)h(Y) = q(Y|X)g(X).$$

Bayes theorem for continuous variables follows immediately, namely

$$q(Y|X) = \frac{p(X|Y)h(Y)}{g(X)}. \tag{2.20}$$

2.3.5. *Bayesian use of Bayes theorem for continuous variables*

In Section 2.2.6 we introduced *Bayes theorem* in terms of discrete random data X_i and discrete hypotheses θ_i. We now illustrate the Bayesian use of the theorem when applied to several continuous variables $\mathbf{X}$, for instance N independent observations of one continuous random variable X_i; and for a continuous range of hypotheses θ, for instance the measurement of a quantity that can take on a continuous range of values like a particle mass.

Let $f_i(X_i|\theta)$ be the p.d.f. of the ith random variable, where θ is a continuous hypothesis representing the possible values of a parameter. The joint density function of the N random variables is given by

$$p(\mathbf{X}|\theta) = \prod_{i=1}^{N} \mathbf{f_i}(\mathbf{X_i}|\theta) \tag{2.21}$$

where one assumes that the same parameter value holds for all variables. In particular, if one has N observations of the same variable X, Eq. (2.21) simplifies to

$$p(\mathbf{X}|\theta) = \prod_{i=1}^{N} \mathbf{f}(\mathbf{X_i}|\theta) \,.$$

Note that the components X_i are still considered as different variables, although the form of the density function is the same.

The question that now arises is the following. Having made N observations X_i^0 from the distribution $f(X|\theta)$, what can one say about the value of θ? Classically, the parameter θ has a true value which is fixed, but unknown. The value can only be estimated using methods we shall discuss in the chapters on estimation (Chapters 7 and 8). In particular, *Bayes theorem* cannot be used to invert X and θ classically.

In the Bayesian methodology, however, one can consider distributions of θ: both the unconditional probability density function $p(\theta)$ and the conditional p.d.f. $p(\theta|\mathbf{X})$ are taken to represent the *degree of belief* in different possible values of θ. *Bayes theorem* then takes the form:

$$p(\theta|\mathbf{X}) = \frac{p(\mathbf{X}|\theta)p(\theta)}{\int p(\mathbf{X}|\theta)p(\theta)d\theta} \,.$$

When we substitute for the random variable $\mathbf{X}$ the particular set of data actually observed $\mathbf{X}^0$, we obtain the form of *Bayes theorem* used in Bayesian parameter estimation:

$$p(\theta|\mathbf{X}^0) = \frac{p(\mathbf{X}^0|\theta)p(\theta)}{\int p(\mathbf{X}^0|\theta)p(\theta)d\theta} \,.$$

Note that although the upper-case P is often used in this formula, we use the small p since none of the factors is a probability. They are respectively:

- $p(\theta|\mathbf{X}^0)$ is a *p.d.f.* and is called the *posterior probability density* for θ.

- $p(\mathbf{X}^0|\theta)$ is the *likelihood function* $L(\theta)$. Note that it is *not* a p.d.f.

- $p(\theta)$ is the *prior probability density* for θ, discussed briefly below. This is a p.d.f. and is required mathematically so that both sides of the equation transform the same way under a change of variables $\theta \to \theta'$.

- the denominator of the r.h.s. is just a normalization factor which can be determined from the condition that the integral over all θ must be one for both sides of the equation.

The prior $p(\theta)$ is the major problem in evaluating the above expression, and we discuss this in Section 7.5, which is a brief summary of the enormous Bayesian literature on this topic.

The posterior distribution $p(\theta|\mathbf{X}^0)$ summarizes all one's knowledge and/or belief concerning θ, including both the prior belief and the information supplied by the experimental data $\mathbf{X}^0$. Extracting the usual statistically meaningful quantities from the p.d.f. is however not entirely obvious. This is treated in Section 9.6.

2.4. Properties of Distributions

In this section we define many of the useful quantities which characterize probability distributions: the *expectation* operator, *mean, variance, skewness, kurtosis, cumulants* and other *moments*. We shall also introduce the tools to find these quantities: the *characteristic function*, the *cumulant generating function* and the *probability generating function*.

2.4.1. *Expectation, mean and variance*

The *probability density function* is used as a weighting function to obtain the average value of a function of random variables. If $g(X)$ is some function of a *random variable* X with density $f(X)$, the *expectation* of $g(X)$ is the number

$$E(g) = \int_{\Omega} g(X)f(X)dX\,, \tag{2.22}$$

where the Ω means that the integration is over the entire space of X.

The *expectation* E is a *linear operator*

$$E[ag(X) + bh(X)] = aE[g(X)] + bE[h(X)]\,. \tag{2.23}$$

Note that $E[g(X)]$ is not a function of X (since X has been integrated between given limits).

The expectation of the random variable X itself is called the *mean* of the density $f(X)$ or the *expected value* of X for the density $f(X)$, and it is denoted by μ (sometimes written $\bar{X}$ or $\langle X \rangle$):

$$\mu = \int Xf(X)dX\,. \tag{2.24}$$

The expectation of the function $(X - \mu)^2$ is called the *variance* $V(X)$ of the density $f(X)$:

$$
\begin{aligned}
V(X) = \sigma^2 &= E[(X - \mu)^2] \\
&= E[X^2 - 2\mu X + \mu^2] \\
&= E(X^2) - \mu^2 \\
&= \int (X - \mu)^2 f(X) dX .
\end{aligned}
\tag{2.25}
$$

Also $V(X)$ is a number and not a function of X. The quantity σ is called the *standard deviation*. This is a mere definition, and no statement can be made here about the relation between probability content and standard deviation. Such statements require $f(X)$ to be specified; this will be done in the chapter on confidence intervals (Chapter 9).

The expected value, $E(X)$ is a measure of the location of the distribution, while the variance $V(X)$ is a measure of the spread of the distribution over the space of X. Other measures of the shape of a distribution will be introduced in Section 2.4.6.

Note that the mean of a distribution does not always exist. An example is given by the density of the *Cauchy distribution* (given in Section 4.2.11):

$$
f(X) = \frac{1}{\pi(1 + X^2)} ,
$$

because the integral

$$
\int_{-\infty}^{\infty} f(X) dX = \frac{1}{\pi} \int_{-\infty}^{\infty} \frac{X dX}{1 + X^2}
$$

is undefined. The variance of X, $V(X)$, is infinite. It should be noted that $f(X)$ is the Breit–Wigner formula often encountered in physics.

However, if we consider the truncated distribution of X, given the condition that $-A \le X \le A$, then the above density function becomes

$$
g(X) = \frac{\pi f(X)}{2 \arctan A} , \qquad -A \le X \le A
$$

$$
g(X) = 0 , \qquad \text{otherwise}
$$

and the expectation and variance of X exist. In fact, we have

$$
E_g(X) = 0 ,
$$

$$
V_g(X) = \frac{A}{\arctan A} - 1 .
$$

2.4.2. *Covariance and correlation*

The *expectation* Eq. (2.22) is easily generalized to several dimensions. In particular, the expectation of a function $g(X, Y)$ of two random variables, given their joint density $f(X, Y)$, is

$$E[g(X, Y)] = \int \int g(X, Y) f(X, Y) dX dY .$$

The *mean* and the *variance* of X and Y are

$$\mu_X = E(X) = \int \int X f(X, Y) dX dY = \int X \int f(X, Y) dY dX$$

$$\mu_Y = E(Y) = \int \int Y f(X, Y) dX dY$$

$$\sigma_X^2 = E[(X - \mu_X)^2]$$

$$\sigma_Y^2 = E[(Y - \mu_Y)^2] .$$

(2.26)

Two further important numerical characteristics of the joint density are the *covariance*, defined by

$$\text{cov}(X, Y) = E[(X - \mu_X)(Y - \mu_Y)] = E(XY) - E(X)E(Y) \qquad (2.27)$$

and the *correlation coefficient*, defined by

$$\text{corr}(X, Y) = \rho(X, Y) = \frac{\text{cov}(X, Y)}{\sigma_X \sigma_Y} . \qquad (2.28)$$

The correlation coefficient lies between -1 and $+1$. To prove this we compute the variance of a linear combination of X and Y. Clearly the variance is a positive quantity since it comes from the integration of a positive function. Thus

$$V(\alpha X + Y) = \alpha^2 V(X) + V(Y) + 2\alpha \, \text{cov}(X, Y) \geq 0 .$$

The inequality holds for all α. Since this quadratic function cannot have two distinct roots, its discriminant must be zero or negative:

$$[\text{cov}(X, Y)]^2 - V(X) \cdot V(Y) \leq 0$$

or

$$-1 \leq \rho \leq 1 .$$

(In fact this is a special case of Schwarz' lemma.)

Random variables $(X_1, \ldots, X_N)$ are said to be *mutually independent* if and only if their joint density $f(\mathbf{X})$ is completely factorizable as

$$f(\mathbf{X}) = f(X_1, \ldots, X_N) = f_1(X_1) f_2(X_2) \cdots f_N(X_N) \,.$$

We have already used this result in Section 2.3.5 in writing down the joint density of a set of observed variables. The expectation of the product of two mutually independent variables X and Y is

$$E(XY) = \int XY f(X, Y) dX dY = \int X f_1(X) dX \int Y f_2(Y) dY$$
$$= E(X) E(Y) \,.$$

Comparing this result with Eqs. (2.27) and (2.28) we see that *the covariance and the correlation vanish for independent variables.* The converse statement is not necessarily true.

Random variables for which $\rho = 0$ are said to be *uncorrelated* (but not necessarily independent).

For example, suppose X is symmetrically distributed about zero, with density function $f(X)$. Let $Y = X^2$. Clearly X and Y are very dependent, being functionally related. But the correlation between X and Y is zero. Thus

$$E(X) = 0 \,,$$

and

$$E(Y) = \int X^2 f(X) dX = V(X) = \sigma^2 \,.$$

The covariance between X and Y is given by

$$\mathrm{cov}(X, Y) = E[X(X^2 - \sigma^2)] = E(X^3 - \sigma^2 E(X)) = 0 \,, \qquad (2.29)$$

since $f(X)$ is symmetrical. It follows that X and Y are uncorrelated.

In everyday usage, one often speaks of "uncorrelated variables" meaning "independent variables". In statistical terms "uncorrelated" is much weaker than "independent", the former being implied by the latter, but not vice versa.

When there are more than two variables, the covariance and the correlation can still be defined for each marginal two-dimensional joint distribution (in X_i and X_j). The matrix with elements $\mathrm{cov}(X_i, X_j)$ is called the *covariance matrix*, or sometimes the *variance matrix* or the *error matrix*. Obviously the diagonal elements are the variances $\sigma_{X_i}^2$. If the covariance matrix is not positive-definite, there is at least one linear relation between the X_i. Consider the correlations $\rho(X_k, Y)$ between the variable X_k and every possible linear combination Y of all the other variables X_i, $i \neq k$. A useful quantity is the *global correlation coefficient* ρ_k, defined as the largest value of $\rho(X_k, Y)$. This quantity is a measure of the total amount of correlation between X_k and all the other variables. If $\rho_k = 0$, the variable X_k is uncorrelated with all others. If $\rho_k = 1$, X_k is completely correlated with at least one linear combination of the other variables.

The global correlation coefficient can be found from the diagonal elements V_{kk} of the covariance matrix, and the diagonal elements $(V^{-1})_{kk}$ of its inverse,

$$\rho_k = \sqrt{1 - [V_{kk} \cdot (V^{-1})_{kk}]^{-1}} \, .$$

2.4.3. *Linear functions of random variables*

In this section we apply the concepts introduced above to the important case of linear functions of random variables. Of particular interest is the average of several observations of a random variable.

The expectation of a *linear function* of several random variables $X_1, \ldots, X_N$ is

$$E\left(\sum_{i=1}^{N} a_i X_i\right) = \sum_{i=1}^{N} a_i E(X_i) = \sum_{i=1}^{N} a_i \mu_{X_i} \qquad (2.30)$$

from Eqs. (2.22) and (2.26), and from the linearity of the expectation operator, Eq. (2.23). The variance is

$$V\left(\sum_{i=1}^{N} a_i X_i\right) = \sum_{i=1}^{N} a_i^2 V(X_i) + 2 \sum_{i=1}^{N} \sum_{j=i+1}^{N} a_i a_j \, \text{cov}\,(X_i, X_j) \, . \qquad (2.31)$$

If the X_i are mutually uncorrelated, Eq. (2.31) takes the simple form

$$V\left(\sum_{i=1}^{N} a_i X_i\right) = \sum_{i=1}^{N} a_i^2 V(X_i) \, . \qquad (2.32)$$

Thus the variance of a linear function is a quadratic function in the expansion coefficients.

To illustrate the importance of Eqs. (2.30) - (2.32), consider the case when the X_i are N different trials of the same experiment, the *average* (often called the *sample mean*) being given by the mean value of the observations X_i,

$$\bar{X} = \frac{1}{N} \sum_{i=1}^{N} X_i \, . \qquad (2.33)$$

The expectation of each trial is assumed to be

$$E(X_i) = \mu, \quad \text{for all } i,$$

and the variance

$$V(X_i) = \sigma^2, \quad \text{for all } i.$$

Note that μ (often called the *population mean*) is not the same as $\bar{X}$.
The *expectation of the average of N observations* is then

$$E(\bar{X}) = E\left(\frac{1}{N}\sum_{i=1}^{N} X_i\right) = \frac{1}{N}\sum_{i=1}^{N} E(X_i) = \frac{1}{N}N\mu = \mu \qquad (2.34)$$

and the *variance of the average*

$$V(\bar{X}) = V\left(\frac{1}{N}\sum_{i=1}^{N} X_i\right)$$

$$= \frac{1}{N^2}\sum_{i=1}^{N} V(X_i) + \frac{2}{N^2}\sum_{i=1}^{N}\sum_{j=i+1}^{N} \text{cov}\,(x_i, X_j)$$

$$= \frac{\sigma^2}{N} + \frac{2}{N^2}\sum_{i=1}^{N}\sum_{j=i+1}^{N} \text{cov}\,(X_i, X_j). \qquad (2.35)$$

If the trials are *independent* $\text{cov}(X_i, X_j) = 0$ for every pair (i, j), and the double sum term drops out. We then have the well-known result

$$\sigma_{\bar{X}} = \sqrt{V(\bar{X})} = \frac{\sigma}{\sqrt{N}}, \qquad (2.36)$$

which says that (under the conditions given) the standard deviation of the mean decreases with $N^{-\frac{1}{2}}$ as N increases.

However, if the measurements are correlated, the standard deviation of the average need not decrease with $N^{-\frac{1}{2}}$ as N increases. This depends on the correlation coefficient. Let us take the case of two random variables of identical means μ and variances σ. Then, if one does not assume $\rho = 0$, one has in general

$$V_{N=2}(\bar{X}) = \frac{1}{2}\sigma^2 + \frac{1}{2}\sigma^2\rho = \frac{1}{2}\sigma^2(1 + \rho).$$

As ρ may take any value between -1 and $+1$, one can get any value for $V(\bar{X})$ between 0 and σ^2. Consider the extreme cases:

Complete positive correlation, $\rho = +1$; then $V(\bar{X}) = \sigma^2$, the average is not better than one trial. This is evident since $\rho = +1$ means that the second measurement is identical with the first one, that is $\bar{X} = X_1 = X_2$.

Complete negative correlation, $\rho = -1$; then $V(\bar{X}) = 0$, so that $\bar{X}$ is no longer random but certain. Evidently, to get $\rho = -1$, one must have (from the definition of ρ) $(X_2 - \mu) = -(X_1 - \mu)$. Thus, $\bar{X} = \mu$.

In general it can be seen from Eq. (2.35) that the variance of the average can be made very small indeed, if it is possible to make the observations X_i in such a way that the observational errors are negatively correlated with one another. (See example in Section 5.5.)

2.4.4. *Ratio of random variables*

Suppose X and Y are independently distributed, with density functions $f(X)$ and $g(Y)$, respectively. Define a change of variable by

$$U = \frac{X}{Y}$$
$$V = Y .$$

It is necessary to make the further restriction that the range of Y does not contain zero. Then the joint density function $h(U, V)$ is given by

$$h(U, V)dU\,dV = f(X)g(Y)dX\,dY$$
$$= f(UV)g(V)\left\|\frac{\partial(X, Y)}{\partial(U, V)}\right\|dU\,dV$$
$$= f(UV)g(V)V\,dU\,dV .$$

The density function of U is

$$p(U) = \int_0^\infty h(U, V)dV = \int_0^\infty f(UV)g(V)V\,dV .$$

If the restriction on the range of Y is relaxed, the joint density function (42) becomes

$$h(U, V) = f(UV)g(V)|V| , \qquad V < 0$$
$$= f(UV)g(V)V , \qquad V \geq 0$$

and the density function of the ratio U is given by

$$p(U) = \int_0^\infty f(UV)g(V)V\,dV - \int_{-\infty}^0 f(UV)g(V)V\,dV . \qquad (2.37)$$

Applying this formula to the case where X and Y are Normal [b] variables with mean zero and variance 1, that is

$$f(X) = g(X) = \frac{1}{\sqrt{2\pi}} e^{-\frac{1}{2}X^2} \,,$$

we find that

$$p(U) = \frac{1}{\pi(1 + U^2)} \,.$$

Thus the distribution of the ratio of two Normally distributed numbers with mean zero, is the Cauchy distribution (see 4.2.11), which has no mean and an infinite variance.

Similarly, by applying Eq. (2.37) to the general case of Normal variables $N(\mu_X, \sigma_X^2)$ and $N(\mu_Y, \sigma_Y^2)$, it can be shown that the probability density function of $Z = X/Y$ is

$$p(Z) = \frac{1}{\pi} \frac{\sigma_X/\sigma_Y}{\left(1 + Z^2 \frac{\sigma_X^2}{\sigma_Y^2}\right)} \exp\left[-\frac{1}{2}\left(\frac{\mu_Y^2}{\sigma_X^2} + \frac{\mu_Y^2}{\sigma_Y^2}\right)\right]\left[1 + be^{\frac{1}{2}b^2}\int_0^b e^{-\frac{1}{2}V^2} dV\right]$$

$$(2.38)$$

where

$$b = \frac{(\mu_X/\sigma_X) + Z(\sigma_X/\sigma_Y)(\mu_Y/\sigma_Y)}{\sqrt{1 + Z^2(\sigma_X/\sigma_Y)^2}} \,.$$

Thus $p(Z)$ takes the form of a Cauchy distribution times $f(Z)$. It can be shown that the variance of this distribution also is infinite, for all finite values of μ_X, μ_Y.

If μ_X/σ_X is sufficiently large so that one may take X to have practically only positive values, then the density function of $Z = Y/X$ becomes

$$p(Z) = \frac{1}{\sqrt{2\pi}} \frac{\mu_X\sigma_Y^2 + \mu_Y\sigma_X^2 Z}{(\sigma_Y^2 + \sigma_X^2 Z^2)^{\frac{3}{2}}} \exp\left[-\frac{1}{2}\frac{(\mu_Y - \mu_X Z)^2}{\sigma_Y^2 + \sigma_X^2 Z^2}\right] \qquad (2.39)$$

From this distribution, it follows that the variable

$$\frac{\mu_Y - \mu_X Z}{\sqrt{(\sigma_Y^2 + \sigma_X^2 Z^2)}}$$

is Normally distributed with zero mean and unit variance [Kendall I, p. 271].

---

[b] The properties of the Normal (Gaussian) distribution will be treated in Section 4.2.1.

2.4.5. *Approximate variance formulae*

In this section we give approximate formulae for the *variance* of a function of random variables. Suppose we have a function $f(X_1, \ldots, X_N)$ where $X_1, \ldots, X_N$ are random variables with means $\mu_1, \ldots, \mu_N$. We also suppose that the function can be expanded about $\mu_1, \ldots, \mu_N$ in a Taylor series. Then we have

$$f(X_1, \ldots, X_N) = f(\mu_1, \ldots, \mu_N) + \sum_{i=1}^{N} (X_i - \mu_i) \frac{\partial f}{\partial \mu_i} + O[(X_i - \mu_i)^2]$$

where

$$\frac{\partial f}{\partial \mu_i} = \frac{\partial f}{\partial X_i} \bigg|_{\mu_i}.$$

Ignoring terms of order greater than 1 in $\Delta X_i = (X_i - \mu_i)$, and taking expectations, we have

$$E(f) \simeq f(\mu_1, \ldots, \mu_N).$$

The variance of f is given by

$$V[f(X_1, \ldots, X_N)] = E[f - E(f)]^2$$

$$\simeq E \left[\sum_{i=1}^{N} (X_i - \mu_i) \frac{\partial f}{\partial \mu_i} \right]^2$$

$$\simeq \sum_{i=1}^{N} \sum_{j=1}^{N} \frac{\partial f}{\partial \mu_i} \frac{\partial f}{\partial \mu_j} E\left[(X_i - \mu_i)(X_j - \mu_j) \right]$$

$$\simeq \sum_{i=1}^{N} \sum_{j=1}^{N} \frac{\partial f}{\partial \mu_i} \frac{\partial f}{\partial \mu_j} \, \text{cov} \, (X_i, X_j). \tag{2.40}$$

Note that in matrix notation Eq. (2.40) takes the simple form $A \, V \, A^T$, where V is the covariance matrix. If the variables X_i are independent, Eq. (2.40) reduces to

$$V[f(X_1, \ldots, X_N)] \simeq \sum_{i=1}^{N} \left(\frac{\partial f}{\partial \mu_i} \right)^2 V(X_i). \tag{2.41}$$

In an exactly similar manner, the covariance between two functions $f(X_1, \ldots, X_N)$ and $g(X_1, \ldots, X_N)$ is given approximately by

$$\text{cov}(f, g) \simeq \sum_{i=1}^{N} \sum_{j=1}^{N} \frac{\partial f}{\partial \mu_i} \frac{\partial g}{\partial \mu_j} \, \text{cov}(X_i, X_j). \tag{2.42}$$

If X_i are independent,

$$\text{cov}(f, g) \simeq \sum_{i=1}^{N} \frac{\partial f}{\partial \mu_i} \frac{\partial g}{\partial \mu_j} V(X_i)$$

Applying Eq. (2.40) to the *ratio of two random variables*, X and Y, we have, approximately,

$$V\left(\frac{X}{Y}\right) \simeq \left(\frac{\mu_X}{\mu_Y}\right)^2 \left[\frac{\sigma_X^2}{\mu_X^2} + \frac{\sigma_Y^2}{\mu_Y^2} - \frac{2\rho_{XY}}{\mu_X} \frac{\sigma_X \sigma_Y}{\mu_Y}\right], \tag{2.43}$$

where μ_X, μ_Y, σ_X^2, σ_Y^2 are the means and variances of X and Y, and ρ is the correlation coefficient.

Note that the above approximation may be expected to be good only under somewhat severe conditions, namely

$$|\mu_X| \gg \sigma_X$$

$$|\mu_Y| \gg \sigma_Y$$

and

$$Y \neq 0.$$

In particular, if X and Y are Normal distributions, $V(X/Y) = \infty$. (See Section 2.4.4.)

2.4.6. *Moments*

The *expectation* of X^n given the density $f(X)$ is called the nth moment of $f(X)$ (or the moment of nth order). More generally

$$\mu_n' = E(X^n) \qquad \text{is the } n^{\text{th}} \text{ } algebraic \text{ } moment,$$

$$\mu_n = E\{[X - E(X)]^n\} \quad \text{is the } n^{\text{th}} \text{ } central \text{ } moment,$$

$$\nu_n' = E(|X|^n) \qquad \text{is the } n^{\text{th}} \text{ } absolute \text{ } moment,$$

$$\nu_n = E[|X - E(X)|^n] \quad \text{is the } n^{\text{th}} \text{ } absolute \text{ } central \text{ } moment,$$

of the density $f(X)$.

In particular, the *mean* μ is the first algebraic moment, and the *variance* is the second central moment of $f(X)$.

The moments of many-dimensional distributions are defined in an analogous way. For instance, the algebraic moment of $f(X, Y)$, of order m in X and n in Y, is

$$\mu_{mn}' = E(X^m Y^n).$$

A distribution which is not symmetrical about its mean may be *skew*. A measure of the skewness of a distribution is given by the coefficient β_1 defined as

$$\beta_1 = \frac{\mu_3^2}{\mu_2^3} .$$

Clearly, if the distribution is symmetrical, μ_3, and therefore β_1, vanishes. (The converse is not necessarily true.)

Another useful coefficient for describing properties of the shape of a distribution is

$$\beta_2 = \frac{\mu_4}{\mu_2^2} .$$

For many purposes, more convenient coefficients are

$$\gamma_1 = \sqrt{\beta_1} = \frac{\mu_3}{\mu_2^{3/2}} , \quad \text{the coefficient of skewness}$$

and

$$\gamma_2 = \beta_2 - 3 , \quad \text{the coefficient of kurtosis} .$$

The number 3 subtracted from β_2 in the definition of γ_2 is conventional, and has the effect of making $\gamma_2 = 0$ for the Normal distribution. (See Section 4.2.1.)

2.5. Characteristic Function

The Fourier transform of the probability density function is called the *characteristic function* of the random variable. It has many useful and important properties. In particular, the distribution of sums of random variables is most easily handled by means of characteristic functions. For many distributions, the moments are also best obtained from the properties of these functions. Special cases of characteristic functions are the probability, moment and cumulant generating functions.

2.5.1. *Definition and properties*

Given a *random variable* X with density $f(X)$, the *characteristic function* is defined as:

$$\phi_X(t) = E(e^{itX}) \qquad (t \text{ real})$$

$$= \int_{-\infty}^{\infty} e^{itX} f(X) dX \qquad (X \text{ continuous})$$

$$= \sum_k p_k e^{itX_k} \qquad (X \text{ discrete}) . \qquad (2.44)$$

The characteristic function $\phi_X(t)$ determines completely the probability distribution of the random variable. In particular, if $F(X)$ is continuous everywhere and $dF(X) = f(X)dX$, then

$$f(X) = \frac{1}{2\pi} \int_{-\infty}^{\infty} \phi_X(t)e^{-iXt}dt \,.$$

The function $\phi(t)$ has the properties

$$\phi(0) = 1 \,; \quad |\phi(t)| \leq 1$$

and it exists for all t.

If a and b are constants, the characteristic function for $aX + b$ is

$$\phi_{aX+b}(t) = e^{ibt}\phi_X(at)$$

as can be proved by taking expectations:

$$E[e^{it(aX+b)}] = E(e^{itb}e^{iatX}) = e^{itb}\phi_X(at) \,.$$

If X and Y are *independent* random variables, with characteristic functions $\phi_X(t)$, $\phi_Y(t)$, then the characteristic function of the sum $(X + Y)$ is

$$\phi_{X+Y}(t) = \phi_X(t) \cdot \phi_Y(t) \,.$$

This we again prove by taking expectations:

$$
\begin{aligned}
\phi_{X+Y}(t) &= E[e^{it(X+Y)}] \\
&= E(e^{itX}e^{itY}) \\
&= E(e^{itX})E(e^{itY}) \\
&= \phi_X(t) \cdot \phi_Y(t) \,.
\end{aligned}
$$

This exemplifies one virtue of using characteristic functions: the characteristic function of a sum of independent variables is much simpler than the corresponding p.d.f. of the sum:

$$f(Z = X + Y) = \int_{-\infty}^{\infty} f_X(Z - t)f_Y(t)dt \,. \tag{2.45}$$

In general, the characteristic function of the sum of N independent random variables is

$$\phi_{X_1+\cdots+X_X}(t) = \prod_{i=1}^{N} \phi_{X_i}(t) \,. \tag{2.46}$$

The relation between the characteristic function and the moments of a distribution can be found from the formal expansion

$$\phi_X(t) = E(e^{itX}) = E\left[\sum_{r=0}^{\infty} \frac{(itX)^r}{r!}\right]$$

$$= \sum_{r=0}^{\infty} \frac{(it)^r}{r!} E(X^r)$$

$$= \sum_{r=0}^{\infty} \frac{(it)^r}{r!} \mu'_r . \qquad (2.47)$$

Thus the moments μ'_r appear as the coefficients of $(it)^r/r!$ in the expansion of $\phi(t)$.

If a distribution has a variance σ^2 and mean μ, then one can find an approximate characteristic function by writing down the Taylor expansion

$$\phi_X(t) = 1 + it\mu + \frac{1}{2}(\sigma^2 + \mu^2)(it)^2 + O(t^2) .$$

If the mean of a distribution is μ, then the transformation

$$e^{-i\mu t}\phi_X(t) = E[e^{it(X-\mu)}] = \phi_{X-\mu}(t) \qquad (2.48)$$

generates the central moments about the mean. Thus one finds the moments

$$\mu'_r = \frac{1}{i^r}\left(\frac{d}{dt}\right)^r \phi_{X-\mu}(t)\Bigg|_{t=0} ,$$

$$\qquad\qquad (2.49)$$

$$\mu_r = \frac{1}{i^r}\left(\frac{d}{dt}\right)^r e^{-i\mu t}\phi_X(t)\Bigg|_{t=0} .$$

We now give two examples of the use of characteristic functions:
(i) Suppose that X is Normally distributed, $N(\mu, \sigma^2)$.[c] The characteristic function is

$$\phi_X(t) = \frac{1}{\sqrt{2\pi}\sigma} \int_{-\infty}^{\infty} \exp\,(iXt)\,\exp\left[-\frac{1}{2}\left(\frac{X-\mu}{\sigma}\right)^2\right] dX$$

$$= e^{i\mu t - \frac{1}{2}t^2\sigma^2} .$$

---

[c]The properties of the Normal (Gaussian) distribution will be treated in Section 4.2.1.

To find the moments we use Equation (2.48):

$$e^{-i\mu t}\phi_X(t) = e^{-\frac{1}{2}t^2\sigma^2}$$

$$= \sum_{r=0}^{\infty} \frac{(it)^{2r}}{r!} \left(\frac{\sigma^2}{2}\right)^r .$$

Comparison with Eq. (2.47) yields

$$\mu_{2r-1} = 0$$

$$\mu_{2r} = \frac{(2r)!}{r!} \left(\frac{\sigma^2}{2}\right)^r .$$

(ii) For the *Cauchy distribution* (treated in Section 4.2.11) the characteristic function is

$$\phi_X(t) = \frac{1}{\pi} \int_{-\infty}^{\infty} \frac{e^{iXt}}{1+X^2} dX = e^{-|t|} .$$

Since $\phi_X(t)$ has no derivatives at $t = 0$, Eq. (2.49) tells us that this distribution has no moments. In particular, the mean of the distribution does not exist.

2.5.2. *Cumulants*

Given a random variable X, with a *characteristic function* $\phi_X(t) = E(e^{itX})$, the *cumulant generating function* (or semi-invariant generating function) is defined as

$$K(t) = \ln \phi_X(t). \qquad (2.50)$$

Expanding $K(t)$ as a power series in (it), we have

$$K(t) = \sum_{r=0}^{\infty} K_r \frac{(it)^r}{r!} . \qquad (2.51)$$

The coefficients, K_r, are called the *cumulants* or semi-invariants of the distribution of X.

The cumulants are a series of constants which are useful for measuring the properties of a distribution, and in some cases, for specifying it. They bear a close relation to the moments of a distribution, as is clear from the definition.

The first cumulants are [Kendall I, p. 67]:

$$K_1 = \mu \equiv \text{ mean},$$

$$K_2 = \mu_2 \equiv \text{ variance},$$

$$K_3 = \mu_3,$$

$$K_4 = \mu_4 - 3\mu_2^2,$$

$$K_5 = \mu_5 - 10\mu_3\mu_2.$$

A general formula for higher cumulants is given at the end of the next section.

For two independent random variables, X, Y, with cumulant generating functions $K_X(t)$ and $K_Y(t)$, the *cumulant generating function* of the sum $X+Y$ is $K_X(t) + K_Y(t)$; that is the cumulants of the sum of independent random variables are the sums of the cumulants.

2.5.3. *Probability generating function*

A characteristic function can also be defined for a discrete random variable r. Let the probability of r be $P(r) = P_r$; then the *characteristic function* according to Eq. (2.44), is

$$\phi_r(t) = \sum_{r=0}^{\infty} p_r e^{itr}.$$

However, it is more convenient to write $Z = e^{it}$. Then we call

$$G(Z) = E(Z^r) = \sum_{r=0}^{\infty} p_r Z^r \qquad (2.52)$$

the *probability generating function* of r. $G(Z)$ is regular for, at least, $|Z| < 1$, since

$$p_r \geq 0 \quad \text{and} \quad \sum_{r=0}^{\infty} p_r = 1.$$

It is easy to see that the derivatives of $G(Z)$ evaluated at $Z = 1$ are related to the moments. Thus the mean μ is obtained as

$$G'(1) = G'(Z)\Big|_{Z=1} = \sum_{r=0}^{\infty} r Z^{r-1} p_r \Big|_{Z=1} = \sum_{r=0}^{\infty} r p_r = E(r) = \mu.$$

Using the second derivative

$$G''(1) = G''(Z)\Big|_{Z=1} = \sum_{r=0}^{\infty} r(r-1)Z^{r-2}p_r \Big|_{Z=1} = \sum_{r=0}^{\infty} r(r-1)p_r$$

$$= E[r(r-1)] = E(r^2) - E(r),$$

and the definition of the variance, Eq. (2.25), one obtains an expression for the variance

$$V(r) = G''(1) + G'(1) - [G'(1)]^2.$$

This procedure is frequently the most direct method of obtaining the mean and variance of a discrete distribution.

The probability generating function can be used to calculate cumulants of any order using:

$$K_n = \frac{d^n \log(G(e^x))}{dx^n}\Big|_{x=0}$$

2.5.4. *Sums of a random number of random variables*

Consider the sum, S_N, of N independent random variables X

$$S_N = \sum_{i=1}^{N} X_i,$$

where N itself is a random variable with probability generating function

$$g(s) = \sum_{i=0}^{\infty} g_i s^i.$$

Suppose that each X has the probability generating function

$$f(s) = \sum_{j=0}^{\infty} f_j s^j.$$

We can write the probability that S_N is j as

$$h_j \equiv P(S_N = j) = \sum_{n=0}^{\infty} P(S_n = j | N = n) P(N = n). \qquad (2.53)$$

For fixed n, the sum S_n has a probability generating function

$$\{f(s)\}^n = \sum_{j=0}^{\infty} f_j^{(n)} s^j \,,$$

that is $P(S_n = j | N = n) = f_j^{(n)}$. Therefore, we can write

$$h_j = \sum_{n=0}^{\infty} f_j^{(n)} g_n \,.$$

Then the compound probability generating function of S_N is

$$h(s) = \sum_{j=0}^{\infty} h_j s^j$$

$$= \sum_{j=0}^{\infty} \sum_{n=0}^{\infty} f_j^{(n)} g_n s^j$$

$$= \sum_{n=0}^{\infty} g_n \{f(s)\}^n = g\{f(s)\} \,. \tag{2.54}$$

Example

Consider the case when N is a Poisson [d] variable of mean λ, and the X_i are also Poisson variables, of mean μ.

The generating function for the Poisson distribution is

$$g(s) = e^{-\lambda + \lambda s} \,.$$

Therefore from Eq. (2.54), the generating function of S_N is given by

$$h(s) = \exp\left(-\lambda + \lambda e^{-\mu + \mu s}\right) \,. \tag{2.55}$$

This is a particular case of the *compound Poisson distribution* which has a probability generating function of the form

$$h(s, t) = e^{-\lambda t + \lambda t f(s)} \,.$$

Considering Eq. (2.55), it is easy to show from the results of the previous section that

$$E(S_N) = \lambda \mu$$

---

[d]The properties of the Poisson distribution will be treated in Secs. 4.1.3 and 4.1.4.

and that

$$V(S_N) = \lambda\mu(1 + \mu).$$

To obtain the probabilities (2.53) we need to calculate

$$h_n \equiv \frac{1}{n!} \frac{d^n h(s)}{ds^n}\bigg|_{s=0}.$$

It turns out that one may write

$$h_n = \left[\sum_{r=0}^{n} v_r^{(n)}\right] \exp\left(-\lambda + \lambda e^{-\mu+\mu_s}\right),$$

where

$$v_r^{(n)} = \frac{r\mu}{n} v_r^{(n-1)} + \frac{\lambda\mu e^{-\mu+\mu_s}}{n} v_{r-1}^{(n-1)}, \quad r = 0, \ldots, n \qquad (2.56)$$

with

$$v_0^{(0)} = 1,$$

and

$$v_r^{(n)} = 0 \quad \text{if} \quad r < 0 \quad \text{or} \quad r > n.$$

The calculation of the probability distribution of S_N for any particular case is therefore relatively straightforward.

2.5.5. *Invariant measures*

Most of the measures discussed in this chapter are based on *mean* values, which are in turn defined by sums. These measures are convenient because the sum (and therefore the *mean*) is a *linear operator*. Measures based on the mean have however an unfortunate property: they are not in general *invariant* under change of variable. This means, for example, that the *expectation* of a square is not equal to the square of the expectation:

$$E(x^2) \neq E^2(x).$$

If however, instead of the *mean expectation* we use the *median expectation*, defined as the point in the distribution for which half the probability (or half the observed values) lie below the point and half above, the new measure is now invariant. It is no longer a linear operator, so the computational complexity is considerably increased, but with modern computers this should not be an overriding concern. As we shall see in Section 8.7, the *median* is also a more *robust* measure of the centre of a distribution than the mean. Related to this robustness, we note also that there are distributions (for example the *Cauchy distribution* and the *Landau distribution*, for which the mean does not exist, but the median is well-defined and finite.

Chapter 3

CONVERGENCE AND THE LAW
OF LARGE NUMBERS

In this chapter, we present the fundamental theorems which form the basis for the practical results to follows. The reader interested only in applications may skip Sections 3.1 and 3.2.

3.1. The Tchebycheff Theorem and Its Corollary

This theorem with its corollary, the Bienaymé–Tchebycheff inequality, is one of the basic tools of convergence theorems.

3.1.1. *Tchebycheff theorem*

Let $h(X)$ be a non-negative function of the random variable X. The *Tchebycheff theorem* then says that an upper limit can be set for the probability that the function $h(X)$ will exceed some value K:

$$P[h(X) \geq K] \leq \frac{E[h(X)]}{K} \tag{3.1}$$

for every $K > 0$, depending only on the expectation of h, and not on its shape.

The proof is quite simple. In the region R where $h(X) \geq K$, the expectation

$$E[h(X)] = \int h(X)f(X)dX$$

is never smaller than

$$K \int_R f(X)dX = K \cdot P[h(X) \geq K].$$

Unfortunately, the theorem is too general to be useful in practical calculations since the upper limit it sets can usually be improved by using more knowledge about the function $h(X)$. It is, however, useful in demonstrating convergence theorems.

3.1.2. *Bienaymé–Tchebycheff inequality*

If in Eq. (3.1) we make the substitutions

$$h(X) \rightarrow [X - E(X)]^2$$

$$K \rightarrow (k\sigma)^2$$

we get

$$P(|X - E(X)| \geq k\sigma) \leq \frac{1}{k^2}. \tag{3.2}$$

This is the *Bienaymé–Tchebycheff inequality*, which gives an upper limit on the probability of exceeding any given number of standard deviations, independent of the shape of the function, provided its variance is known. Like the Tchebycheff theorem, it is too general to be really useful outside the purely theoretical domain, but it can have some practical applications when very little is known about the function under consideration.

With more knowledge about the p.d.f., one can set stronger limits. In the case, for example, of a continuous p.d.f. with a single maximum at X_0, the limit is

$$P(|X - X_0| \geq k\tau) \leq \frac{4}{9k^2}, \tag{3.3}$$

where

$$\tau^2 = \sigma^2 + (X_0 - \mu)^2.$$

The inequality (3.3) can be rewritten as

$$P(|X - \mu| \geq k\sigma) \leq \frac{4}{9} \frac{1 + s^2}{(k - |s|)^2},$$

where

$$s = \frac{\mu_1 - X_0}{\sigma}.$$

Similarly, if a higher even order moment is known, say μ_4, it can be shown [Cramér, p. 256] that the limit is

$$P(|X - \mu| \geq k\sigma) \leq \frac{\gamma_2 + 2}{(k^2 - 1)^2 + \gamma_2 + 2}. \tag{3.4}$$

This inequality has been obtained with the following substitutions in Eq. (3.1):

$$h(X) = 1 + \frac{\sigma^2(k^2 - 1)[(X - \mu)^2 - k^2\sigma^2]}{\mu_4 + k^4\sigma^4 - 2k^2\sigma^4},$$

$$k = 1,$$

$$\gamma_2 = \frac{\mu_4}{\sigma^4} - 3.$$

Applying the above formulae to the Normal distribution (treated in Section 4.2.1) gives the results shown in Table 3.1.

Table 3.1. Limits for $P(|X - E(X)| \leq k\sigma)$ for the Normal distribution.

k	1	2	3	4
Bienaymé–Tchebycheff (3.2)	1	1/4	1/9	1/16
Unimodal inequality (3.3)	4/9	1/9	4/81	1/36
Stronger inequality (3.4)	1	2/11	2/66	2/227
Exact probability	0.317	0.0555	0.0027	0.000063

3.2. Convergence

Physicists usually feel that they know what they mean by convergence and limiting processes from their familiarity with ordinary calculus where this concept is important. In statistics, however, convergence is a more complicated idea since one has in addition to cope with random fluctuations and probabilities. Because of this, it is necessary to define different kinds of convergence [Tortrat].

3.2.1. *Convergence in distribution*

This is the weakest kind of convergence. Consider a sequence $(X_1, \ldots, X_n)$ of random variables with cumulative distribution functions $F_1(X), \ldots, F_n(X)$. Then it is said that the sequence X_n *converges in distribution* (when $n \to \infty$) to X, of cumulative distribution F, if, *for every point where F is continuous*, we have

$$\lim_{n \to \infty} F_n(X) = F(X). \tag{3.5}$$

For example, let X_n be distributed as $N(0, 1/n)$: then X_n converges in distribution to a p.d.f. which is a Dirac delta-function at zero. It is easy to show that for $X \neq 0$, $F_n(X) \to F(X)$, where

$$F(X) = 0, \quad X \leq 0$$

$$F(X) = 1, \quad X > 0.$$

At the point $X = 0$, Eq. (3.5) does not hold, since $F_n(0) = 1/2$ and $F(0) = 0$. But $F(X)$ is not continuous at this point, and our definition does therefore indeed hold.

3.2.2. *The Paul Levy theorem*

Convergence properties are often demonstrated most easily using the *characteristic function* ϕ, of a random variable. The theorem which allows us to do this was formulated by Paul Levy:

If $\{\phi_n(t)\}$ is a sequence of characteristic functions, corresponding to distribution functions $F_n(X)$, if $\phi_n(t)$ converges to some function $\phi(t)$, and if the real part of $\phi(t)$ is continuous at $t = 0$ [remember that $\phi(t)$ is a complex function], then

    (i) $\phi(t)$ is a characteristic function.
    (ii) F_n converges to F, the distribution function corresponding to ϕ.

3.2.3. *Convergence in probability*

The sequence $\{X_1, \ldots, X_n\}$ is said to *converge in probability* to X if for any $\varepsilon > 0$ and any $\eta > 0$, a value of N can be found such that

$$P(|X_n - X| > \varepsilon) < \eta$$

for all $n \geq N$. Convergence in distribution is weaker than convergence in probability since the former says nothing about the "distance" of X_n from X. In fact it can be shown that convergence in probability implies convergence in distribution, whereas the reverse is not generally true.

One should note here that both X_n and X are random variables. It is clear that the X_i are in general strongly correlated. In fact, one can show that, if the X_i are independent (and therefore uncorrelated), then X must be a *certain* variable (i.e. a random variable with a p.d.f. which is a δ-function), and convergence in law and in probability are equivalent. In the general case

(X not certain), convergence in probability implies convergence in law and some strong correlation between the X_i.

3.2.4. *Stronger types of convergence*

Other kinds of convergence encountered in mathematical statistics are known as *almost certain convergence* and *convergence in quadratic mean*. They both imply convergence in probability. Although they are important theoretically, they are of no interest practically in this work and will not be discussed here. The relationships between the different types of convergence are illustrated in the diagram below [Tortrat].

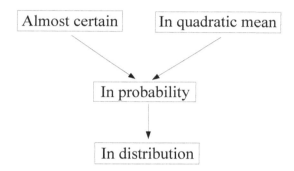

3.3. The Law of Large Numbers

The most important applications of convergence theorems are the laws of large numbers, which concern the convergence of the average (sample mean). There are in fact two laws of large numbers, known as the *weak law* and the *strong law*, corresponding to different types of convergence. The distinction between the two does not interest us here, and we simply state the general result.

Let $\{X_1, \ldots, X_N\}$ be a sequence of *independent* random variables, each having the same mean, and variances σ_i^2. Recall the definition of the average

$$\bar{X} = \frac{1}{N} \sum_{i=1}^{N} X_i .$$

If the mean μ exists, then the *weak law* of large numbers states that, if

$$\lim_{N \to \infty} \left[\frac{1}{N^2} \sum_{i=1}^{N} \sigma_i^2 \right] = 0 ,$$

then $\bar{X}$ converges to μ in *quadratic mean*. The *strong law* states that, if

$$\lim_{N \to \infty} \left[\sum_{i=1}^{N} \left(\frac{\sigma_i}{i} \right)^2 \right]$$

is finite, then $\bar{X}$ converges *almost certainly* to μ. Both laws imply convergence in probability [Fourgeaud]. These results are easily generalised to the case of different means.

3.3.1. *Monte Carlo integration*

Perhaps the most common example of the use of the *law of large numbers* is the calculation of an integral by the *Monte Carlo method*. By the definition of expectation we have

$$E(g) = \int_a^b g(u)f(u)du \,. \tag{3.6}$$

We take for $f(u)$ the normalized uniform distribution[b]

$$f(u) = \frac{1}{b-a} \quad a \le u \le b \,.$$

Now from the *weak law* of large numbers we know that if the u_i are chosen uniformly and independently at random over the interval (a, b), the corresponding function values g_i will satisfy

$$\frac{1}{N} \sum_{i=1}^{N} g_i \to E(g) \quad \text{as} \quad N \to \infty \,. \tag{3.7}$$

Combining Eqs. (3.6) and (3.7) gives us the usual formula for *Monte Carlo integration*

$$I = \frac{b-a}{N} \sum_{i=1}^{N} g(u_i) \to \int_a^b g(u)du \,,$$

with u_i chosen uniformly from a to b.

Note that the above method is quite different from usual numerical methods such as the trapezoidal rule. Under certain general conditions, all these quadrature methods converge to the correct value of the integral, but the error on the integral for finite values of N tends to increase exponentially with the

---

[b]We shall discuss the properties of the uniform distribution in Section 4.2.7.

number of dimensions. On the other hand, the the variance of the Monte Carlo estimate I is simply

$$V(I) = \frac{V(g)}{N},$$

independent of the dimensionality.

A complete discussion of Monte Carlo methods, although closely related to statistical problems in physics, is beyond the scope of this work, and the reader is referred to the bibliography for further reference [James 1980].

3.3.2. *The Central Limit theorem*

This powerful theorem is of central importance in both theoretical and practical problems in statistics. In fact it has already been used earlier, but we take the time here to state it fully and to discuss it in some detail.

We just recall the result from Section 2.4.3 that if we have a sequence of independent random variables X_i, each from a distribution with mean μ_i and variance σ_i^2, then the distribution of the sum $S = \sum X_i$ will have a mean $\sum \mu_i$ and a variance $\sum \sigma_i^2$. This holds for any distributions provided that the X_i are independent, the individual means and variances exist and do not increase too rapidly with i. Nothing is said about the distribution of the sum except for its mean and variance.

Under the same conditions as above, the Central Limit theorem states how the sum is distributed in the limit of large N (still independent of the original sub-distributions). In particular it says that as $N \to \infty$,

$$\frac{S - \sum_{i=1}^{N} \mu_i}{\sqrt{\sum_{i=1}^{N} \sigma_i^2}} \quad \to \quad N(0,1). \tag{3.8}$$

The theorem holds in the above general case, but we will demonstrate it for the usual case where all $\mu_i = \mu$ and all $\sigma_i = \sigma$. Let $X_1, \ldots, X_N$ be independent identically distributed random variables with $E(X) = \mu$, $V(X) = \sigma^2 < \infty$, and also with finite third moments. Define

$$Y_N = \frac{X_1 + \cdots + X_N}{N}$$

and a standardized variable

$$Z_N = \frac{Y_N - \mu}{\sigma/\sqrt{N}},$$

then the expectation and variance of Z_N are

$$E(Z_N) = 0, \quad V(Z_N) = 1.$$

The Central Limit theorem states that Z_N converges in distribution to the standard Normal distribution (see 4.2.1.).

$$\lim_{N \to \infty} [p(Z_N < a)] = \frac{1}{\sqrt{2\pi}} \int_{-\infty}^{a} e^{-\frac{1}{2}X^2} dX.$$

Without loss of generality, suppose that $\mu = 0$. Then, the *characteristic function* $\phi(t)$ of X has the expansion

$$\phi(t) = E(e^{itX}) = 1 + \sigma^2 \frac{(it)^2}{2} + O(t^3).$$

It follows that the characteristic function of NY_N is $[\phi(t)]^N$ and that that of Z_N is

$$\psi_{Z_N}(t) = \left[\phi\left(\frac{t}{\sigma\sqrt{N}} \right) \right]^N.$$

Taking logarithms we obtain

$$\ln \psi_{Z_N}(t) = N \ln \phi\left(\frac{t}{\sigma\sqrt{N}} \right)$$

$$= N \ln \left[1 + \frac{\sigma^2}{2} \frac{(it)^2}{N\sigma^2} + O\left(\frac{t^3}{N^{\frac{3}{2}}} \right) \right]$$

$$= \frac{(it^2)^2}{2} + O\left(\frac{1}{\sqrt{N}} \right)$$

for any fixed t. Thus we have proved that in the limit

$$\psi_{Z_N}(t) \to e^{-\frac{1}{2}(t^2)}$$

the *characteristic function* of the standard Normal distribution is obtained. One gets the same result if X_i have different variances σ_i^2, provided they are all finite, or at least do not approach infinity as fast as i. Refined conditions may be found in [Kendall I, p. 194] or [Feller].

For the Normal distribution, both the *skewness* and *kurtosis* (defined in Section 2.4.6) are zero. For other distributions, these quantities are a measure of how far the distribution is from Normal.

For example, the χ^2 distribution[d] with N degrees of freedom, has the characteristic function

$$\phi(t) = (1 - 2it)^{N/2}$$

Taking logarithms we obtain

$$\ln \phi(t) = -\frac{N}{2} \ln(1 - 2it)$$

$$= \frac{N}{2}\left[2it + 4\frac{(it)^2}{2} + 2^3\frac{(it)^3}{3} + 2^4\frac{(it)^4}{4} + \cdots\right].$$

Hence the moments are

$$\mu = N,$$

$$\sigma^2 = 2N,$$

$$\mu_3 = 8N,$$

$$\mu_4 = 3\mu_2^2 = 48N,$$

the coefficient of skewness is

$$\gamma_1 = \frac{8N}{(2N)^{3/2}} = 2\sqrt{2/N},$$

and the coefficient of kurtosis is

$$\gamma_2 = \frac{48N}{4N^2} = \frac{12}{N}.$$

It is generally accepted that χ^2 with $N > 30$ is reasonably Normal (γ_1 and γ_2 less than 0.5). Note, however, that this is not enough to assume Normality, since higher order moments may be important (see Fig. 4.6b).

3.3.3. *Example: Gaussian (Normal) random number generator*

All Monte Carlo calculations require a sequence of random numbers. Most commonly, they are uniformly distributed between zero and one, but it is sometimes more useful to have a generator producing random numbers which are distributed not uniformly but Normally, say as $N(0, 1)$. Using the Central Limit theorem one can easily make such a Gaussian generator if one already

---

[d]We shall discuss the properties of the χ^2 distribution in Section 4.2.3.

has a uniform generator. If u_i is the ith number from a uniform generator, use a sequence of N numbers u_i to form each number g using

$$g = \frac{\displaystyle\sum_{i=1}^{N} u_i - \frac{N}{2}}{\sqrt{\dfrac{N}{12}}}. \tag{3.9}$$

From Eq. (3.8) and the fact that $\mu = \frac{1}{2}$ and $\sigma^2 = 1/12$ for the uniform distribution, it is immediately seen that Eq. (3.9) becomes $N(0,1)$ as $N \to \infty$. In practice, g is already very close to Normal when $N = 10$ for instance. Compared with an exact Gaussian distribution, the tails of the distribution of the values of Eq. (3.9) are truncated:

$$-\sqrt{3N} \le g \le \sqrt{3N}\,,$$

which may or may not be a desirable property. A common and convenient choice for N is $N = 12$, which produces values as far out as six standard deviations, and yields the simple formula

$$g = \sum_{i=1}^{12} u_i - 6\,.$$

Note however, that exact methods, based on change of variables, are more efficient.

The exact p.d.f. of the average, $\bar{u}$, of N uniformly distributed random numbers u_i is

$$p(\bar{u}) = \frac{N^N}{(N-1)!} \sum_{i=1}^{k} (-1)^i \binom{N}{i} \left(\bar{u} - \frac{i}{N} \right)^{N-1}, \tag{3.10}$$

$$\frac{k}{N} \le \bar{u} \le \frac{k+1}{N}, k = 0, \ldots, N-1\,.$$

This function is interesting, since it consists of N polynomial arcs of degree $N-1$ in $\bar{u}$, joined at $N-1$ points with their first $N-2$ derivatives continuous. Thus the function (3.10) is a form of *spline function*.

Chapter 4

PROBABILITY DISTRIBUTIONS

The probability densities and cumulative distributions encountered in real life are often well approximated by a few mathematical functions. In Sections 4.1 and 4.2 we shall therefore present several such "ideal" distributions, for discrete and continuous random variables, respectively. In Section 4.3 we shall come back to real life distributions, and several methods for handling them.

4.1. Discrete Distributions

4.1.1. *Binomial distribution*

Variable	r, positive integer $\leq N$.
Parameters	N, positive integer.
	p, $0 \leq p \leq 1$.
Probability function	$P(r) = \binom{N}{r} p^r (1-p)^{N-r}$,
	$r = 0, 1, \ldots, N$. $\qquad (4.1)$
Expected value	$E(r) = Np$. $\qquad\qquad (4.2)$
Variance	$V(r) = Np(1-p)$. $\qquad (4.3)$

Skewness
$$\gamma_1 = \frac{1 - 2p}{[Np(1 - p)]^{1/2}}.$$

Kurtosis
$$\gamma_2 = \frac{1 - 6p(1 - p)}{Np(1 - p)}.$$

Probability generating function $G(Z) = [pZ + (1 - p)]^N$.

The binomial distribution gives the probability of finding exactly r successes in N trials, when the probability of success in each single trial is a constant, p. The distribution of the number of events in a single bin of a histogram is binomial (if the bin contents are independent).

If p is unknown, an unbiased estimate of the variance is given by

$$V(r) = \frac{N}{N - 1} N\left(\frac{r}{N}\right)\left(1 - \frac{r}{N}\right).$$

Examples

(i) Suppose that the outcome A corresponds to getting an event in a histogram bin j, and $\bar{A}$ corresponds to getting an event in any other histogram bin. The probability p of getting a success A is actually the integral of the p.d.f. over the bin (sometimes it is well approximated by the product of the bin width and the value of the p.d.f. at the middle of the bin). The probability of getting r events in the bin j and $N - r$ events outside j is given by Eq. (4.1). The expected number of events in j is Np from Eq. (4.2) and the variance is $Np(1 - p)$ from Eq. (4.3).

(ii) Consider a study of forward-backward asymmetry. Suppose that one wants to end an experiment when N events have been collected. (N is not a random variable!) Let F and B denote the numbers of events in the forward and backward hemispheres, respectively:

$$N = F + B.$$

Then F is distributed according to the binomial law

$$P(F) = \binom{N}{F} p^F (1 - p)^B$$

with mean Np and variance $Np(1 - p)$. For large N the variance is well approximated by $F(1 - p)$ and the standard deviation by $\sqrt{F(1 - p)}$ (not by $\sqrt{F}$).

Usually the forward-backward asymmetry is defined as

$$R = \frac{F - B}{F + B} = \frac{2F}{N} - 1 \, .$$

It follows that the variance $V(R)$ is given by $V(R) = 4p(1 - p)/N$. For large N this variance is well approximated by

$$V(R) \simeq \frac{4FB}{N^3} = \frac{4FB}{(F + B)^3} \tag{4.4}$$

and thus the standard deviation becomes approximately

$$\sigma \sim \frac{2}{(F + B)} \sqrt{\frac{FB}{(F + B)}} \, .$$

(iii) A desired event has a probability p of occurring in an experimental trial. Suppose that one wants to know how many trials must be carried out in order that the probability of at least one event occurring is α. Let X be the number of events in N trials, N being unknown. The problem is to find N such that

$$P(X \geq 1) \geq \alpha \, ,$$

or

$$1 - P(X = 0) \geq \alpha \, ,$$

or

$$P(X = 0) \leq 1 - \alpha \, .$$

Expressing the left-hand side of the inequality by means of Eq. (4.1) with $r = 0$, one has

$$(1 - p)^N \leq 1 - \alpha \, .$$

Taking logarithms, and using the fact that

$$(1 - p) < 1 \, ,$$

one has

$$N \geq \, \log \, (1 - \alpha)/\log \, (1 - p) \, .$$

4.1.2. *Multinomial distribution*

Variable $r_i, i = 1, 2, \ldots, k,$ positive integers $\leq N$.

Parameters N,   positive integer.

k,   positive integer.

$p_1 \geq 0, p_2 \geq 0, \ldots, p_k \geq 0$

with

$$\sum_{i=1}^{k} p_i = 1 \,.$$

Probability function

$$P(r_1, r_2, \ldots, r_k) = \frac{N!}{r_1! r_2! \cdots r_k!} p_1^{r_1} p_2^{r_2} \cdots p_k^{r_k} \,. \tag{4.5}$$

Expected values $E(r_i) = N p_i \,.$

Variances $V(r_i) = N p_i (1 - p_i) \,.$ $\qquad$ (4.6)

Covariances $\mathrm{cov}(r_i, r_j) = -N p_i p_j, \quad i \neq j \,.$ $\qquad$ (4.7)

Probability generating function

$$G(Z_2, \ldots, Z_k) = (p_1 + p_2 Z_2 + \cdots + p_k Z_k)^N \,.$$

The generalization of the binomial distribution to the case of more than two possible outcomes of an experiment is called the multinomial distribution. It gives the probability [Eq. (4.5)] of exactly r_i outcomes of type i in N independent trials, where the probability of outcome i in a single trial is p_i, $i = 1, 2, \ldots, k$.

An example of the multinomial distribution is a *histogram* containing N events distributed in k bins, with r_i events in the ith bin. The random variable is a k-dimensional vector, but the range of values is restricted to a $(k-1)$-dimensional space, since

$$\sum_{i=1}^{k} r_i = N \,.$$

Since N is fixed (see below), one can use Eq. (4.7) to demonstrate that the numbers of events in any two bins i and j are *negatively correlated*, with correlation coefficient

$$\text{corr}(r_i, r_j) = -\sqrt{p_i p_j / (1 - p_i)(1 - p_j)}. \qquad (4.8)$$

In many experimental situations, the observations r_j are independent, so that also the total $N = \sum r_i$ may be considered as the observation of a random variable. The data may then be treated in two ways:

(i) Unnormalized: If it is useful to retain N as a random variable, consider the $k+1$ variables (r_i and N) of which the r_i are mutually uncorrelated, but N is correlated with the r_i with covariance matrix of rank k. The distribution of events in bins is then *not multinomial* (but Poisson, Section 4.1.3).

(ii) Normalized: If it is not useful to treat N as a random variable (for example if there is no theory which predicts the expectation of N), the values r_i must be considered as conditional on the fixed observed value of N. In this case the multinomial distribution is applicable: the k variables r_i are all correlated with correlations given by Eq. (4.8), and the rank of the covariance matrix is simple $k - 1$.

When $p_i \ll 1$ (many bins), Eq. (4.6) can be approximated by

$$V(r_i) \sim N_{p_i} \sim r_i$$

and the standard deviation of the number of events in a bin becomes

$$\sigma_i \sim \sqrt{r_i}.$$

4.1.3. *Poisson distribution*

Variable $\qquad\qquad\qquad\qquad r$, positive integer.

Parameter $\qquad\qquad\qquad\quad \mu$, positive real number.

Probability function

$$P(r) = \frac{\mu^r e^{-\mu}}{r!}. \qquad (4.9)$$

Expected value $\qquad\qquad\qquad E(r) = \mu.$

Variance $\qquad\qquad\qquad\qquad V(r) = \mu.$

Skewness $\qquad\qquad\qquad\qquad \gamma_1 = \frac{1}{\sqrt{\mu}}.$

Kurtosis $$\gamma_2 = \frac{1}{\mu}.$$

Probability generating function

$$G(Z) = e^{\mu(Z-1)}.$$

The Poisson distribution gives the probability of finding exactly r events in a given length of time, if the events occur independently, at a constant rate. It is a limiting case of the binomial distribution for $p \to 0$ and $N \to \infty$, when $Np = \mu$, a finite constant.

As $\mu \to \infty$, the Poisson distribution converges to the Normal distribution.

An important relationship exists between the cumulative Poisson distribution and the cumulative χ^2 distribution (see Section 4.2.3), namely

$$P(r \leq N_0 \,|\, \mu) = 1 - P[\chi^2(2N_0 + 2) < 2\mu],$$

$$\text{or} \quad P(r > N_0 \,|\, \mu) = P[\chi^2(2N_0 + 2) < 2\mu],$$

$$\text{or} \quad P(r \geq N_0 \,|\, \mu) = P[\chi^2(2N_0) < 2\mu].$$

If events occur randomly and independently in time, so that the number of events occuring in a fixed time interval is Poisson-distributed, then the time between two successive events is exponentially distributed (see Section 4.2.9).

Examples

(i) Suppose that particles are emitted from a radioactive source at an average rate of ν particles per unit time, in such a way that the probability of emission in δt is $\nu\,\delta t$, and the probability of more than one emission in δt is $O(\delta t^2)$. Then the distribution of the number X of particles, emitted in a fixed time interval t, is Poisson, with mean νt:

$$P(X = r) = \frac{(\nu t)^r e^{-\nu t}}{r!}.$$

(ii) Many phenomena involving interactions of particles with matter produce events (sparks, showers, bubbles, droplets, etc.) which are distributed as either Poisson or compound Poisson (see below, Section 4.1.4) variables.

(iii) Consider the ratio of particles emitted in the forward and backward directions. Suppose that the numbers F and B are independent random variables distributed according to the Poisson law.

Then F has mean μ_F and variance μ_F, which can be approximated by F. The asymmetry is defined by

$$R = \frac{F - B}{F + B}.$$

Assuming that the error on F and B is small (large number of events), one may use a differential formula to obtain an approximation to the variance of R:

$$\frac{dR}{R} = \frac{d(F - B)}{(F - B)} - \frac{d(F + B)}{(F + B)} = 2\frac{BdF - FdB}{(F - B)(F + B)}.$$

Taking expectations of the squares of both sides,

$$\frac{E(dR^2)}{R^2} \simeq 4\frac{B^2 E(dF^2) + F^2 E(dB)^2}{(F - B)^2(F + B)^2} \simeq \frac{4BF}{(F - B)^2(F + B)}.$$

one finds the variance

$$V(R) \simeq \frac{4BF}{(F + B)^3}. \tag{4.10}$$

This is the same result as in example (ii) of the binomial distribution in Section (5.1.1) and Eq. (4.4). Note, however, the following differences:

— Equation (4.10) is an approximation (we used differentiation) whereas the binomial calculation leading to Eq. (4.4) is rigorous. In fact, Eq. (4.10) would have looked different from Eq. (4.4) had we not replaced μ_p by F and μ_B by B.
— In the binomial case the number of events was fixed *a priori*. When N is not fixed, then the correct distribution is a Poisson law $e^{-\nu}\nu^N/N!$ for the total number of events N, and a binomial law for F and B conditional on N.

Therefore the probability of finding N events, of which F are forward and B backward is given by

$$f(N, F, B) = e^{-\nu}\frac{\nu^N}{N!}\binom{N}{F}p^F(1 - p)^B = e^{-\nu}\frac{(\nu p)^F [\nu(1 - p)]^B}{F!B!}$$

which is the *product of the two Poisson laws* for F and B.

4.1.4. *Compound Poisson distribution*

Variable r, positive integer.

Parameters μ, λ, positive real numbers.

Probability function

$$P(r) = \sum_{N=0}^{N} \left[\frac{(N\mu)^r e^{-N\mu}}{r!} \cdot \frac{\lambda^N e^{-\lambda}}{N!} \right]. \tag{4.11}$$

Expected value $E(r) = \lambda\mu.$

Variance $V(r) = \lambda\mu(1 + \mu).$

Probability generating function

$$G(Z) = \exp(-\lambda + \lambda e^{-\mu + \mu Z}).$$

The compound Poisson distribution, sometimes known as the distribution of a *branching process*, is the distribution of the sum of N Poisson variables n_i all of mean μ, where N is also a Poisson variable of mean λ. A more transparent way to express this is to define

$$r \equiv \sum_{i=1}^{N} n_i.$$

Then the n_i have the distributions

$$P(n_i) = \frac{\mu^{n_i} e^{-\mu}}{n_i!}$$

and N has the distribution

$$P(N) = \frac{\lambda^N e^{-\lambda}}{N!}.$$

Using the fundamental relationship

$$P(r) = \sum_N P(r|N)P(N)$$

and the fact that a sum of N Poisson variables of mean μ is itself a Poisson variable of mean $N\mu$, the probability density function (4.11) can be evaluated easily by the method of Section 2.5.4.

A more general case of the compound Poisson distribution occurs when the variable r is the sum of N random observations n_i from any distribution, where N is a Poisson variable of mean λt. Then the generating function becomes (see Section 2.5.4):

$$G(Z, t) = \exp[-\lambda t + \lambda t f(Z)]$$

where $f(Z)$ is the probability generating function of the n_i.

This distribution has the remarkable factorization property

$$G(Z, t_1 + t_2) = G(Z, t_1)G(Z, t_2). \tag{4.12}$$

The random variable associated with $G(Z, t)$ may be called the contribution from the period t. For the compound Poisson process, the result (4.12) shows that the contributions from non-overlapping periods are independent random variables, also with compound Poisson distributions. This property is *unique* to the compound Poisson process.

The distribution has very wide applications in ecology, nuclear chain reactions, genetics, and queueing theory. We illustrate its application to the results of an experimental search for quarks.

Example: A quark experiment

As an example of both simple and compound Poisson distributions we consider the formations of droplets by charged particles traversing a cloud chamber. We take as a concrete case a "quark search" experiment carried out using this technique [McCusker].

When fast (relativistic) charged particles traverse a cloud chamber, the resulting ionization causes the formation of droplets which are observed as tracks. It is assumed that the probability of formation of a droplet per unit track length is constant and proportional to the square of the charge of the particle. Since nearly all the particles traversing the chamber have a single unit of charge, one could establish the existence of particles with less than one unit of charge by finding a track with *significantly* fewer droplets per unit length than the average track. The object of the experiment is to establish the existence of the "quark", a particle of charge 2/3 of the unit (proton) charge.

From the above assumptions, the distribution of the number of droplets in a given length of track should follow the Poisson law, with the mean μ determined to be equal to 229 from counting droplets on normal tracks over a certain fixed length. Over the same fixed length, one track was observed with only 110 droplets. The statistical problem is to calculate the probability that a particle of unit charge would form only 110 (or less) droplets, and to compare this with $1/55{,}000 \approx 2 \times 10^{-5}$, since a total of 55,000 tracks were examined in the course of the experiment. This probability is

$$P(r \leq 110) = \sum_{i=0}^{110} \frac{229^i e^{-229}}{i!} \approx 1.6 \times 10^{-18}\,.$$

One can certainly say that the result is significant (i.e. highly improbable if only particles of unit charge exist).

However, in a subsequent article [Adair] the physical assumptions under-lying the experiment were challenged, and in particular the mechanism of formation of droplets. It was pointed out that the phenomenon which is Poisson-distributed is that of elementary scattering events, each of which pro-duces about four droplets. Assuming that each "Poisson" event makes exactly four droplets, and rounding off to even multiples of four, one can now recalcu-late the probability of making only 110 (112), or less, droplets:

$$P(r' \leq 28) = \sum_{i=0}^{28} \frac{57^i e^{-57}}{i!} \approx 6.7 \times 10^{-6}$$

which is much less significant, and demonstrates how sensitive these probabil-ities can be to the exact assumptions involved.

A more refined model for the droplet-formation mechanism would take account of the fact that the elementary scattering events do not each make the same number of droplets, but in fact this number of droplets per event should also be Poisson-distributed. Then one uses not the simple but the compound Poisson distribution. Taking $\lambda\mu = 229$ and $\mu = 4$, one obtains, from the result (2.56)

$$P(r \leq 110) = \sum_{i=0}^{110} \sum_{N=0}^{\infty} \left[\frac{(N\mu)^i e^{-N\mu}}{i!} \cdot \frac{\lambda^N e^{-\lambda}}{N!} \right] \approx 4.7 \times 10^{-5}$$

thus reducing still further the significance of the observation.

4.1.5. *Geometric distribution*

Variable	r, positive integer.
Parameters	p, $0 \leq p \leq 1$.

Probability function

$$P(r) = p(1-p)^{r-1} .$$ (4.13)

Expected value $$E(r) = \frac{1}{p} .$$

Variance $$V(r) = \frac{(1-p)}{p^2} .$$

Skewness $$\gamma_1 = \frac{(2-p)}{(1-p)^{\frac{1}{2}}} .$$

Probability generating function

$$G(Z) = \frac{pZ}{1-(1-p)Z} .$$

Kurtosis $$\gamma_2 = \frac{p^2 - 6p + 6}{1-p} .$$

The geometric distribution gives the probability of having to wait exactly r trials before the occurrence of the first successful event, when the probability of the occurrence of a successful event in a single trial is p.

4.1.6. *Negative binomial distribution*

Variable $\qquad\qquad r,\quad$ positive integer$\geq m$.

Parameters $\qquad\qquad m$, positive integer.

$\qquad\qquad\qquad\qquad p,\quad 0 \leq p \leq 1$.

Probability function

$$P(r) = \binom{r-1}{m-1} p^m (1-p)^{r-m} .$$ (4.14)

Expected value

$$E(r) = \frac{m}{p} .$$

Variance
$$V(r) = \frac{m(1-p)}{p^2}.$$

Skewness
$$\gamma_1 = \frac{2-p}{[m(1-p)]^{\frac{1}{2}}}.$$

Kurtosis
$$\gamma_2 = \frac{p^2 - 6p + 6}{m(1-p)}.$$

Probability generating function

$$G(Z) = \left(\frac{pZ}{1-(1-p)Z}\right)^m.$$

The negative binomial distribution gives the probability of having to wait r trials until m successes have occurred, when p is the probability of the occurrence of a success in a single trial.

The distribution is sometimes used to give the probability of the number of failures occurring while waiting for m successes. In this case the probability function becomes

$$P(s) = \binom{s+m-1}{s} p^m (1-p)^s \qquad (4.15)$$

with

$$E(s) = \frac{m(1-p)}{p}.$$

4.2. Continuous Distributions

4.2.1. *Normal one-dimensional (univariate Gaussian)*

Parameters
$\mu,$ real.

$\sigma,$ positive real number.

Probability density function

$$f(X) = N(\mu, \sigma^2) = \frac{1}{\sigma\sqrt{2\pi}} \exp\left[-\frac{1}{2}\frac{(X-\mu)^2}{\sigma^2}\right]. \qquad (4.16)$$

Cumulative distribution

$$F(X) = \phi\left(\frac{X-\mu}{\sigma}\right) \qquad \text{where} \qquad \phi(Z) = \frac{1}{\sqrt{2\pi}}\int_{-\infty}^{Z} e^{-\frac{1}{2}x^2}\,dx. \qquad (4.17)$$

Expected value

$$E(X) = \mu .$$

Variance

$$V(X) = \sigma^2 .$$

Skewness

$$\gamma_1 = 0 .$$

Kurtosis

$$\gamma_2 = 0 .$$

Characteristic function

$$\phi(t) = \exp\left[it\mu - \frac{1}{2}t^2\sigma^2 \right] .$$

Moments

$$\mu_{2r} = \frac{(2r)!}{2^r (r)!}\sigma^{2r} , \quad \mu_{2r+1} = 0 , \quad r \geq 1 .$$

The most important theoretical distribution in statistics is the Normal probability density function, or Gaussian, usually abbreviated $N(\mu, \sigma^2)$. Its cumulative distribution, Eq. (4.17), is called the *Normal probability integral* or *error function*. One may find in the literature several variations of the definition of the error function.

Note that the standard deviation σ is not the width of the p.d.f. at half the height. The half-width at half-height is 1.176σ. The probability content of various intervals is given below:

$$P\left(-1.64 \leq \frac{X - \mu}{\sigma} \leq 1.64 \right) = 0.90$$

$$P\left(-1.96 \leq \frac{X - \mu}{\sigma} \leq 1.96 \right) = 0.95$$

$$P\left(-2.58 \leq \frac{X - \mu}{\sigma} \leq 2.58 \right) = 0.99$$

$$P\left(-3.29 \leq \frac{X - \mu}{\sigma} \leq 3.29 \right) = 0.999 .$$

The function $N(0, 1)$ is called the *standard* [a] Normal density, and its cumulative function,

$$\phi(X) = \frac{1}{\sqrt{2\pi}} \int_{-\infty}^{X} e^{-\frac{1}{2}t^2} dt \,,$$

is called the *standard Normal distribution*.

A number of important results can be derived for random variables X_i, which are independent and Normally distributed:

(a) Any linear combination of the X_i is also Normal. Suppose that X_1 and X_2 have means μ_1, μ_2 and variances σ_1^2, σ_2^2, respectively. Then aX_1 and bX_2 are independent Normal variables, with characteristic functions

$$\phi_{aX_1}(t) = \exp\left(ita\mu_1 - \frac{1}{2}t^2 a^2 \sigma_1^2 \right)$$

and

$$\phi_{bX_2}(t) = \exp\left(itb\mu_2 - \frac{1}{2}t^2 b^2 \sigma_2^2 \right) .$$

The characteristic function of $Z = aX_1 + bX_2$ is given by

$$\phi_Z(t) = \phi_{aX_1}(t) \cdot \phi_{bX_2}(t) = \exp\left[it(a\mu_1 + b\mu_2) - \frac{1}{2}t^2(a^2\sigma_1^2 + b^2\sigma_2^2) \right].$$

Thus Z is also Normal, with mean $a\mu_1 + b\mu_2$ and variance $a^2\sigma_1^2 + b^2\sigma_2^2$. In the same way one can show that any linear combination

$$Z = \sum_{i=1}^{N} a_i X_i$$

is Normal.

---

[a] A random variable is said to be *standardized* when its expectation is zero and its variance is unity. Thus if X is a random variable with mean μ and variance σ^2, then the standardized form of X is $\dfrac{X - \mu}{\sigma}$.

(b) The *average* or *sample mean*

$$\bar{X} = N^{-1} \sum_{i=1}^{N} X_i \,,$$

and the *sample variance*

$$S^2 = N^{-1} \sum_{i=1}^{N} (X_i - \bar{X})^2 \tag{4.18}$$

are independent if, and only if the X_i have the same Normal distribution (same μ and same σ). This property is unique to the Normal distribution.

(c) If the X_i are *standard* Normal, the probability density is constant on the hypersphere

$$\sum_{i=1}^{N} X_i^2 = \text{constant} \,,$$

since the density is a function of $\sum X^2$ only. This property of radial symmetry is not possessed by any other distribution.

(d) If $\mathbf{X}$ is a vector with components which are independent random variables, not necessarily identically distributed, and if a non-trivial transformation $\mathbf{X} = \mathcal{C}\,\mathbf{Y}$ gives a vector $\mathbf{Y}$ of independent random variables, then each X_i is Normally distributed, the transformation is orthogonal, and each Y_i is Normal.

4.2.2. *Normal many-dimensional (multivariate Gaussian)*

Variables $\mathbf{X}$, k-dimensional real vector.

Parameters $\boldsymbol{\mu}$, k-dimensional real vector.

$\underset{\sim}{V}$, $k \times k$ matrix, positive semi-definite.

Probability density function

$$f(\mathbf{X}) = \frac{1}{(2\pi)^{k/2}|\underset{\sim}{V}|^{\frac{1}{2}}} \exp\left[-\frac{1}{2}(\mathbf{X} - \boldsymbol{\mu})^{\mathrm{T}} \underset{\sim}{V}^{-1} (\mathbf{X} - \boldsymbol{\mu}) \right]. \tag{4.19}$$

Expected values $E(\mathbf{X}) = \boldsymbol{\mu}\,.$

Covariances $\qquad\qquad\qquad\qquad \mathrm{cov}(\mathbf{X}) = \underset{\sim}{V}$

$$V(X_i) = V_{ii}$$

$$\mathrm{cov}(X_i X_j) = V_{ij}, \quad \text{the } (i,j)^{\text{th}} \text{ element of } \underset{\sim}{V} .$$

Characteristic function

$$\phi(\mathbf{t}) = \exp\left[it\boldsymbol{\mu} - \frac{1}{2}\mathbf{t}^{\mathrm{T}} \underset{\sim}{V} \mathbf{t} \right].$$

In generalizing the Normal distribution to many dimensions, it is natural to look for a density function which has, as its exponent, a quadratic form in the component variables:

$$f(\mathbf{X}) \propto \exp\left[-\frac{1}{2}\sum_{i=1}^{k}\sum_{j=1}^{k} a_{ij} \left(\frac{X_i - \mu_i}{\sigma_i} \right)\left(\frac{X_j - \mu_j}{\sigma_j} \right) \right].$$

As in the one-dimensional case, μ_i and σ_i are location and scale parameters which may be removed by a standardizing transformation. Let us use matrix notation, and write this density function as

$$f(X) \propto \exp\left[-\frac{1}{2}(\mathbf{X} - \boldsymbol{\mu})^{\mathrm{T}} \underset{\sim}{A} (\mathbf{X} - \boldsymbol{\mu}) \right].$$

The condition that this function $f(\mathbf{x})$ is a density function imposes the condition that $\underset{\sim}{A}$ be positive-definite. Determining the constant of proportionality from the condition

$$\int f(\mathbf{X})d\mathbf{X} = 1,$$

the probability density Eq. (4.19) follows.

The *covariance matrix* of $\mathbf{X}$ is then conveniently defined as

$$\underset{\sim}{V} = E\left[(\mathbf{X} - \boldsymbol{\mu})(\mathbf{X} - \boldsymbol{\mu})^{\mathrm{T}} \right] = \underset{\sim}{A}^{-1}$$

$$= \begin{bmatrix} \sigma_1^2 & \rho_{12}\sigma_1\sigma_2 \cdots & \rho_{1k}\sigma_1\sigma_k \\ \rho_{12}\sigma_1\sigma_2 & \sigma_2^2 \cdots & \vdots \\ \vdots & \vdots & \vdots \\ \rho_{1k}\sigma_1\sigma_k & \cdots & \sigma_k^2 \end{bmatrix}$$

where ρ_{ij} is the correlation coefficient between components X_i and X_j

$$\rho_{ij} = E\left[(X_i - \mu_i)(X_j - \mu_j)\right]/\sigma_i\sigma_j .$$

In two dimensions, X and Y, the density (4.19) becomes

$$f(X,Y) = \frac{1}{2\pi\sigma_X\sigma_Y\sqrt{(1-\rho^2)}}$$

$$\times \exp\left[-\frac{1}{2(1-\rho^2)}\left\{\frac{(X-\mu_X)^2}{\sigma_X{}^2} - 2\rho\frac{(X-\mu_X)(Y-\mu_Y)}{\sigma_X\sigma_Y} + \frac{(Y-\mu_Y)^2}{\sigma_Y{}^2}\right\}\right].$$

(4.20)

Equation (4.20) is explicitly parametrized in terms of its expectations μ_X, μ_Y, standard deviations σ_X, σ_Y and correlation ρ. It follows that the *conditional density*

$$f(X|Y) = \frac{1}{\sqrt{2\pi}\sigma_X\sqrt{1-\rho^2}}\exp\left[-\frac{1}{2\sigma_X{}^2(1-\rho^2)}\left\{X - \left(\mu_X + \frac{\rho\sigma_X}{\sigma_Y}(Y-\mu_Y)\right)\right\}^2\right]$$

is Normal, with mean $\mu_X + \rho(\sigma_X/\sigma_Y)(Y - \mu_Y)$ and variance $\sigma_X^2(1 - \rho^2)$.

If **X** is Normal and $\underset{\sim}{V}$ is non-singular, the quantity

$$(\mathbf{X} - \boldsymbol{\mu})^{\mathrm{T}}\underset{\sim}{V}^{-1}(\mathbf{X} - \boldsymbol{\mu})$$

is called the *covariance form* of **X**. This has a χ^2-distribution with k degrees of freedom.

The many-dimensional Normal distribution is interesting for many reasons:

(a) it is a function only of the means, variances, and two-variable correlations;
(b) the p.d.f. has a many-dimensional "bell-shaped" form;
(c) contours of constant probability density are given by the quadratic form,

$$(\mathbf{X} - \boldsymbol{\mu})^{\mathrm{T}}\underset{\sim}{V}^{-1}(\mathbf{X} - \boldsymbol{\mu}) = \text{constant};$$

(d) any section through the distribution, say at $X_i = $ constant, gives again a Normal distribution in $k-1$ dimensions of the form (4.19), with covariance matrix $\underset{\sim}{V}_{k-1}$ obtained by removing the ith row and column of $\underset{\sim}{V}^{-1}$ and inverting the resultant submatrix;
(e) any projection into a lower space gives a marginal distribution which is again Normal, of the form (4.19), with covariance matrix obtained by deleting appropriate rows and columns of $\underset{\sim}{V}$. In particular, the marginal distribution of X_i is $N(\mu_i, \sigma_i^2)$;

(f) a set of variables, each of which is a linear function of a set of Normal variables, has itself a many-dimensional Normal distribution;

(g) the average vector $\bar{\mathbf{X}}$, and *sample covariance* s_{ij} of a set of N independent observations $\mathbf{X}$ are defined as

$$\bar{X}_i = \frac{1}{N} \sum_{l=1}^{N} X_{il} \tag{4.21}$$

and

$$s_{ij} = \frac{1}{N-1} \sum_{l=1}^{N} \left(X_{il} - \bar{X}_i \right) \left(X_{jl} - \bar{X}_j \right),$$

where $X_{i\ell}$ is the ith component of the ℓth vector observed. $\bar{\mathbf{X}}$ and the matrix (s_{ij}) are independently distributed if, and only if, the parent distribution of the $\mathbf{X}_k$ is many-dimensional Normal.

4.2.3. *Chi-square distribution*

Variable X, positive real number.

Parameter N, positive integer ["degrees of freedom"].

Probability density function $f(X) = \dfrac{\dfrac{1}{2}\left(\dfrac{X}{2}\right)^{(N/2)-1} e^{-X/2}}{\Gamma\left(\dfrac{N}{2}\right)}.$ $\tag{4.22}$

Expected value $E(X) = N.$

Variance $V(X) = 2N.$

Skewness $\gamma_1 = 2\sqrt{\dfrac{2}{N}}.$

Kurtosis $\gamma_2 = \dfrac{12}{N}.$

Characteristic function

$$\phi(t) = (1 - 2it)^{-N/2}. \tag{4.23}$$

Illustration   Fig. 4.1.

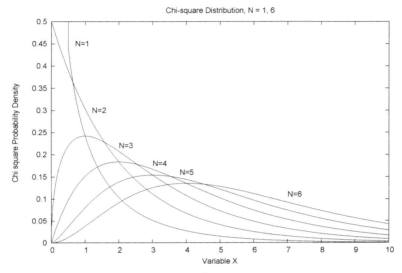

(a) Chi-square distributions $\chi^2(N)$ with $N = 1, 2, 3, 4, 5, 6$.

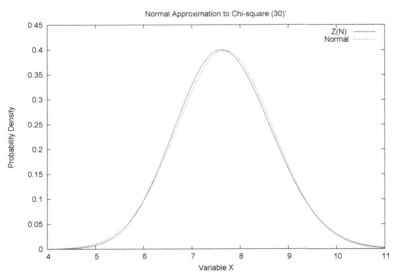

(b) Solid line: $\sqrt{2x}\chi_x^2(30)$ versus $\sqrt{2x}$;   Dotted line: $N(\sqrt{2 \times 30 - 1}, 1)$.

Fig. 4.1.

Suppose that $X_1, \ldots, X_N$ are independent, *standard Normal* variables, $N(0,1)$. Then the sum of squares

$$X_{(N)}^2 = \sum_{i=1}^{N} X_i^2 \qquad (4.24)$$

is said to have a *chi-square distribution* $\chi^2(N)$, with N degrees of freedom.

To derive the distribution, consider the characteristic function of the square of a standard Normal variable, X.

$$E\left(e^{itX^2}\right) = \int_{-\infty}^{\infty} e^{itX^2} \frac{1}{\sqrt{2\pi}} e^{-\frac{1}{2}X^2} dX$$

$$= \int_{-\infty}^{\infty} \frac{1}{\sqrt{2\pi}} e^{-\frac{1}{2}X^2(1-2it)} dX = \frac{1}{\sqrt{(1-2it)}}. \qquad (4.25)$$

From Eqs. (2.46), (4.24) and (4.25) it follows that the characteristic function of $X_{(N)}^2$ is given by Eq. (4.23), which has the p.d.f. given in Eq. (4.22).

If $X_{(N)}^2$ and $X_{(M)}^2$ have independent χ^2 distributions with N and M degrees of freedom, respectively, then the sum

$$X_{(K)}^2 = X_{(N)}^2 + X_{(M)}^2$$

has a χ^2 distribution with $K = N + M$ degrees of freedom. This is obvious from the definition (4.24).

The function

$$R_N = \sqrt{X_{(N)}^2}$$

has the $\chi(N)$ distribution, with density function

$$P(R_N) = \frac{\left(\frac{1}{2}\right)^{N/2-1} R_N^{N-1} e^{-\frac{1}{2}R_N^2}}{\Gamma\left(\frac{1}{2}N\right)}.$$

Asymptotically, the $\chi^2(N)$ and $\chi(N)$ distributions are both approximately Normal for $N > 30$, see Fig. 4.1b.

More specifically it can be shown that for large N the quantities

$$Z_N = \frac{X_{(N)}^2 - N}{\sqrt{2N}}$$

and

$$Z'_N = \sqrt{2X^2_{(N)}} - \sqrt{2N - 1}$$

are standard Normal, $N(0, 1)$.

When the independent variables X_i are Normal, $N(\mu_i, 1)$ (not standard Normal, as in the previous case), then $X^2_{(N)}$ of Eq. (4.24) has a *non-central* χ^2 *distribution* with N degrees of freedom, denoted $\chi^2(N, \Delta)$ where

$$\Delta = \sum_{i=1}^{N} \mu_i^2$$

is called the non-centrality parameter.

The characteristic function of $\chi^2(N, \Delta)$ is

$$E\left(e^{itX_{(N)}}\right) = \frac{\exp\left[\dfrac{it\Delta}{(1 - 2it)}\right]}{(1 - 2it)^{N/2}}.$$

Note that $\mu_1, \ldots, \mu_N$ enter the distribution only through Δ. If $\Delta = 0$ (i.e. $\mu_i = 0$, $i = 1, \ldots, N$), we recover the central χ^2 distribution.

One can approximate $\chi^2(N, \Delta)$ by a variable proportional to a central chi square, say $Y = \beta X^2_{(M)}$, choosing β and M so that Y has the same mean and variance as $\chi^2(N, \Delta)$. This is true for

$$\beta = 1 + \frac{\Delta}{N + \Delta},$$

$$M = N + \frac{\Delta^2}{N + 2\Delta} = \frac{(N + \Delta)^2}{N + 2\Delta}.$$

4.2.4. *Student's t-distribution*

Variable t, real number.

Parameter N, positive integer.

Probability density function

$$f(t) = \frac{\Gamma\left(\dfrac{N + 1)}{2}\right)}{\sqrt{N\pi}\,\Gamma\left(\dfrac{N}{2}\right)} \cdot \frac{1}{\left(1 + \dfrac{t^2}{N}\right)^{(N+1)/2}}. \tag{4.26}$$

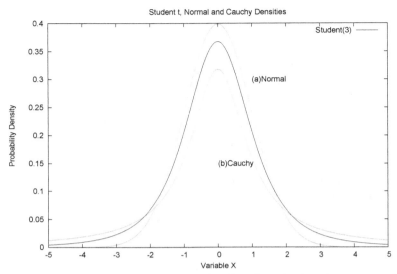

Fig. 4.2. Student's distribution for $N = 3$ (solid line). The dotted curves are (a) the standard Normal distribution $N(0,1)$, or Student's distribution for $N = \infty$, and (b) the Cauchy distribution, or Student's distribution for $N = 1$.

Expected value
$$E(t) = 0\,.$$

Variance
$$V(t) = \frac{N}{N-2}\,,\quad N > 2\,.$$

Skewness
$$\gamma_1 = 0\,.$$

Kurtosis
$$\gamma_2 = \frac{6}{N-4}\,,\quad N > 4\,.$$

Moments
$$\mu_{2r} = \frac{N^r \Gamma\left(r + \frac{1}{2}\right)\Gamma\left(\frac{N}{2} - r\right)}{\Gamma\left(\frac{1}{2}\right)\Gamma\left(\frac{N}{2}\right)}\,,\quad 2r < N$$

$$\mu_{2r+1} = 0\,,\quad 2r < N\,.$$

Moments of order greater than N do not exist.

Illustration
Fig. 4.2.

In Section 4.2.1, we have seen that if the measurements $X_1, X_2, \ldots, X_N$, are Normal $N(\mu, \sigma^2)$, then $\sqrt{N}(\bar{X} - \mu)$ is also Normal, $N(0, \sigma^2)$. Further

$$s^2 = \frac{1}{N-1} \sum_{i=1}^{N} (X_i - \bar{X})^2 \tag{4.27}$$

is distributed independently of $\bar{X}$. It can be shown that $(N-1)s^2/\sigma^2$ is distributed as a χ^2 variable with $(N-1)$ degrees of freedom.

Frequently one wants to test whether an observed mean $\bar{X}$ is consistent with a theoretical value, μ. The correct *statistic* (as defined in chapter 10) to use for this test is

$$t = \frac{\sqrt{N}(\bar{X} - \mu)}{s} \tag{4.28}$$

which has a *Student's t-distribution*, with $(N-1)$ degrees of freedom.

More generally, a variable Y^2 is said to be a *Normal theory estimate* of σ^2, with k degrees of freedom, if kY^2/σ^2 is distributed as a χ^2 variable with k degrees of freedom. If X is $N(0, \sigma^2)$, independent of Y, then

$$t_k = X/Y$$

has Student's t-distribution with k degrees of freedom.

The Student's distribution is symmetrical around $t = 0$. For $N = 1$ it reduces to the Cauchy distribution (Section 5.2.11) and for $N \to \infty$ it approaches the standard Normal distribution $N(0, 1)$.

From the p.d.f. one can construct confidence limits for μ, or tests of the hypothesis that $\mu = \mu_0$ (See Chapters 9 – 11).

If $X_1, \ldots, X_M$ and $Y_1, \ldots, Y_N$ are distributed as $N(\mu_X, \sigma^2)$ and $N(\mu_Y, \sigma^2)$, respectively, all observations being independent, then the quantity

$$(\bar{X} - \bar{Y}) - (\mu_X - \mu_Y)$$

has the distribution

$$N\left[0, \sigma^2\left(\frac{1}{M} + \frac{1}{N}\right)\right].$$

The *pooled estimate* of σ^2,

$$s^2 = \left[\sum_{i=1}^{M}(X_i - \bar{X})^2 + \sum_{i=1}^{N}(Y_i - \bar{Y})^2\right]/(N + M - 2) \tag{4.29}$$

is a *Normal theory estimate* of σ^2 with $M + N - 2$ degrees of freedom. Now s^2 and $\bar{X} - \bar{Y}$ are independent [since neither sum in Eq. (4.29) depends on both $\bar{X}$ and $\bar{Y}$]. Therefore the quantity

$$t = \frac{(\bar{X} - \bar{Y}) - (\mu_X - \mu_Y)}{\sqrt{\dfrac{1}{M} + \dfrac{1}{N}} \; s} \tag{4.30}$$

has a t_{M+N-2} distribution. The statistic (4.30) provides a test of whether the observed difference in mean,

$$\bar{X} - \bar{Y}$$

is consistent with a theoretical difference

$$\delta \equiv \mu_X - \mu_Y \, .$$

In particular, if $\delta = 0$, one can test if $\mu_X = \mu_Y$.

This test is explicitly dependent on the assumption above, that the two distributions have the same variance, σ^2. Without the assumption the test becomes more complicated.

If X is $N(\Delta\sigma, \sigma^2)$, and kY^2/σ^2 is $\chi^2(k)$ (i.e. Y^2 is a *Normal theory estimate* of σ^2), then $t' = X/Y$ has a *non-central t-distribution* with k degrees of freedom and non-centrality parameter Δ, provided X, Y are independent.

Clearly

$$P\left(t' = \frac{X}{Y} < q\right) = P(X - qY < 0) \, . \tag{4.31}$$

Now X was defined to be Normal, and Y is fairly Normal, if $k > 20$. Also, the expectation and variance of $X - qY$ are approximately

$$E(X - qY) \simeq \sigma(\Delta - q)$$

$$V(X - qY) \simeq \sigma^2 \left(1 + \frac{q^2}{k}\right) \, .$$

Hence the probability in Eq. (4.31) is approximately

$$P\left(\frac{X}{Y} < q\right) \simeq \Phi\left(\frac{q - \Delta}{\sqrt{1 + \dfrac{q^2}{k}}}\right)$$

where Φ denotes the Normal probability integral.

4.2.5. *Fisher–Snedecor F and Z distributions*

The F-distribution:

Variable F, positive real number.

Parameters ν_1, ν_2, positive integers (degrees of freedom).

Probability density function

$$f(F) = \frac{\nu_1^{\nu_1/2} \nu_2^{\nu_2/2} \Gamma\left(\dfrac{\nu_1 + \nu_2}{2}\right)}{\Gamma(\nu_1/2)\Gamma(\nu_2/2)} \cdot \frac{F^{\frac{1}{2}\nu_1 - 1}}{(\nu_1 + \nu_2 F)^{\frac{1}{2}(\nu_1 + \nu_2)}} \cdot \qquad (4.32)$$

Expected value

$$E(F) = \frac{\nu_2}{\nu_2 - 2}, \quad \nu_2 > 2.$$

Variance

$$V(F) = \frac{2\nu_2^2(\nu_1 + \nu_2 - 2)}{\nu_1(\nu_2 - 2)^2(\nu_2 - 4)}, \quad \nu_2 > 4.$$

Let s_1^2, s_2^2 be independent and Normal theory estimates of σ_1^2 and σ_2^2, with ν_1 and ν_2 degrees of freedom, respectively, that is

$$\frac{\nu_1 s_1^2}{\sigma_1^2} \qquad \text{is distributed as } \chi^2(\nu_1),$$

and

$$\frac{\nu_2 s_2^2}{\sigma_2^2} \qquad \text{is distributed as } \chi^2(\nu_2),$$

and s_1^2 and s_2^2 are defined analogously to s^2 in Eq. (4.27). Then the quantity

$$F = \frac{s_1^2}{\sigma_1^2} \cdot \frac{\sigma_2^2}{s_2^2}$$

is said to have the F distribution with (ν_1, ν_2) degrees of freedom (or variance-ratio distribution). The distribution is positively skew, and tends to Normality as $\nu_1, \nu_2 \to \infty$, but only slowly $(\nu_1, \nu_2 > 50)$.

The Z-distribution:

The variable $Z = \frac{1}{2}\ln F$ has a distribution much closer to Normal. Its properties are:

Variable Z, real number.

Parameters ν_1, ν_2, positive integers.

Probability density function

$$f(Z) = \frac{2\nu_1^{\frac{1}{2}\nu_1}\nu_2^{\frac{1}{2}\nu_2}}{B\left(\frac{1}{2}\nu_1, \frac{1}{2}\nu_2\right)}\frac{e^{\nu_1 Z}}{\left[\nu_1 e^{2Z} + \nu_2\right]^{\frac{1}{2}(\nu_1+\nu_2)}}\,. \tag{4.33}$$

Expected value

$$E(Z) = \frac{1}{2}\left[\frac{1}{\nu_2} - \frac{1}{\nu_1}\right] - \frac{1}{6}\left[\frac{1}{\nu_1^2} - \frac{1}{\nu_2^2}\right]\,.$$

Variance

$$V(Z) = \frac{1}{2}\left[\frac{1}{\nu_1} + \frac{1}{\nu_2}\right] + \frac{1}{2}\left[\frac{1}{\nu_1^2} + \frac{1}{\nu_2^2}\right] + \frac{1}{3}\left[\frac{1}{\nu_1^3} + \frac{1}{\nu_2^3}\right]\,.$$

Characteristic function

$$\phi(t) = \left(\frac{\nu_2}{\nu_1}\right)^{\frac{1}{2}it}\frac{\Gamma\left[\frac{1}{2}(\nu_2 - it)\right]\Gamma\left[\frac{1}{2}(\nu_1 + it)\right]}{\Gamma\left(\frac{1}{2}\nu_1\right)\Gamma\left(\frac{1}{2}\nu_2\right)}\,.$$

Apart from the question of approximation by the Normal distribution, modern practice is in favour of using the simpler statistic F.

With $\nu_1 = 1$, the F distribution specializes to $F = t_{\nu_2}^2$. As $\nu_2 \to \infty$, F approaches

$$\frac{1}{\nu_1}\chi^2(\nu_1)\,.$$

A function equivalent to F for tests is the function

$$U = \frac{\nu_1 F}{\nu_2 + \nu_1 F}\,,$$

which is a monotonic function of F, and has a beta distribution [Section 4.2.8].

4.2.6. *Uniform distribution*

Variable X,   real number.

Parameters a, b, $a < b$.

Probability density function

$$f(X) = \frac{1}{b - a}, \quad a \le X \le b \tag{4.34}$$

$$f(X) = 0, \qquad \text{otherwise.}$$

Expected value $E(X) = \dfrac{a + b}{2}$.

Variance $V(X) = \dfrac{(b - a)^2}{12}$.

Skewness $\gamma_1 = 0$.

Kurtosis $\gamma_2 = -1.2$.

Characteristic function

$$\phi(t) = \frac{e^{itb} - e^{ita}}{it(b - a)}.$$

The rounding-off errors in arithmetical calculations are *uniformly* distributed.

4.2.7. *Triangular distribution*

Variable X,   real number.

Parameters $\Gamma > 0, \mu$,   real numbers.

Probability density function

$$f(X) = \frac{1 - |X - \mu|}{\Gamma}, \quad \mu - \Gamma < X < \mu + \Gamma,$$

$$= 0, \qquad \text{otherwise.} \tag{4.35}$$

Expected value	$E(X) = \mu$.
Variance	$V(X) = \Gamma^2/6$.
Skewness	$\gamma_1 = 0$.
Kurtosis	$\gamma_2 = -0.6$.

Characteristic function

$$\phi(t) = \frac{2e^{it\mu}}{t^2\Gamma^2}(1 - \cos t\Gamma).$$

Examples

(i) The sum of two numbers, each selected independently from a uniform distribution, will have a triangular distribution of mean and variance equal respectively to twice the mean and variance of the uniform distribution.

(ii) The momentum distribution of secondary particles from a synchrotron beam is often approximately *triangular* with central momentum μ and full width Γ at half-height. The height of the distribution is $1/\Gamma$, and its base extends from $\mu - \Gamma$ to $\mu + \Gamma$.

4.2.8. *Beta distribution*

Variable	X, real number.
Parameters	n, m, positive integers.

Probability density function

$$f(X) = \frac{\Gamma(n + m)}{\Gamma(n)\Gamma(m)} X^{m-1}(1 - X)^{n-1}, \quad 0 \le X \le 1.$$

$$= 0, \quad \text{otherwise} \tag{4.36}$$

Expected value

$$E(X) = \frac{m}{m + n}.$$

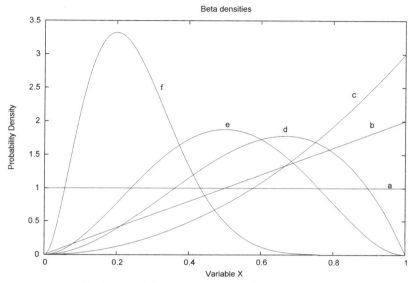

Fig. 4.3. Beta distributions with (a) $n = 1$, $m = 1$, (b) $n = 1$, $m = 2$, (c) $n = 1$, $m = 3$, (d) $n = 2$, $m = 3$, (e) $n = 3$, $m = 3$, (f) $n = 9$, $m = 3$.

Variance
$$V(X) = \frac{mn}{(m+n)^2(m+n+1)}.$$

Skewness
$$\gamma_1 = \frac{2(n-m)\sqrt{(m+n+1)}}{(m+n+2)\sqrt{mn}}.$$

Kurtosis
$$\gamma_2 = \frac{3(m+n+1)\left[2(m+n)^2 + mn(m+n-6)\right]}{mn(m+n+2)(m+n+3)} - 3.$$

Illustration   Fig. 4.3

The *beta distribution* is a basic distribution in statistics for variables bounded at both sides, for example, $0 \leq X \leq 1$. Examples of its use are as the distribution of the proportion of a population located between the lowest and highest values in a sample, and the description of elapsed times to task completion in PERT analyses.

4.2.9. *Exponential distribution*

Variable X, positive real number.

Parameter λ, positive real number.

Probability density function

$$f(X) = \lambda e^{-\lambda X} . \tag{4.37}$$

Expected value

$$E(X) = \frac{1}{\lambda} .$$

Variance

$$V(X) = \frac{1}{\lambda^2} .$$

Skewness

$$\gamma_1 = 2 .$$

Kurtosis

$$\gamma_2 = 6 .$$

Characteristic function

$$\phi(t) = \left(1 - \frac{it}{\lambda} \right)^{-1} .$$

Consider events occurring randomly in time, with an average of λ events per unit time. From the Poisson distribution, the probability of N events occurring in a time interval t is

$$P(N|t) = \frac{1}{N!} (\lambda t)^N e^{-\lambda t} .$$

Then the probability of no events in time t is the *exponential distribution* $e^{-\lambda t}$.

Consider the time interval, Z, between two successive events. In this interval, no events occur. Therefore for fixed X,

$$P(Z > X) = e^{-\lambda X} ,$$

with the cumulative exponential distribution

$$F(X) = P(Z \leq X) = 1 - e^{-\lambda X} .$$

The exponential distribution has "no memory": if no event has occurred up to time y, the probability of no event in a subsequent time x is independent of y. For fixed y

$$P(X > x + y | X > y) = \frac{e^{-\lambda(x,y)}}{e^{-\lambda y}}$$

$$= e^{-\lambda x} = P(X > x).$$

The exponential distribution is applicable to the time between arrival of particles at a counter.

Distributions closely related to the exponential distribution are the *hyperexponential* and the *Erlangian* distributions. The hyperexponential distribution gives the time between events in a process where the events are generated with probability p from one exponential distribution rate λ_1, and with probability $(1 - p_1)$ from another exponential distribution rate λ_2. The probability density function is then

$$f(X) = p_1 \lambda_1 e^{-\lambda_1 X} + (1 - p_1)\lambda_2 e^{-\lambda_2 X}.$$

Thus the hyperexponential distribution is applicable when there is a mixture of exponential processes.

The Erlangian k-distribution gives the time between every k^{th} event of an exponential process; it is the sum distribution of k random variables exponentially distributed with rate $k\mu$. The probability density function is

$$f(X) = \frac{k\mu}{(k-1)!}(k\mu X)^{k-1} e^{-k\mu X}.$$

The hyperexponential distributions correspond to having exponential processes operation in parallel, while the Erlangian distributions give the results of having exponential distributions in series.

4.2.10. *Gamma distribution*

Variable X, real positive number.

Parameters a, b, real positive numbers.

Probability density function

$$f(X) = \frac{a(aX)^{b-1}e^{-aX}}{\Gamma(b)}. \tag{4.38}$$

Expected value

$$E(X) = \frac{b}{a} .$$

Variance

$$V(X) = \frac{b}{a^2} .$$

Skewness

$$\gamma_1 = \frac{2}{\sqrt{b}} .$$

Kurtosis

$$\gamma_2 = \frac{6}{b} .$$

Characteristic function

$$\phi(t) = \left(1 - \frac{it}{a} \right)^{-b} .$$

The *gamma distribution* is a basic statistical tool for describing variables bounded at one side, for example $0 \le X < \infty$. Note that a is a scale parameter only.

The gamma distribution is applicable to the time between recalibration of an instrument that needs recalibration after k uses, and to the time to failure for a system with stand-by components.

The exponential, Erlangian, and chi square distributions are all special cases of the gamma distribution.

4.2.11. *Cauchy, or Breit–Wigner, distribution*

Variable X,   real number.

Parameters   None, except for possible location and scale parameters.

Probability density function

$$f(X) = \frac{1}{\pi} \frac{1}{1 + X^2} . \tag{4.39}$$

Expected value $E(X)$ is undefined.

Variance, Skewness and Kurtosis

are divergent.

Characteristic function

$$\phi(t) = e^{-|t|} .$$

The *Cauchy distribution* represents a pathological case, since, as we have noted in Section 2.4.1, the expectation is undefined, and all other moments diverge. The distribution is identical to the physically important Breit–Wigner distribution, usually written as

$$f(X) = \frac{1}{\pi}\left(\frac{\Gamma}{\Gamma^2 + (X - X_0)^2}\right).\tag{4.40}$$

The parameters X_0 and Γ represent location and scale parameters respectively, being the mode and half-width at half-height respectively. However it should be noted that the mean and moments of the distribution are still undefined.

We shall see in Section 4.3.3 how the Cauchy distribution becomes manageable by truncation.

Note that the properties of the distribution (4.40) may change completely if Γ has some X-dependence.

4.2.12. *Log-Normal distribution*

Variable X, positive real number.

Parameters μ, real number.

                  σ, positive real number.

Probability density function

$$f(X) = \frac{1}{\sqrt{2\pi}\sigma} \cdot \frac{1}{X} \exp\left[-\frac{1}{2\sigma^2}(\ln X - \mu)^2\right].\tag{4.41}$$

Expected value

$$E(X) = e^{(\mu - \frac{1}{2}\sigma^2)}.$$

Variance

$$V(X) = e^{(2\mu - \sigma^2)}(e^{\sigma^2} - 1).$$

Skewness

$$\gamma_1 = \sqrt{(e^{\sigma^2} - 1)}(e^{\sigma^2} + 2).$$

Kurtosis

$$\gamma_2 = (e^{\sigma^2} - 1)(e^{3\sigma^2} + 3e^{2\sigma^2} + 6e^{\sigma^2} + 6).$$

Illustration   Fig. 4.4.

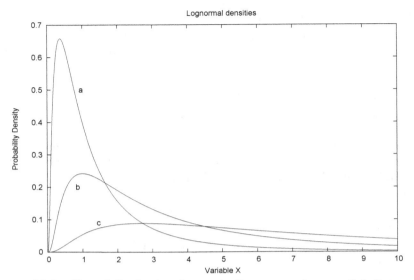

(a) Log-Normal distributions for $\sigma = 1$: For curves a, b, c, $\mu = 0,\ 1,\ 2$.

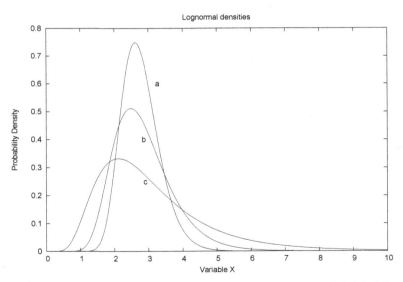

(b) Log-Normal distributions for $\mu = 1$: For curves a, b, c, $\sigma = 0.2,\ 0.3,\ 0.5$.

Fig. 4.4.

The *log-Normal distribution* represents a random variable whose logarithm follows a Normal distribution. It provides a model for the error of a process involving many small multiplicative errors (from the Central Limit Theorem). It is also appropriate when the value of an observed variable is a random proportion of the previous observation.

4.2.13. *Extreme value distribution*

Variable X, real number.

Parameters μ, real number.

               σ, positive real number.

Probability density function

$$f(X) = \frac{1}{\sigma} \exp\left[\pm\frac{\mu - X}{\sigma} - e^{\pm(\mu-X)/\sigma}\right]. \tag{4.42}$$

Cumulative distribution

$$F(X) = \exp\left[-e^{\pm(\mu-X)/\sigma}\right].$$

Expected value

$$E(X) = \mu \pm 0.5776\sigma.$$

Variance

$$V(X) = 1.645\sigma^2.$$

Skewness

$$\gamma_1 = \pm 1.14.$$

Kurtosis

$$\gamma_2 = 2.4.$$

Illustration   Fig. 4.5.

The *extreme value distribution* gives the limiting distribution for the largest (+ sign) or smallest (− sign) element of a set of independent observations from a distribution of exponential type (Normal, gamma, exponential, etc. [NBS]).

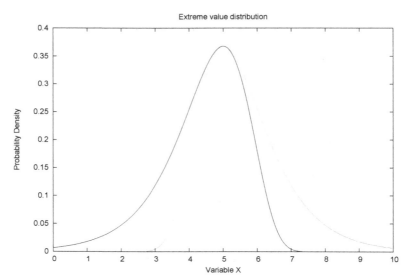

Fig. 4.5. Extreme value distributions for $\mu = 5$, $\sigma = 1$. The dotted curve (solid curve) gives the distribution of the largest (smallest) element.

Consider, as an example, an experiment which results in a large number N of independent histograms. Suppose that one describes the deviation of each histogram from a known theory by one number, such as the mean square deviation of the bins. Let us denote this number by χ^2. A method to estimate the significance of the occurrence of one particularly large value of χ^2 would be to consider the probability distribution of the extreme value of N χ^2 distributions.

Assume, for simplicity, that all the histograms have the same number of bins, and that the degree of freedom r of the N $\chi^2(r)$ distributions is large enough ($r \geq 30$) to be approximated by a Normal law:

$$X = \sqrt{2\chi^2} - \sqrt{2r - 1}.$$

The random variable X is Normal with mean zero, and the largest value of X is then described by the distribution (4.12), with

$$\mu = \sqrt{2 \log N}, \quad \sigma = \mu^{-1}.$$

[Kendall, I, p. 334]. The advantage of this test over the more sophisticated tests in Chapter 11, is that one uses only one value and thus it is suitable for desk calculations.

4.2.14. *Weibull distribution*

Variable X, positive real number.

Parameters η, σ, positive real numbers.

Probability density function

$$f(X) = \frac{\eta}{\sigma}\left(\frac{X}{\sigma}\right)^{\eta-1} \exp\left[-\left(\frac{X}{\sigma}\right)^{\eta}\right], \quad X \geq 0. \tag{4.43}$$

Expected value

$$E(X) = \sigma\Gamma\left(\frac{1}{\eta} + 1\right).$$

Variance

$$V(X) = \sigma^2\left\{\Gamma\left(\frac{2}{\eta} + 1\right) - \left[\Gamma\left(\frac{1}{\eta} + 1\right)\right]^2\right\}.$$

The *Weibull distribution* describes the time to failure of a wide diversity of complex mechanisms. The exponential distribution is a special case, when the probability of failure at time t is independent of t.

The Weibull form also gives the distribution of the minimum of a set of observations from distributions bounded below.

4.2.15. *Double exponential distribution*

Variable X, real number.

Parameters μ, real number.

                  λ, positive real number.

Probability density function

$$f(X) = \frac{\lambda}{2}e^{-\lambda|X-\mu|}. \tag{4.44}$$

Expected value

$$E(X) = \mu.$$

Variance

$$V(X) = 2/\lambda^2.$$

Skewness

$$\gamma_1 = 0.$$

Kurtosis

$$\gamma_2 = 3.$$

Characteristic function

$$\phi(t) = it\mu + \frac{\lambda^2}{\lambda^2 + t^2}.$$

The *double exponential distribution*, sometimes called the *Laplace distribution*, is a symmetric function whose tails fall off less sharply than the Gaussian, but faster than the Cauchy. Unlike the latter two functions, however, the double exponential has a cusp (discontinuous first derivative) at $X = \mu$.

The distribution is interesting from the point of view of parameter estimation since it is the symmetric function for which the sample median is the best estimator (in fact a sufficient statistic — see Section 5.3) for the mean μ.

4.2.16. *Asymptotic relationships between distributions*

Some of the distributions discussed earlier converge asymptotically to certain other types of distributions under various limiting conditions. Rather than discuss each case in detail, we summarize these asymptotic properties in the figure below.

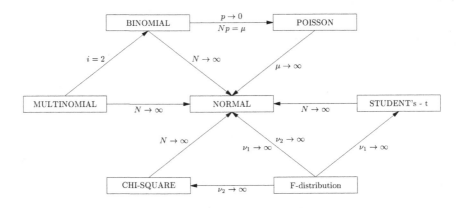

Fig. 4.6. The limiting conditions on the distribution parameters under which the indicated convergence occurs are shown on the arrows.

4.3. Handling of Real Life Distributions

In the preceding paragraphs, the most important ideal distributions have been presented with some useful properties. It is highly dangerous to assume that all physical distributions have some idealized shape. Although ideal distributions are physically realizable under certain conditions, real life is unfortunately not so simple, and one more often has to do with combinations or distortions of these ideal distributions. In this section we show how some typical "real distributions" can be handled using the technique developed above for ideal distributions.

4.3.1. *General applicability of the Normal distribution*

The question naturally arises as to what extent a series of measurements can be considered *a priori* to be Normally distributed. It is often assumed, for example, that a person making repeated measurements of the distance between two fixed points will obtain a set of measurements that is Normally distributed with mean equal to the "true" distance and width given by the precision of the method used.

Mathematicians sometimes assume that measurement errors have been observed in practice, and have been found to be Normally distributed. On the other hand, experimentalists often assume that the distribution of measurement errors is mathematically proved to be Normally distributed, without checking whether the conditions of the Central Limit Theorem are satisfied. In fact, there are suggestions from both sides (mathematical and experimental) that Normality is at least a good approximation to the actual distribution. Therefore, Normality is assumed because it is simple to use (many mathematically simple properties can be demonstrated only for Normal distributions) and because it is, to some extent, empirically supported. Two important cases are cited here, one of which is certainly Normal and the other certainly not Normal.

(i) The sum or the mean value of many independent observations. In this case the Central Limit Theorem can be invoked (see Section 3.3.2) and one obtains the result that the sum or mean of several measurements will be (asymptotically) Normally distributed, even if the parent distribution is not Normal (they need not even have the same parent distribution). Suppose for example that in a study of the π mesons emerging from a target placed in a high-energy beam one wishes to measure the mean value of the π momentum. It is known that in general a π coming from a given reaction will not have a Gaussian momentum spectrum, and furthermore the total momentum spectrum of π from all reactions will not have a Normal distribution either. However, taking the mean value of momentum for N consecutive π as the random variable, repeated observations of this kind will yield a distribution of mean π momentum that becomes Normal as N becomes large.

(ii) A set of Normally-distributed measurements, each having the same mean but different variances. In this case, even though each individual measurement is a Normally distributed random variable, the over-all distribution is the sum of Normal distributions of different widths, which is never Normal. An example is the measurement in a bubble chamber experiment of the mass of a narrow resonance or stable particle. If the mass determination from one event is assumed to be Normally distributed with standard error ΔM, then the over-all experimental mass spectrum obtained will be Normal only if events are selected so that they all have the same ΔM. In the more usual case where there is a spread of ΔM values (often a large spread) the over-all distribution may be quite far from Normal. This important case is described in more detail in Section 4.3.4 below.

To summarize the two examples above, one can say that under certain general conditions the distribution of the sum of independent random variables is Normal, but the sum of Normal distributions of different variances (an ideogram) is not Normal.

4.3.2. *Johnson empirical distributions*

It is often useful to represent data by an empirical distribution. For this purpose, a free-hand fit may be adequate, but greater objectivity, possible automatic analysis, and parametrization are useful benefits from fitting a mathematical form to observed data.

The Johnson families of distributions provide a wide range of possible shapes. They take the form of transformations of the Normal distribution. Let X be the random variable whose distribution is to be represented empirically. The general form of the transformation is then given by

$$Z = \gamma + \eta g(X; \varepsilon, \lambda),$$

where Z is a standard Normal variable, and $\eta, \lambda > 0$, $-\infty < \gamma, \varepsilon < \infty$. g is an arbitrary function.

Johnson proposed three alternative forms for the function g, covering the cases where the variable X is unbounded, bounded below, and bounded above and below. These are especially useful because the properties of these distributions, as functions of the parameters, have been studied analytically. For further details the reader is referred to the literature [Hahn].

4.3.3. *Truncation*

The most obvious modification of ideal distributions comes from the fact that the range of possible measurements is never infinite. Usually one measures the variable X between finite limits A and B, so that p.d.f. is modified as follows:

$$f(X)dX \rightarrow \frac{f(X)dX}{\int_A^B f(X)dX} = \frac{f(X)dX}{F(B) - F(A)} \, .$$

Although this is usually a complicating effect, it is sometimes very useful. This happens for the Cauchy or Breit–Wigner distribution. We have seen above that this distribution has no moments if considered over the full infinite range, but if it is truncated at $+A$ and $-A$, one obtains finite normalization, mean, and variance:

$$g(X) = \frac{f(X)}{\int_{-A}^A \frac{1}{\pi}\frac{1}{1+X^2}dX}$$

$$= \frac{1}{2\arctan A} \cdot \frac{1}{1+X^2}$$

$$E(X) = \frac{1}{2\arctan A}\int_{-A}^A \frac{X}{1+X^2}dX = 0$$

$$V(X) = \frac{1}{2\arctan A}\int_{-A}^A \frac{X^2}{1+X^2}dX$$

$$= \frac{A}{\arctan A} - 1 \, .$$

Taking the limit of this variance for $A \rightarrow \infty$, one sees why the expectation does not exist:

$$\lim_{A\to\infty} V(X) = \infty \, .$$

The tails of the Cauchy distribution fall off so slowly that an arbitrarily large variance can be obtained by considering the function far away from the origin. (This is not the case for the Gaussian distribution.)

Truncation is actually only a special case of a more general class of distortions which may be called *detection efficiency*. When all events do not have the same probability of being observed by the apparatus, the ideal p.d.f., $f(X)$, is distorted and becomes

$$g(X) = \frac{\int_{\Omega_Y} f(X)P(Y|X)e(X,Y)dY}{\int_{\Omega_X}\int_{\Omega_Y} f(X)P(Y|X)e(X,Y)dYdX}$$

where Y denotes a set of ancillary variables and $e(X,Y)$ is the probability density for observing an event at (X,Y). We shall defer a detailed discussion of this problem to Section 8.5.

4.3.4. *Experimental resolution*

Another frequent source of distortion of ideal distributions is that of experimental uncertainty in measurements. That is, an event having a true value X will yield a measured value X' with a probability density, sometimes called the resolution function:

$$r(X, X')\,.$$

Then if the true density of X is $f(X)$, the measured density will be

$$g(X') = \int_{\Omega} r(X, X')f(X)dX\,. \tag{4.45}$$

This is fundamentally different from the case of detection efficiency for now the "true" variable X has been integrated out and replaced by a measured variable X'. This can give rise, for example, to a measured value at a point where the ideal density is zero. In such cases, it is clearly essential to take account of this distortion.

As described above in Section 4.3.1 it is often assumed that the resolution function $r(X, X')$ is a Normal distribution

$$r(X, X') = \frac{1}{\sqrt{2\pi}\sigma} \exp\left[-\frac{(X'-X)^2}{2\sigma^2}\right]\,. \tag{4.46}$$

Two important results from the application of this ideal resolution function may be demonstrated easily. If the distribution being measured is $N(\mu, \tau^2)$ and the resolution function is Eq. (4.46) then the resulting measured distribution is

$$g(X') = \frac{1}{\sqrt{2\pi}\sqrt{\sigma^2+\tau^2}} \exp\left[-\frac{(X'-\mu)^2}{2(\sigma^2+\tau^2)}\right] = N(\mu, \sigma^2+\tau^2)\,.$$

That is, the variance of the final Gaussian distribution is the sum of the variances of the original Gaussian distribution and the Gaussian resolution function.

One often encounters the case when the original distribution $f(X)$ is a Cauchy or Breit–Wigner distribution, and the resolution function is normal. The integral (4.45) can then be expressed in terms of the *error function for complex argument* [Abramovitz].

Let $r(X, X')$ be given by Eq. (4.46) and $f(X)$ by the Breit–Wigner distribution

$$f(X) = \frac{1}{\pi} \frac{\Gamma \sqrt{X_0}}{(X - X_0)^2 + \Gamma^2 X_0} ,$$

and make the substitutions

$$x = (X' - X_0)/\sigma\sqrt{2} ,$$
$$y = \Gamma\sqrt{X_0}/\sigma\sqrt{2} ,$$
$$z = x + iy ,$$
$$t = (X' - X)/\sigma\sqrt{2} .$$

Then the integral (4.45) can be written

$$g(X') = \frac{y}{\sqrt{\pi}} \int_{-\infty}^{\infty} dt \frac{y \exp(-t^2)}{(x - t)^2 + y^2} = y\sqrt{\pi} \operatorname{Re} w(z) ,$$

where $w(z)$ is the error function for complex argument

$$w(z) = 2e^{-z^2} \phi(-i\sqrt{2}z) .$$

In the limiting case when σ^2 is small compared to Γ, the resulting shape, $g(X')$, is approximately given by the Breit–Wigner alone. When the opposite is true the resulting distribution is nearly $N(\mu, \sigma^2)$, where μ is the expectation of the truncated Breit–Wigner.

4.3.5. *Examples of variable experimental resolution*

If the *resolution function* for a given measurement is Gaussian but the variance σ^2 is not the same for all measurements in an experiment, then the total resolution function for the experiment is not Gaussian. Let us consider three examples of *variable resolution*.

(i) If the true distribution $f(X)$ is uniform or slowly varying, the effect of variable resolution will be to create peaks at the points where the resolution is best. Although this may be obvious, it is often overlooked since the most usual effect of resolution is to broaden or smooth the measured distribution with respect to the true distribution.

(ii) Suppose that σ depends on some ancillary variables (such as the position of an event in a track chamber) and its distribution can be described by a density function $q(\sigma^2)$. Then the resolution function will be

$$r(X, X') = \int_0^\infty q(\sigma^2) \frac{1}{\sqrt{2\pi\sigma^2}} \exp\left(\frac{(X' - X)^2}{2\sigma^2}\right) d\sigma . \qquad (4.47)$$

In this case, r will be Gaussian only if $q(\sigma^2)$ is a delta-function (i.e. σ^2 is constant). More usually, the apparatus will produce events with different precision, depending on quantities like track length, which may be nearly independent of the quantity we want to measure. Let us first consider a functional form

$$q(\sigma) = \frac{2}{\sqrt{2\pi}\tau\sigma^2} \exp\left(-\frac{1}{2\tau^2\sigma^2}\right) . \qquad (4.48)$$

For large values of σ, the above distribution goes to zero as $1/\sigma^2$, and for small values of σ, there is a reasonably sharp cut-off [$q(\sigma)$ and all its derivatives are zero at $\sigma = 0$]. The resolution function corresponding to the above error distribution is found as follows

$$r(X, X') = \int_0^\infty \frac{2}{2\pi\tau\sigma^3} \exp\left(-\frac{1}{2\tau^2\sigma^2}\right) \exp\left[-\frac{1}{2}\frac{(X - X')}{\sigma^2}\right] d\sigma$$

$$= \frac{1}{\pi} \frac{\tau}{1 + \tau^2(X - X')^2} .$$

This is just the Cauchy or Breit–Wigner distribution, which we have seen above to have no mean and infinite variance if it is not truncated. Since, in practice, this distribution is always truncated, one need not worry about the long tails which give rise to the infinite variance, and it is sufficient to consider whether the behaviour of $q(\sigma)$ is reasonable at small values of σ. If Eq. (4.48) is approximately valid, then the resolution function takes on a Breit–Wigner shape, and this allows one to fit both wide and narrow resonances using the same general shape. Conversely, if a Breit–Wigner shape is used to fit a resonance that is narrow compared with experimental resolution, then the error distribution $q(\sigma)$ is implicitly assumed to be valid for small σ.

(iii) Suppose that $q(\sigma^2)$ in Eq. (4.47) is given by the uniform distribution in $1/\sigma$ between $A \leq 1/\sigma \leq B$.

Let $t = \frac{1}{\sigma}$, then $q(t) = \frac{1}{B-A}\big|_A^B$ and

$$r(X, X') = \frac{1}{(B-A)\sqrt{2\pi}} \int_A^B \exp\left[-\frac{(X-X')^2 t^2}{2}\right] dt$$

$$= \frac{1}{(B-A)\sqrt{2\pi}(X-X')^2}\{\exp[-(X-X')^2 A^2/2] - \exp[-(X-X')^2 B^2/2]\}$$

Setting $X' = 0$, one can find the expectation and variance of $r(X,0)$:

$$E(X) = 0; \quad V(X) = \frac{1}{AB}$$

Thus the variance of this resolution function can be arbitrarily large if the lower limit A becomes arbitrarily small.

Suppose that one wishes to estimate $E(X)$ by taking an average (mean) of several observations from this distribution, with a fixed maximum precision $1/\sigma = B$, but the possibility of discarding the worst measurements with precision in the range $A.le.1/\sigma.lt.A'$. This clearly improves the average precision, but it also reduces the size of the sample (N). The resulting variance of the average will be

$$V(\overline{X}) = \frac{1}{N}V(X) \propto \frac{1}{A'B(B-A')}.$$

The minimum variance then results when $A' = B/2$. Therefore, with this error distribution, if it is desired to estimate the mean by an unweighted average, the minimum variance is obtained by not using any events with precision $(1/\sigma)$ less than half of the maximum precision.

The seeming paradox here is resolved by noting that a simple unweighted average is not the best estimator of the mean. This will be discussed later under estimation theory.

Chapter 5

INFORMATION

There are different meanings of the word *information*, usually distinguished by attaching the name of the person who introduced the particular definition in question. From the properties we require for information, we will be led to the definition of R.A. Fisher. We require that:

(i) The information in a set of observations should *increase with the number of observations*. Doubling the number of events should double the amount of information available, if the events are independent.

(ii) Information should be *conditional on what we want to learn* from the experiment. Data which is irrelevant to the hypothesis under test (or the parameters to be measured) should contain no information. Of course, the same data may contain information relative to other parameters or other tests.

(iii) Information should be *related to precision*; the greater the information, the better the precision of the experiment.

A quantity with the above properties will clearly be a valuable tool in *data reduction*. It is rare to find an experiment which has so little raw data that it can all be published, or even recorded. Thus a criterion is needed for the efficient rejection or reduction of data. We shall see that a good aim is to look for the maximum data reduction consistent with minimum loss of information. In particular we shall consider in detail when the information loss can be made zero.

Other definitions of information, which will not concern us here, are the information of Kullback [Kullback], which is primarily of theoretical interest, and that of Shannon [Shannon], much used in communications theory [Wiener].

5.1. Basic Concepts

We will discuss a real random variable (or a set of N real random variables) $\mathbf{X}$, with p.d.f. $f(\mathbf{X}|\boldsymbol{\theta})$, where $\boldsymbol{\theta}$ is a real parameter (or a set of k real parameters). The set of all allowed values of $\mathbf{X}$ (the range of $\mathbf{X}$) will be denoted by Ω_θ, where the subscript denotes the possible dependence of Ω on the parameter $\boldsymbol{\theta}$.

5.1.1. *Likelihood function*

Consider a set of N independent observations of X, say $X_1, \ldots, X_N$. These could be N events found in an experiment, an event being the measurement of p quantities.

The joint p.d.f. of $\mathbf{X}$ is, by independence,

$$P(\mathbf{X}|\boldsymbol{\theta}) = P(X_1, \ldots, X_N|\boldsymbol{\theta}) = \prod_{i=1}^{N} f(X_i|\boldsymbol{\theta}) \,. \tag{5.1}$$

When the variable $\mathbf{X}$ is replaced by the observed data $\mathbf{X}^0$, then P is no longer a p.d.f., and it is usual to denote it by L and call $L(\mathbf{X}^0|\boldsymbol{\theta})$ the *likelihood function*, which is now a function of $\boldsymbol{\theta}$ only:

$$L(\boldsymbol{\theta}) = P(\mathbf{X}^0|\boldsymbol{\theta}) \,.$$

We shall use it in this form in Chapters 7–9.

For the purposes of this chapter however, it will be convenient to keep $\mathbf{X}$ as a variable, but nevertheless to use the notation $L(\mathbf{X}|\boldsymbol{\theta})$, since we will be primarily interested in it as a function of $\boldsymbol{\theta}$. And indeed the dependence on X will soon disappear as we will always take expectations with respect to X.

5.1.2. *Statistic*

Suppose a new random variable, T, is defined by $T = T(X_1, \ldots, X_N)$. Any such function T is called a *statistic*. For example, the average $\overline{X}$ [Eq. (2.33)] is a statistic.

The word *statistic* is easily confused with the word *statistics*, the latter being used loosely by physicists to indicate an amount of data (*e.g.*, a high-statistics experiment). We will always use *statistic* in the technical sense defined above.

5.2. Information of R.A. Fisher

5.2.1. *Definition of information*

The amount of information given by an observation X about the parameter θ is defined by the following expression (if it exists)

$$I_X(\theta) = E\left[\left(\frac{\partial \ln L(X|\theta)}{\partial \theta}\right)^2\right]$$

$$= \int_{\Omega_\theta} \left(\frac{\partial \ln L(X|\theta)}{\partial \theta}\right)^2 L(X|\theta)dX. \qquad (5.2)$$

If $\boldsymbol{\theta}$ has k dimensions, the definition becomes

$$[\mathcal{I}_X(\boldsymbol{\theta})]_{ij} = E\left[\frac{\partial \ln L(X|\boldsymbol{\theta})}{\partial \theta_i} \cdot \frac{\partial \ln L(X|\boldsymbol{\theta})}{\partial \theta_j}\right]$$

$$= \int_{\Omega_\theta} \left[\frac{\partial \ln L(X|\boldsymbol{\theta})}{\partial \theta_i} \cdot \frac{\partial \ln L(X|\boldsymbol{\theta})}{\partial \theta_j}\right] L(X|\boldsymbol{\theta})dX. \qquad (5.3)$$

Thus, in general, $\mathcal{I}_X(\boldsymbol{\theta})$ is a $k \times k$ matrix.

There are two justifications of this apparently arbitrary definition: firstly it has the properties given at the beginning of this chapter, which we require of information; and secondly it is closely related to the maximum attainable precision in estimating the value of θ, given X (shown in Section 7.4.3).

5.2.2. *Properties of information*

In order to make information a useful concept, the definition (5.2) must be supplemented with the following two conditions:

(i) Ω_θ is independent of $\boldsymbol{\theta}$.
(ii) $L(\mathbf{X}|\boldsymbol{\theta})$ is regular enough to allow the operators $\partial^2/\partial\theta_i\,\partial\theta_j$ and $\int dX$ to commute.

If condition (i) holds, condition (ii) is expected to hold for all distributions met in physics. However, condition (i) is not always fulfilled. For example, suppose one wants to study the mass of the K^0 meson from its $\pi^+\pi^-$ decay and its line of flight. It is clear that the transverse momentum of each π takes a value between 0 and p^*, the total momentum of each pion, which depends on the K^0 mass. Thus Ω_θ [0 to p^*] depends on θ [K^0 mass].

Assuming that conditions (i) and (ii) hold, it is simple to show that the definition (5.3) of information can be written

$$\left[\mathcal{I}_X \left(\boldsymbol{\theta} \right) \right]_{ij} = -E \left[\frac{\partial^2}{\partial \theta_i \partial \theta_j} \ln L(X|\boldsymbol{\theta}) \right].$$

The *additivity property* now follows immediately from condition (ii). Information from different experiments, about one set of parameters, can be added. In particular, the information I_1, based on one event, is related to the information from N events by

$$I_N(\boldsymbol{\theta}) = N I_1(\boldsymbol{\theta}). \tag{5.4}$$

This nice property can be made quite useful via the relation between information and precision, Sec. 7.4. It can be illustrated here by a simple example.

Let X be Normally distributed with *known* variance σ^2, and unknown mean μ. Then the information from a single observation X is

$$I_1(\mu) = -\int_{-\infty}^{\infty} \frac{\partial^2}{\partial \mu^2} \left(-\frac{(X-\mu)}{2\sigma^2} - \ln \sigma \sqrt{2\pi} \right) \frac{1}{\sqrt{2\pi}\,\sigma} \exp\left[-\frac{1}{2} \frac{(X-\mu)^2}{\sigma^2} \right] dX$$

$$= +\int_{-\infty}^{\infty} \frac{1}{\sigma^2} \frac{1}{\sqrt{2\pi}\,\sigma} \exp\left[-\frac{1}{2} \frac{(X-\mu)^2}{\sigma^2} \right] dX = \frac{1}{\sigma^2}.$$

For N observations X_i, one has

$$I_N(\mu) = -\int_{-\infty}^{\infty} \cdots \int_{-\infty}^{\infty} \frac{\partial^2}{\partial \mu^2} \left[\sum_{i=1}^{N} \left(-\frac{(X_i-\mu)^2}{\sigma^2} \right) - \ln \sigma \sqrt{2\pi} \right]$$

$$\times \prod_{i=1}^{N} \frac{1}{\sqrt{2\pi}\,\sigma} \exp\left[-\frac{1}{2} \frac{(X_i-\mu)^2}{2\sigma^2} \right] dX_i.$$

$$= \frac{N}{\sigma^2} = N I_1(\mu). \tag{5.5}$$

Thus $I_N(\mu)$ is equal to the reciprocal of the variance of $\overline{X}$, and $\overline{X}$ is the usual estimate of the mean μ of the population. It will be shown later that Eq. (5.5) gives the minimum variance attainable by *any* estimate of μ in this case.

Let us now go to the other desirable feature of information mentioned in the introduction to this chapter, namely its applicability to data reduction problems. For that purpose we introduce the notion of the *sufficient statistic*.

5.3. Sufficient Statistics

5.3.1. *Sufficiency*

A statistic $T = T(\mathbf{X})$ is said to be *sufficient* for θ if the conditional density function of $\mathbf{X}$, given T, $f(\mathbf{X}|T)$ is independent of θ. T and θ may be multidimensional, and of *different* dimensions.

If T is a *sufficient statistic* for θ, then any strictly monotonic function of T is also a sufficient statistic for θ. We shall use this property later.

The importance of sufficiency for data reduction is evident from its definition; there is as much information about θ in T as in the original data $\mathbf{X}$. In other words, no other function of the data can give any further information about θ. [See Sec. 5.4.]

The set $T = \mathbf{X}$ is evidently sufficient for θ, since it carries all the initial information. A set of sufficient statistics will be useful only if it gives rise to a *data reduction*.

From its definition, one can test for sufficiency, but it is preferable (for practical reasons) to rely on the following result.

A necessary and sufficient condition for $T(\mathbf{X})$ to be a sufficient statistic for θ is that the likelihood factorizes as

$$L(\mathbf{X}|\theta) = g(T, \theta)h(\mathbf{X}),\tag{5.6}$$

where

(i) $h(\mathbf{X})$ does not depend on θ, and

(ii) $g(T, \theta) \propto A(T|\theta)$, the *conditional probability density* for T, given θ.

In many cases, condition (ii) is easily verified. Thus we can compute

$$A(T|\theta) = \int_{\Omega_\theta} L(\mathbf{X}|\theta)d\mathbf{X} \ T = \text{constant}$$

$$= \int_{\Omega_\theta} g(T, \theta)h(\mathbf{X})d\mathbf{X} \ T = \text{constant}$$

$$= g(T, \theta) \int_{\Omega_\theta} h(\mathbf{X})d\mathbf{X} \ T = \text{constant}.$$

Therefore, *where the range Ω_θ of $\mathbf{X}$ does not depend on θ*, condition (ii) follows from condition (i). Otherwise, both conditions must be checked before $T(\mathbf{X})$ is taken as a sufficient statistic.

5.3.2. *Examples*

(i) Suppose that $\mathbf{X}$ follows a Normal law $N(\mu, \sigma^2)$. Then the likelihood function is given by

$$
L(\mathbf{X}|\mu, \sigma^2) = (2\pi\sigma^2)^{-N/2} \exp\left[-\frac{1}{2\sigma^2} \sum_{i=1}^{N} (X_i - \mu)^2\right]
$$

$$
= (2\pi\sigma^2)^{-N/2} \exp\left[-\frac{N}{2\sigma^2}(\overline{X} - \mu)^2\right] \exp\left[-\frac{1}{2\sigma^2} \sum_{i=1}^{N} (X_i - \overline{X})^2\right],
$$

$$(5.7)$$

with the average $\overline{X}$ given by Eq. (2.33).

When σ^2 is known, $\overline{X}$ follows a Normal law $N(\mu, \sigma^2/N)$, and the likelihood function (5.7) factorizes as Eq. (5.6), with

$$
g(\overline{X}|\mu) = \left(\frac{N}{2\pi\sigma^2}\right)^{1/2} \exp\left[-\frac{N}{2\sigma^2}(\overline{X} - \mu)^2\right].
$$

Clearly the conditions (i) and (ii) of Sec. 5.3.1 are both verified. Therefore $\overline{X}$ is a sufficient statistic for μ.

When, on the other hand, μ is known and σ^2 is unknown, it is easy to show that

$$
\hat{\sigma}^2 = \frac{1}{N} \sum_{i=1}^{N} (X_i - \mu)^2
$$

is sufficient for σ^2. The conditional p.d.f. of $\hat{\sigma}^2$, given σ^2, is

$$
g(\hat{\sigma}^2|\sigma^2) = \frac{1}{\Gamma(\frac{N}{2})} \left(\frac{N}{2\sigma^2}\right)^{N/2} (\hat{\sigma}^2)^{(N/2)-1} \exp\left(-\frac{N}{2\sigma^2}\hat{\sigma}^2\right).
$$

(ii) We now give an example where one cannot find a sufficient statistic. Suppose the density function is

$$
f(X|\theta) = \frac{\theta e^{\theta X}}{e^{\theta^2} - 1},
$$

where

$$
0 \le X \le \theta.
$$

Here Ω_θ does explicitly depend on θ. One can write the likelihood function, given observations $X_1, \ldots, X_N$, as

$$L(\mathbf{X}|\theta) = \left(\frac{\theta}{e^{\theta^2} - 1}\right)^N e^{\theta T}, \quad \text{with } T = \sum_{i=1}^{N} X_i.$$

The only obvious factorization, with $h(\mathbf{X}) = 1$, leads to $g(T|\theta) = L(\mathbf{X}, \theta)$ which is *not* the conditional p.d.f. of T, given θ. Thus T cannot be sufficient for θ. This does not exclude the possibility of some other statistic being sufficient.

5.3.3. *Minimal sufficient statistics*

It is clear that the original data set $(X_1, \ldots, X_N)$ is a set of sufficient statistics. One can show that, if θ has k dimensions, one may find S sufficient statistics for a set of N data, where evidently $k \leq N$, $S \leq N$, but S may be smaller than, equal to, or greater than k. A set of sufficient statistics having the smallest S is said to be minimal sufficient. It is a function of all other sets of sufficient statistics. It is clearly the most useful set, since it gives the largest *data reduction*.

The data reduction achieved by a set of S sufficient statistics may be such that S is proportional to N; that is, the number of sufficient statistics depends on the number of observations. However, the existence of sufficient statistics becomes most useful when S is fixed for a fixed k, and is *independent* of N. Distributions yielding such a set are defined by the Darmois theorem (Sec. 5.3.4).

In particular, the Normal p.d.f. $N(\mu, \sigma^2)$ has either
(a) a single sufficient statistic for one parameter μ or σ^2 with the other one fixed (this we have seen in Sec. 5.3.), or
(b) a pair of jointly sufficient statistics for the two parameters μ, σ^2 when they are both unknown.

In fact, whatever the p.d.f. one can prove that (a) implies (b). Let us choose, for example, the uniform distribution:

$$f(X|\theta) = \frac{1}{\theta}, \quad \frac{-\theta}{2} \leq X \leq \frac{\theta}{2}.$$

Then one can show that $[\min X_i, \max X_i]$ form a pair of sufficient statistics for θ which is minimal. Note that this is a case of $S > k$. There does not exist a *single* sufficient statistic for θ. Or let us take

$$\prod_{i=1}^{N} f(X_i|\theta) = \frac{1}{\theta(2\pi)^{N/2}} \exp\left\{-\frac{1}{2}\left[\frac{N^2}{\theta^2}(\overline{X} - \mu)^2 + \sum_{i=2}^{N} X_i^2\right]\right\}.$$

Then one can show that the average, $\overline{X}$, is a single sufficient (minimal of course) statistic for *both* μ and θ. Here $S < k$. The problem is to find what functions of $\overline{X}$ best estimate μ and θ. [Kendall II, p. 193].

5.3.4. *Darmois theorem*

This theorem proves that only a very restricted class of probability density functions admits a number of sufficient statistics independent of the number of observations. The results of the theorem are stated as follows for the one dimensional case.

(i) Whatever Ω_θ, if there exists a number $N > 1$ such that the set $(X_1, X_2, \ldots, X_N)$ admits a sufficient statistic for θ, then the probability density function is of the "exponential form"

$$f(X|\theta) = \exp[\alpha(X)a(\theta) + \beta(X) + c(\theta)]. \tag{5.8}$$

(ii) Inversely, $(X_1, X_2, \ldots, X_N)$ admits a sufficient statistic for all $N > 1$ (but only if Ω_θ does not depend on θ), if $f(X|\theta)$ has the exponential form, and if the mapping

$$(X_1, X_2, \ldots, X_N) \Rightarrow (R, X_2, \ldots, X_N),$$

with

$$R = \sum_{i=1}^{N} \alpha(X_i),$$

is one-to-one and continuously differentiable for all $\mathbf{X}$. R is sufficient for θ, and so is any monotonic function of R.

The multidimensional "exponential form" is

$$f(\mathbf{X}|\boldsymbol{\theta}) = \exp\left[\sum_{j=1}^{S} \alpha_j(\mathbf{X})a_j(\boldsymbol{\theta}) + \beta(X) + c(\boldsymbol{\theta})\right], \tag{5.9}$$

and

$$R_j = \sum_{i=1}^{N} \alpha_j(X_i), \quad j = 1, \ldots, S$$

is one possible set of S jointly sufficient statistics for $\boldsymbol{\theta}$.

Examples

In the notation of Eq. (5.8) a Normal law $N(\mu, \sigma^2)$ can be written as

$$\ln f = X\frac{\mu}{\sigma^2} - \frac{X}{2\sigma^2} - \frac{\mu}{2\sigma^2} - \ln\left(\sigma\sqrt{2\pi}\right).$$

(i) Assuming σ^2 known and μ unknown, then the conditions of Darmois theorem are satisfied with the choice

$$a = \mu, \quad \alpha = \frac{X}{\sigma^2}, \quad c = -\frac{\mu}{2\sigma^2}, \quad \beta = -\frac{X^2}{2\sigma^2} - \ln(\sigma\sqrt{2\pi}).$$

A sufficient statistic for μ is given by

$$A = \sum_{i=1}^{N} \alpha_i = \frac{N}{\sigma^2}\overline{X}.$$

Alternatively, $\overline{X}$ is a sufficient statistic.

(ii) Assuming μ known and σ^2 unknown, one chooses

$$a = \frac{1}{\sigma^2}, \quad \alpha = X\left(\mu - \frac{X}{2}\right), \quad c = -\frac{\mu^2}{2\sigma^2} - \ln(\sigma\sqrt{2\pi}), \quad \beta = 0.$$

The sufficient statistic is given by

$$A = \sum_{i=1}^{N} \alpha_i = \mu\sum_{i=1}^{N} X_i - \frac{1}{2}\sum_{i=1}^{N} X_i^2.$$

Then the statistic

$$A' \equiv N\hat{\sigma}^2 = \sum_{i=1}^{N}(X_i - \mu)^2 = 2A - N\mu^2$$

is also sufficient for σ^2.

(iii) Assuming μ and σ^2 both unknown, one chooses

$$a_1 = \frac{\mu}{\sigma^2}, \quad \alpha_1 = X, \quad c = -\frac{\mu}{2\sigma^2} - \ln(\sigma\sqrt{2\pi}),$$

$$a_2 = \frac{1}{\sigma^2}, \quad \alpha_2 = -\frac{X^2}{2}, \quad \beta = 0.$$

The joint sufficient statistics for μ and σ^2 are

$$A_1 = \sum_{i=1}^{N} X_i \quad \text{and} \quad A_2 = -\sum_{i=1}^{N} \frac{X_i^2}{2}.$$

(iv) Assuming again μ and σ^2 unknown, but that one is interested only in one parameter, then one can choose

$$a = \mu, \quad \alpha = \frac{X}{\sigma^2}, \quad c = -\frac{\mu^2}{2\sigma^2}, \quad \beta = -\frac{X^2}{2\sigma^2} - \ln(\sigma\sqrt{2\pi}),$$

so that

$$R = \sum_{i=1}^{N} \frac{X_i}{\sigma^2}$$

is sufficient for μ. Then $\overline{X}$ is also sufficient, independently of σ^2.

On the other hand, it turns out to be impossible to find a statistic for σ^2, independent of μ, because the only possible expressions for a, α, c, β are those given in example (ii) above.

5.4. Information and Sufficiency

As we have already noted, a statistic T sufficient for θ contains *all the information* on θ from the observations $X_1, \ldots, X_N$. Thus one achieves data reduction without loss of information if one can find a sufficient statistic. If the p.d.f. is not of the exponential form (5.9), it is impossible to find a sufficient statistic. Then any *data reduction* to a fixed number of statistics, independent of the number of observations, must involve some loss of information. Note, however, that it may be possible to reduce the number of statistics from N to αN, where $\alpha < 1$, with no loss of information. Such a situation is possible when there are $1/\alpha$ observations (e.g. momentum components) at each of αN events, and from each event we are interested only in a single quantity (e.g. momentum). If, *for a given event* a sufficient statistic exists for the required quantity, then some data reduction is possible with no loss of information.

Choosing between several insufficient statistics, that T is preferred which contains the maximum of information $I_T(\theta)$. Clearly $I_T(\theta)$ is bound by $I_X(\theta)$, the total information on θ in the set of data X,

$$I_T(\theta) \leq I_X(\theta). \tag{5.10}$$

The proof of Eq. (5.10) is quite simple. Let us factorize $L(X|\theta)$ into

$$L(X|\theta) = g(T|\theta) \cdot h(X,\theta), \tag{5.11}$$

which is always possible since h is explicitly a function also of θ; where $g(T|\theta)$ is the conditional distribution of T, given θ. Insertion of Eq. (5.11) into Eq. (5.2) yields

$$I_X(\theta) = E\left\{\left[\frac{\partial}{\partial\theta}\ln g(T|\theta)\right]^2\right\} + E\left\{\left[\frac{\partial}{\partial\theta}\ln h(\overline{X},\theta)\right]^2\right\}$$

$$= I_T(\theta) + E\left\{\left[\frac{\partial}{\partial\theta}\ln h(X,\theta)\right]^2\right\}.$$

The last term is necessarily positive, thus Eq. (5.10) is proved. The equality sign in Eq. (5.10) obviously holds only when h does not depend on θ. But then T is sufficient, by the definition (5.6).

The generalization of Eq. (5.10) to the case when there are k parameters θ, is

$$\tilde{U}\, \mathcal{L}_T\, U \leq \tilde{U}\, \mathcal{L}_X\, U\,.$$

where $\mathcal{L}_T$ and $\mathcal{L}_X$ are $k \times k$ matrices and U is any k-dimensional vector.

This result is very important. It shows that if, for instance one diagonalizes $\mathcal{L}_T$ by a unitary transformation, each diagonal component of $\mathcal{L}_T$ is smaller than the corresponding diagonal element in the (non-diagonal) matrix $\mathcal{L}_X$. In general if one has found a set of statistics T which contain the amount $\mathcal{L}_T\,(\boldsymbol{\theta})$ of information on k parameters $\boldsymbol{\theta}$, they contain the same amount of information on any k linear combinations of the parameters $\boldsymbol{\theta}$.

5.5. Example of Experimental Design

Given the data, one can compare the information in different possible statistics as we have shown. But one may as well, in view of the measurement of a set of parameters, design the experiment in order to get a maximum of information with respect to the different possibilities.

Consider the problem of weighing four objects with a balance. The simple-minded approach is to weigh each object separately, totalling four measurements. (If the precision of the balance is not good enough, one can repeat the measurements until the required precision is achieved.) The question arises, however, of whether it is possible to improve precision of four measurements by weighing four suitable combinations of the objects rather than by weighing them separately [Lindley, I p. 131, II p. 119].

Let us choose four new variables X_i which are linear combinations of the individual weights θ_j, $j = 1,\ldots,4$ of the four objects. Assume that the X_i follow a Normal law, $N(\Sigma_j a_{ij}\theta_j, \sigma^2)$, where the a_{ij} are the coefficients of the linear combinations, and σ is the uncertainty of the weighings. We assume explicitly that σ is always the same, independent of the linear combination. The possible combinations are defined by $a_{ij} = -1$, 0, or $+1$ (i.e. in the ith weighing the jth object is in the left scale, does not participate in the measurement, or is in the right scale, respectively). The likelihood of the measurement is

$$L(\mathbf{X}|\boldsymbol{\theta}) = \left(\frac{1}{\sigma\sqrt{2\pi}}\right)^4 \exp\left[-\frac{1}{2\sigma^2}\sum_{i=1}^{4}\left(X_i - \sum_{j=1}^{4} a_{ij}\theta_j\right)^2\right].$$

From this one can calculate the information matrix

$$-\frac{\partial^2 \ln L(\mathbf{X}, \boldsymbol{\theta})}{\partial \theta_j \partial \theta_k} = \sum_{i=1}^{4} \frac{a_{ij} a_{ik}}{\sigma^2}$$

$$= E\left(-\frac{\partial^2 \ln L}{\partial \theta_j \partial \theta_k}\right)$$

$$= [\mathcal{I}(\boldsymbol{\theta})]_{jk} .$$

Let us construct a matrix $\mathcal{Q}$ with the coefficients a_{ij}, thus

$$\mathcal{I}(\boldsymbol{\theta}) = \frac{1}{\sigma^2} \, \mathcal{Q}^{\mathrm{T}} \cdot \mathcal{Q} .$$

Now we can compare the two methods. In the first one, all four objects are weighed independently, thus $a_{ij} = \delta_{ij}$. Then we have, evidently,

$$[\mathcal{I}_1(\boldsymbol{\theta})]_{ij} = \delta_{ij}/\sigma^2 .$$

In the second method one again carries out four weighings, but all the four objects participate with coefficients a_{ij} being -1 or $+1$. Moreover the combinations chosen must be linearly independent (in order to enable one to solve four equations for the four unknowns θ_j). One possible set of coefficients is

$$\mathcal{Q} = \begin{bmatrix} 1 & 1 & -1 & -1 \\ 1 & -1 & 1 & -1 \\ 1 & -1 & -1 & 1 \\ 1 & 1 & 1 & 1 \end{bmatrix} .$$

Then it follows immediately that the information matrix is

$$[\mathcal{I}_2(\boldsymbol{\theta})]_{ij} = 4\delta_{ij}/\sigma^2 = 4[\mathcal{I}_1(\boldsymbol{\theta})]_{ij} .$$

Obviously the second method is preferable since it yields four times as much information. Thus it leads to an uncertainty in the determination of the θ_j two times smaller.

Chapter 6

DECISION THEORY

Classical statistics helps the experimental physicist to summarize his experiment with minimum loss of information. It may be, however, that we want to use the experimental data in order to reach a *decision*. A decision is something that has different consequences depending on what the true state of nature turns out to be. For example, one has to make decisions about the design of the detector, about how much time to spend on different activities, or when to publish an important result. The physicist, therefore, needs a *decision theory*, closely tied to the powerful methods of statistics. In this chapter we shall briefly outline such a theory.

The basic approach to decision theory can best be illustrated by a simple example. Suppose one is running a particle detector, with a maximum daily counting rate $\theta_{\max}$, but which may drift from its optimal setting. Each morning, one has to decide whether or not to reset the detector, the data available being the previous day's observation, t, and the fact that readjustment takes some known fraction, u, of the day, $0 \leq u \leq 1$.

In order to make a rational choice between the two possible decisions, one needs to introduce a measure of the loss incurred in making each decision. This measure is given by a *loss function*, $L(\theta, d)$, giving the loss incurred in making decision d when the true counting rate is θ. A possible function could be

$$L(\theta, d_1) = \theta_{\max} - \theta$$
$$L(\theta, d_2) = u\theta \,,$$

where d_1 is the decision "Do not reset" and d_2 is the decision "Reset".

If θ is known, the choice of decision is obvious. The detector is reset if $\theta \leq (1 - u)\theta_{\max}$. However, in general, θ is not known, only an observation t is known from a distribution depending on θ. Decision theory is concerned with formulating a technique for choosing the best decision on the basis of the observations, the essence of this technique being to minimize the loss resulting from making a wrong decision.

6.1. Basic Concepts in Decision Theory

6.1.1. *Subjective probability, Bayesian approach*

In Secs. 2.2.5 and 2.3.5, we have seen how *Bayes theorem* may be used to update one's *prior knowledge* about some unknown parameter, when new observations are given. In order to do this practically, one needs a quantitative expression of prior knowledge in the form of a probability distribution.

Suppose that one is interested in knowing θ, the parameter of a distribution $f(X, \theta)$ of an observable random variable X. The Bayesian approach is to consider θ to have a prior distribution $\pi(\theta)$, which describes one's *degree of belief* in the different possible values of θ.

Let us show how it may be possible to convert such degrees of belief into numbers. Consider a parameter which can have only two possible values, θ_1 and θ_2. One now asks the question: "On the basis of my present information, what are the minimum *odds* at which I would be prepared to gamble that θ was θ_1 rather than θ_2?"

If the answer is

$$\text{"minimum odds} = b = 4\text{"} ,$$

this can be interpreted by saying that one's belief in θ_1 is four times greater than in θ_2. Let us denote the degrees of belief in θ_1 and θ_2 by $\pi(\theta_1)$, $\pi(\theta_2)$. Since $\pi(\theta_1)$, $\pi(\theta_2)$ are determined only relatively,

$$b = \frac{\pi(\theta_1)}{\pi(\theta_2)} ,$$

we may choose to express them in normalized units

$$\sum_{i=1}^{2} \pi(\theta_i) = 1 .$$

Thus one has

$$\pi(\theta_1) = \frac{b}{b + 1} , \quad \pi(\theta_2) = \frac{1}{b + 1} .$$

This argument can be generalized to give a distribution $\pi(\theta)$ of degrees of belief over a range of values of θ.

It can be shown [Lindley, II, p. 33] that such degrees of belief satisfy the axioms of probability, and the distribution $\pi(\theta)$ can, therefore, be handled as a probability distribution. Thus the *subjective probability* distribution $\pi(\theta)$ summarizes one's knowledge of θ.

Using the prior distribution $\pi(\theta)$, the observations $\mathbf{X}$ with probability $p(\mathbf{X}|\theta)$, and *Bayes theorem*, one obtains the *posterior density*

$$p(\theta|\mathbf{X}) \propto p(\mathbf{X}|\theta)\pi(\theta) \, .$$

This can be interpreted as representing one's knowledge of θ, given the observations $\mathbf{X}$.

In decision theory a technique is developed for calculating an optimum decision, given $\mathbf{X}$. The posterior distribution provides the link between the observations $\mathbf{X}$, and the parameter θ about which the decision is made.

6.1.2. *Definitions and terminology*

Decision theory deals with three different spaces:

An *observable space* χ, in which all possible observations $\mathbf{X} = (X_1, \ldots, X_N)$ fall.

A *parameter space* Ω contains all possible values of the parameter θ, or the parameters $\boldsymbol{\theta} = (\theta_1, \ldots, \theta_p)$. The possible values of $\boldsymbol{\theta}$ are often called *states of nature*.

A *decision space* $\mathcal{D}$, which contains all possible decisions d.

A *decision rule* δ, alternatively *decision procedure*, *decision function*, specifies what decision d is to be taken given the observation $\mathbf{X}$, that is

$$d = \delta(\mathbf{X}) \, .$$

We shall limit ourselves to *non-random decision rules* δ, which define d completely when $\mathbf{X}$ is given.

To choose between decision rules one needs a *loss function*, $L(\theta, d)$, defined as the loss incurred in taking the decision d, when θ is assumed to be the true value of the parameter.

The introduction of a loss function is an essential part of decision theory. Any rational choice between decisions must be based on a calculation of the loss (such as the *cost*) or the *gain* incurred by each decision. Consider the function $L[\theta, \delta(\mathbf{X})]$, which is a random variable, being a function of $\mathbf{X}$. The *risk function* for the decision rule δ is defined as

$$R_\delta(\theta) \equiv E\{L[\theta, \delta(\mathbf{X})]\} = \int L[\theta, \delta(\mathbf{X})] f(\mathbf{X}|\theta) d\mathbf{X} . \tag{6.1}$$

Thus $R_\delta(\theta)$ gives the average loss over all possible observations.

From the Bayesian point of view, θ can be treated as a random variable. The expected risk over θ

$$r_\pi(\delta) = \int R_\delta(\theta) \pi(\theta) d\theta \tag{6.2}$$

is called the *posterior risk* of using decision rule δ, given the prior density $\pi(\theta)$. Using Eq. (6.1), the posterior risk (6.2) can be written

$$r_\pi(\delta) = E_\theta\{E_\mathbf{X}\{L[\theta, \delta(\mathbf{X})]|\theta\}\}$$
$$= E_\mathbf{X}\{E_\theta\{L[\theta, \delta(\mathbf{X})]|\mathbf{X}\}\} ,$$

where the subscript of the expectation operator refers to the variable over which the average is taken. The quantity

$$E_\theta\{L[\theta, \delta(\mathbf{X})]|\mathbf{X}\} , \tag{6.3}$$

is called the *posterior loss*, given the observations $\mathbf{X}$. It is the average loss over the posterior density $p(\theta|\mathbf{X})$ incurred by using the decision $\delta(\mathbf{X})$.

6.2. Choice of Decision Rules

6.2.1. *Classical choice: pre-ordering rules*

The classical choice of a *decision rule* δ is based on the *risk function* $R_\delta(\theta)$, Eq. (6.1). The best decision rule is the one which gives the smallest risk. Thus if δ and δ' are two possible rules, and if

$$R_\delta(\theta) < R_{\delta'}(\theta) , \quad \text{for all } \theta \tag{6.4}$$

then δ is a better decision rule than δ'. We shall use the notation

$$\delta > \delta'$$

to indicate that δ is *preferable* to δ'.

Consider a situation with several decision rules δ_i, the risk functions being given in Fig. 6.1. Since $R_{\delta_1}(\theta) < R_{\delta_2}(\theta)$ for all values of θ, one can say that rule δ_1 is always better than rule δ_2. However, rule δ_1 is not better than all other rules. For some values of θ, δ_3 and δ_4 are better rules.

It is, therefore, possible to *pre-order decision rules*. In general, it will not be possible to find a single best rule for all θ, but for every value of θ there will be an optimal rule. The basic uncertainty in the choice is that the true value of θ is unknown.

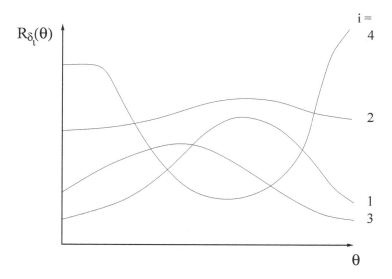

Fig. 6.1. Example of four risk functions $R_{\delta i}(\theta)$, $i = 1, \ldots, 4$.

A decision rule δ' is an *admissible decision rule* if there is no other rule $\delta \neq \delta'$ that has a lower risk than δ' for all values of θ.

Clearly one has to choose decision rules among the admissible class. The two problems of decision theory are to find admissible solutions, and to choose between them. A useful tool for this purpose is the Bayesian family, discussed in the next section.

6.2.2. *Bayesian choice*

The Bayesian choice of a *decision rule* δ, given a *prior density* $\pi(\theta)$, is based on the *posterior risk* function $r_\pi(\delta)$, Eq. (6.2). The *Bayesian decision rule* δ, corresponding to the prior $\pi(\theta)$, is the one which gives the smallest posterior risk, that is

$$r_\pi(\delta) \le r_\pi(\delta') \quad \text{for any } \delta' \,. \tag{6.5}$$

Thus the Bayesian approach is to locate the uncertainty about the true value θ in a prior distribution of beliefs, $\pi(\theta)$. By averaging the risk over the posterior density, the basic uncertainty of the classical approach can be avoided.

If a decision δ exists, which minimizes the *posterior loss* (6.3) for all $\mathbf{X}$, it obviously also minimizes the posterior risk $r_\pi(\delta)$, and it is therefore a Bayesian solution. Similarly, if there is a decision which minimizes the risk $R_\delta(\theta)$ for all θ, it is a Bayesian solution.

Under fairly general conditions (practically only that R_δ is bounded for all δ), it can be shown that *all admissible solutions are Bayesian solutions*. Thus, if δ is an admissible decision rule, then there exists some prior density $\pi(\theta)$ such that δ is the Bayesian solution for π [Wald 1950], [Fourgeaud].

Conversely, given any decision rule, there exists some Bayesian rule which is equivalent or preferable. The class of Bayesian rules is called a *complete class*.

If δ_B is a Bayesian rule, there is no rule which is strictly better than δ_B for all θ. On the other hand, δ_B may not always be admissible, for example when the corresponding prior distribution $\pi_\text{B}(\theta)$ is zero for some θ.

In the particular case when Ω is a finite space of discrete values $\theta_1, \ldots, \theta_r$, the class of Bayesian solutions for the prior distribution $\pi(\theta_i)$, $i = 1, 2, \ldots, r$, coincides with the class of admissible solutions, provided that

$$\pi(\theta_i) \ne 0 \,, \quad i = 1, \ldots, r \,.$$

6.2.3. *Minimax decisions*

Among other procedures which have been proposed for choosing between decision rules, the most important is the *minimax* method of von Neumann. The minimax method is to choose that decision rule which minimizes the maximum risk.

By definition, the minimax rule is admissible, if it exists. One can show that it does exist, under the general conditions of the previous section. It follows that the *minimax decision rule*, although a classical tool, must be a Bayesian solution corresponding to some particular prior distribution $\pi_0(\theta)$.

It can be shown that $\pi_0(\theta)$ is the most unfavourable prior distribution that can be chosen, in the sense that

$$\min\left[r_{\pi_0}(\delta)\right]_{\text{for all }\delta} \geq \min\left[r_\pi(\delta)\right]_{\text{for all }\delta}$$

for any prior distribution $\pi(\theta)$. In other words, the Bayesian decision rule for π_0 has a higher posterior risk than the Bayesian rule for any other prior distribution.

Thus the minimax rule leads to the most pessimistic, or conservative, decision.

6.3. Decision-theoretic Approach to Classical Problems

Estimation and testing are vast subjects in Statistics which will be treated in all the remaining chapters, essentially from the point of view of obtaining minimum loss of information and optimal frequentist properties. The alternative approach, based on decision theory, is outlined in this section. It will be seen that this approach has many interesting features, but it requires subjective input.

6.3.1. *Point estimation*

The decision problem in point estimation is what value $\hat{\theta}$ to choose for a parameter θ, given N observations $X_1, X_2, \ldots, X_N$ from a density $f(\mathbf{X}, \theta)$. The decision space $\mathcal{D}$ is in one-to-one correspondence with the parameter space Ω: corresponding to each possible value θ, there is a decision d to assign this value to θ.

Let $\delta(\mathbf{X}) = \hat{\theta}$ be the decision function, which maps the observable space $(\chi)^N$ onto $\mathcal{D} = \Omega$.

The loss is a function $L(\theta - \hat{\theta})$ of the distance between the estimate $\hat{\theta}$ and the true value θ. The loss function is taken to have a minimum at $\theta = \hat{\theta}$. Since a regular function may be approximated at a minimum by a quadratic form, one often chooses[a]

$$L(\theta - \hat{\theta}) = (\theta - \hat{\theta})^T W (\theta - \hat{\theta}).$$
$$= \omega(\theta - \hat{\theta})^2 \qquad\qquad (6.6)$$

---

[a]Note that if the range of θ is not bounded, this loss function is not bounded, and some care should be taken when applying the results of Sec. 6.2.

in the one-dimensional case considered here. With this loss function, one can write down the posterior loss as

$$E[L(\theta - \hat{\theta})|\mathbf{X}] = E[\omega(\theta - \hat{\theta})^2|\mathbf{X}]$$
$$= E\{\omega[\theta - E(\theta|\mathbf{X})]^2|\mathbf{X}\} + \omega[\hat{\theta} - E(\theta|\mathbf{X})]^2$$
$$= \omega\{V(\theta|\mathbf{X}) + [\hat{\theta} - E(\theta|\mathbf{X})]^2\}\,.$$

Clearly, the posterior loss is minimal when the decision rule is $\hat{\theta} = E(\theta|\mathbf{X})$.

For the quadratic loss function (6.6), the Bayesian point estimate is the mean of the posterior density of θ. Obviously, for a different loss function, reflecting a different situation (such as preference for underestimation rather than overestimation), one will find a different decision rule.

6.3.2. *Interval estimation*

Often a point estimate $\hat{\theta}$ is of little use, unless related to some interval (as we shall see in Chapter 9). A well-known example is the interval $\hat{\theta} \pm \Delta\hat{\theta}$, where $\Delta\hat{\theta}$ is the standard deviation of the estimate $\hat{\theta}$.

Since the posterior density $p(\theta|X)$ summarizes one's knowledge of θ, the decision problem may be to choose an interval (a, b) in the range of θ which best describes $p(\theta|X)$. The loss function is then a function $L(\theta; a, b)$ of θ and the interval. A possible loss function is

$$L(\theta; a, b) = \begin{cases} \omega_1(b-a)^2 & \text{if} \quad \theta \in (a, b)\,. \\ \omega_2(\theta-a)^2 & \text{if} \quad \theta < a\,. \\ \omega_3(\theta-b)^2 & \text{if} \quad \theta > b\,. \end{cases}$$

The posterior loss now becomes

$$E[L(\theta; a, b)|X] = \omega_1 \int_a^b (b-a)^2 p(\theta|X)d\theta + \omega_2 \int_{-\infty}^a (\theta-a)^2 p(\theta|X)d\theta$$
$$+ \omega_3 \int_b^\infty (\theta-b)^2 p(\theta|X)d\theta\,.$$

The solution of the decision problem is to choose the interval (a, b) to minimize this posterior loss, which depends obviously on the values assigned to ω_1, ω_2 and ω_3.

6.3.3. *Tests of hypotheses*

Suppose that one has to decide between two hypotheses, H_0 and H_1, which uniquely correspond to values 0 and 1, respectively, of a parameter θ. Usually,

Table 6.1. Loss function.

Decision State of nature	Choose $\theta = 0$	Choose $\theta = 1$
H_0 true	0	ℓ_0
H_1 true	ℓ_1	0

Table 6.2. Probabilities of decisions, $P(H_i|\theta)$.

Decision State of nature	Choose H_0	Choose H_1
H_0 true	$1 - \alpha(\delta)$	$\alpha(\delta)$
H_1 true	$\beta(\delta)$	$1 - \beta(\delta)$

one chooses a loss function which has the value 0 (no loss) for the right decision. Such a loss function is the one defined by the values in Table 6.1.

In the Bayesian approach, one attributes prior probabilities μ and $(1-\mu)$ to H_0 and H_1, respectively. Given some decision rule δ, there is a finite probability $\alpha(\delta)$ of choosing H_1 when H_0 is true, and $\beta(\delta)$ of choosing H_0 when H_1 is true, as shown in Table 6.2.

The Bayesian solution is to choose the decision rule δ to minimize

$$r_\mu(\delta) = \alpha(\delta)\ell_0\mu + \beta(\delta)\ell_1(1 - \mu).$$

The situation may be seen graphically by plotting the possible points $[\alpha(\delta), \beta(\delta)]$. It can be shown that the accessible region is convex, and will have the general shape as illustrated in Fig. 6.2. Bayesian decision rules correspond to points on the lower boundary of this region. For $0 < \mu < 1$, the Bayesian family is identical to the set of admissible decision rules. It is clear that the minimum risk will be obtained at the decision corresponding to the point B where the line

$$r = \ell_0\mu\alpha + \ell_1(1 - \mu)\beta$$

is tangential to the region of possible points. This decision is found by considering the posterior loss, given the observations $\mathbf{X}$. Thus the expected loss in choosing H_0 is given by

$$\ell_1 P(H_1|\mathbf{X}) = \ell_1(1 - \mu)P(\mathbf{X}|H_1),$$

where $P(H_1|\mathbf{X})$ is the posterior probability that H_1 is true, (i.e. $\theta = 1$). The expected loss in choosing H_1, on the other hand, is given by

$$\ell_0\mu P(\mathbf{X}|H_0).$$

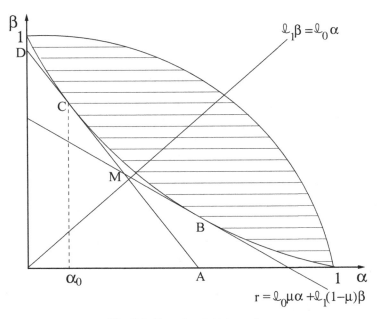

Fig. 6.2. Bayesian decision rules.

Thus the *minimum risk decision rule* (point B in Fig. 6.2) is:
   choose H_0 if

$$\ell_1(1 - \mu)P(\mathbf{X}|H_1) < \ell_0\mu P(\mathbf{X}|H_0)$$

   or

$$\frac{P(\mathbf{X}|H_1)}{P(\mathbf{X}|H_0)} < \frac{\ell_0\mu}{\ell_1(1 - \mu)},$$

otherwise choose H_1.

Classically, one would not speak of loss ℓ_0 and ℓ_1, and one would choose the "significance level"[b] α_0 arbitrarily. One would then minimize β, and obtain point C. It seems obvious, and it may be shown rigorously [Fourgeaud, Ch. 16], that this procedure corresponds to a Bayesian solution with a particular choice of the ratio

$$\ell_0\mu/\ell_1(1 - \mu), \tag{6.7}$$

---

[b]We shall discuss significance levels in more detail in Chapters 10–11.

namely, the ratio of distances OD/OA in Fig. 6.2. Similarly, the minimax decision rule corresponds to the point M, at the intersection with the line

$$\ell_1 \beta = \ell_0 \alpha \,.$$

Intuitively, a small α means that one would choose H_1 only when the observations make H_0 quite improbable. This line of reasoning, which makes no sense in the classical approach, is justified in the Bayesian interpretation by the fact that the absolute value of the slope (6.7) at B is a decreasing function of α; this is a consequence of the convexity of the domain. The smaller α is, then, the bigger should the prior probability for H_0 (or the loss ℓ_0) be, in order to tolerate unlikely observations without taking the decision to choose H_1. Therefore, the choice of a small α is equivalent to giving preference *a priori* to the hypothesis H_0.

A small confidence level α can be interpreted as the assumption that H_0 most probably is true, and that one is not prepared to commit the error of disproving a true theory in more than $\alpha\%$ of one's professional lifetime. On the other hand, when one decides to fix α rather than β, it implies that one considers it unlikely to commit the error of accepting H_0 as true when in fact H_1 is true (and perhaps less dangerous for one's professional career).

Obviously this is a very particular way of formulating a decision problem, and certainly it is not convenient in all situations.

6.4. Examples: Adjustment of an Apparatus

We now turn to less classical examples, hinting how decision theory can be used in "real life".

6.4.1. *Adjustment given an estimate of the apparatus performance*

Let us return to the example of the introduction to this chapter, where one had two choices

$$d_1 \quad \text{no adjustment}$$

$$d_2 \quad \text{adjustment}\,.$$

θ is the mean number of events that can be recorded by the apparatus per day. The losses if θ is the true state of nature are, respectively,

$$L(\theta, d_1) = (\theta_{\max} - \theta)$$

$$L(\theta, d_2) = \theta_{\max}\, p\,, \quad 0 < p < 1$$

where $\theta_{\max}$ is the maximum counting rate.

Suppose that one has recorded t events the previous day. One may then use this number as an estimate of the performance θ of the apparatus.

Let the distribution of t around θ be

$$f(t,\theta) = \frac{\theta^t}{t!}e^{-\theta}$$

(the Poisson distribution).

Suppose also that the prior distribution of θ is chosen to be

$$-(\theta) = \begin{cases} \dfrac{1}{\theta_{\max}} & 0 \le \theta \le \theta_{\max} \\ 0 & \theta < 0 \quad \text{or} \quad \theta > \theta_{\max} \end{cases}.$$

uniform between 0 and $\theta_{\max}$. One may have chosen this, for instance, because one knows from previous experience that there is no accumulation of θ around any point.

Then the posterior distribution of θ, given t, is

$$P(\theta|t)d\theta \propto \frac{1}{\theta_{\max}}\frac{\theta^t}{t!}e^{-\theta}d\theta.$$

Making the change of variable $u = 2\theta$,

$$P(u=2\theta|t)du \propto \left(\frac{u}{2}\right)^t e^{-\frac{u}{2}}du, \quad 0 < u < 2\theta_{\max}.$$

This is a χ^2 law of $n = 2(1+t)$ degrees of freedom.

Given d_1, the posterior loss when t is observed is

$$E_\theta\{L(d_1)|t\} = \int_0^{\theta_{\max}} (\theta_{\max} - \theta)P(\theta|t)d\theta$$

$$= \theta_{\max}\int_0^{2\theta_{\max}} P(u|t)du - \frac{1}{2}\int_0^{2\theta_{\max}} uP(u|t)du \qquad (6.8)$$

where we have used the fact $P(u)du = P(\theta)d\theta$. After some rearrangement, Eq. (6.8) may be written

$$E_\theta\{L(d_1)|t\} = \theta_{\max}P\{\chi^2(2+2t) < 2\theta_{\max}\} - (1+t)P\{\chi^2(3+2t) < 2\theta_{\max}\}.$$

Given now the decision d_2, the posterior loss is

$$E_\theta\{L(d_2)|t\} = p\theta_{\max}.$$

The Bayesian decision rule is, therefore:

Choose d_1 if
$$E_\theta\{L(d_1)|t\} \leq E_\theta\{L(d_2)|t\}.$$
Choose d_2 otherwise.

When $\theta_{\max}$ is large, the expressions above simplify and one will reset the apparatus when $t < (1-p)\theta_{\max} - 1$.

6.4.2. *Adjustment with estimation of the optimum adjustment*

Suppose that observations are taken of a random variable X, distributed as $N(\theta, \tau^2)$ where θ is unknown, τ^2 is known. θ is the deviation from some target value, taken to be zero. The setting of the machine can be deterministically adjusted by any amount. The decisions are (for different values of y):

d_y : adjust setting to decrease the deviation by y.

Two possible loss functions may be

$$\text{(i)} \quad L(d_y, \theta) = a(y - \theta)^2, \quad a > 0$$

$$\text{(ii)} \quad L(d_y, \theta) = \begin{cases} a(y - \theta)^2 + b, & y \neq 0 \\ a\theta^2, & y = 0. \end{cases}$$

In case (ii), a premium is placed on *not* making an adjustment. The prior density of θ is assumed to be $N(\mu, \sigma^2)$.

The posterior density of θ is then proportional to $P(X|\theta)P(\theta)$

$$= \exp\left\{ -\frac{1}{2}\left[\frac{1}{\tau^2} + \frac{1}{\sigma^2}\right]\left[\theta - \frac{\frac{X}{\tau^2} + \frac{\mu}{\sigma^2}}{\frac{1}{\tau^2} + \frac{1}{\sigma^2}}\right]^2 \right\}.$$

Thus $P(\theta|X)$ is

$$N\left(\frac{\frac{X}{\tau^2} + \frac{\mu}{\sigma^2}}{\frac{1}{\tau^2} + \frac{1}{\sigma^2}}, \frac{1}{\frac{1}{\tau^2} + \frac{1}{\sigma^2}} \right).$$

If $\sigma^2 \gg \tau^2$, the posterior density is $N(X, \tau^2)$. For the loss function (i) the posterior loss is

$$E[(\theta - Y)^2|X] = a\big(V[(\theta - Y)|X] + \{E[(\theta - Y)|X]\}^2\big)$$
$$= a\big(V[(\theta - Y)|X] + \{E(\theta|X) - Y\}^2\big).$$

Thus, to get minimum posterior loss, the decision is to choose

$$Y = E(\theta|X) = \frac{\dfrac{X}{\tau^2} + \dfrac{\mu}{\sigma^2}}{\dfrac{1}{\tau^2} + \dfrac{1}{\sigma^2}}$$

with a posterior loss

$$a \left[\frac{1}{\dfrac{1}{\tau^2} + \dfrac{1}{\sigma^2}} \right] .$$

In case (ii), one can show similarly that it is optimal to adjust, if and only if

$$b < a \left(\frac{\dfrac{X}{\tau^2} + \dfrac{\mu}{\sigma^2}}{\dfrac{1}{\tau^2} + \dfrac{1}{\sigma^2}} \right)^2 ,$$

and if an adjustment is needed, then this adjustment is

$$Y = E(\theta|X) = \frac{\dfrac{X}{\tau^2} + \dfrac{\mu}{\sigma^2}}{\dfrac{1}{\tau^2} + \dfrac{1}{\sigma^2}} .$$

6.5. Conclusion: Indeterminacy in Classical and Bayesian Decisions

In the classical approach, one must choose among different possible decision rules, each of which is optimal for some unknown value of the parameters. The Bayesian approach centralizes the indeterminacy of the problem into a subjective prior distribution over the possible values of the parameters, and then has a single decision rule.

In the classical case, the fundamental indeterminacy lies in the many possible decision rules from which a choice must be made. There is only one decision rule for any individual Bayesian, but then the indeterminacy lies in the possible prior distributions which may be chosen by different individuals. Asymptotically, when the number of observations becomes large both approaches lead to consistent results.

In fact, it turns out that any optimal classical decision rule is also some Bayesian rule. In other words, even if the decision maker is not a Bayesian, he will behave as if he were!

A problem of philosophical nature remains, namely how this basic indeterminacy can be tolerated by a scientist who aims to behave objectively. The scientist concerned with describing his results in such a way as to convey a maximum of information about the unknown parameter, can attain his aim of objectivity by careful use of the frequentist methods described in the other chapters of this book. However, when making a decision, he or she must recognize that this necessarily involves some subjectivity.

Chapter 7

THEORY OF ESTIMATORS

Estimation may be considered as the measurement of a parameter (which is assumed to have some fixed but unknown value) based on a limited number of experimental observations. Given these observations, *point estimation* consists in determining a single value as close as possible to the true value, and *interval estimation* determines a range of values most likely to include the true parameter value. The main subject here is the exact sense in which "close" and "likely to include" are to be understood.

There exists a close relationship between estimation and information. In particular, it is clear that an estimate of a parameter is a *statistic* (i.e. it is a function of the data) and the properties of statistics discussed earlier apply also to estimates.

We have attempted to divide the vast subject of *point estimation* into a theoretical chapter, the present one, and an applied chapter, the next one. Still the theoretical part contains practical examples and the applied parts cannot altogether avoid theory. The reader interested mainly in applications should, however, be able to skip all of this chapter, except for Section 7.1.

7.1. Basic Concepts in Estimation

To estimate a parameter, one first chooses a function of the observations (i.e. a method for proceeding from the observations to the estimate) which is called the *estimator*. The numerical value yielded by the estimator for a particular set of observations is the *estimate*.

Having chosen an estimator, one can then discuss its goodness in terms of four important desirable properties:

(i) *consistency*,
(ii) *unbiasedness*,
(iii) *information content* or *efficiency*,
(iv) *robustness*.

The first two properties above are discussed in this section. The third is treated at length in Chapters 7 and 8. The fourth property, although it may be of considerable importance in practical cases, is not relevant to the basic theory of estimation and will not be discussed until Section 8.7.

7.1.1. *Consistency and convergence*

An estimator is called a *consistent estimator* if its estimates converge toward the true value of the parameter as the number of observations increases. Different kinds of consistency can be defined using different kinds of convergence (Chapter 3). In the following, we shall always imply *consistency in probability*, defined using convergence in probability (Section 3.2.3). Let $\hat{\theta}_n$ be an estimator of the parameter θ based on n observations. Given any $\varepsilon > 0$ and any $\eta > 0$, $\hat{\theta}_n$ is a consistent estimator of θ if an N exists such that

$$P(|\hat{\theta}_n - \theta| > \varepsilon) < \eta$$

for all $n > N$. This says that $\hat{\theta}_n$ converges (in probability) toward θ as n increases.

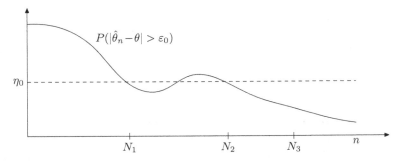

Fig. 7.1. Consistency in probability. For any $\eta > 0$ and any $\varepsilon > 0$, an N can be found such that the curve stays below the value of η for all $n > N$. For the values ε_0 and η_0 shown here, N_2 and N_3 satisfy the definition, but N_1 is not big enough.

The law of large numbers (Section 3.3) is equivalent to the statement that the sample mean[a] is a consistent estimator of the parent mean[a] (subject to a finiteness condition on the variance).

Since consistency is very much an asymptotic property, it does not necessarily imply that the precision is a monotonic function of n. That is, even if an estimator is consistent, adding some observations need not always increase the precision. This situation is illustrated in Fig. 7.1.

7.1.2. *Bias and consistency*

Let $\hat{\theta}$ be an estimator of parameter θ, based on N observations. We define the *bias* b of the estimator $\hat{\theta}$ as the deviation of the expectation of $\hat{\theta}$ from the true value θ_0,

$$b_N(\hat{\theta}) = E(\hat{\theta}) - \theta_0 = E(\hat{\theta} - \theta_0). \tag{7.1}$$

Thus, an estimator is *unbiased* if, for all N and θ_0,

$$b_N(\hat{\theta}) = 0$$

or

$$E(\hat{\theta}) = \theta_0 .$$

There is some arbitrariness in this definition, since there exist other measures of the centre of a distribution which could have been used. The expectation is conventionally chosen to be the arithmetic mean, probably because it is a linear operator, and because of its properties for the Normal distribution. Note that if an estimator is unbiased, this does not necessarily mean that on the average half the estimates will lie above θ_0 and half below. This would have been true if the median had been chosen instead of the arithmetic mean.

It may seem that unbiasedness and consistency are related, but one can easily show that neither one implies the other. Suppose that N observations of a given estimator yield estimates $\hat{\theta}$, distributed with probability density function $f(\hat{\theta}, \theta_0)$ around the true value θ_0. In Fig. 7.2 we give examples of four different functions $f(\hat{\theta}, \theta_0)$, using estimators with different combinations of bias and consistency. In each subfigure of Fig. 7.2 the family of narrowing curves represent the behaviour of $f(\hat{\theta})$ as N increases. In (1), each $f(\hat{\theta})$ is centred at θ_0, and the narrowest peaks converge toward θ_0, representing both unbiasedness and consistency. In (2), each distribution $f(\hat{\theta})$ is biased to the right of θ_0, but this estimator is nevertheless consistent since it converges

---

[a]Recall that the sample mean is the mean of the observations at hand, whereas the parent mean is the mean of the underlying distribution from which the sample has been drawn.

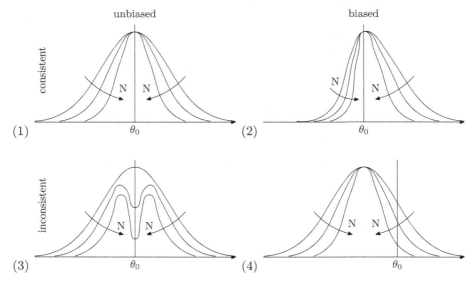

Fig. 7.2. Examples of probability density functions $f(\hat{\theta}, \theta_0)$ (not normalized) with different combinations of consistency and bias. The arrows show the direction of increasing N.

to the true value θ_0. In (3), each $f(\hat{\theta})$ is unbiased, but there is no convergence to θ_0. Finally, (4) shows an estimator which is both biased and inconsistent.

If t is an unbiased estimator of θ, one might think that t^2 would be an unbiased estimator of θ^2. Unfortunately, this is not the case, as bias, being a linear operator, is not invariant under change of variables. It would have been invariant if the definition of bias had been *median bias* instead of *mean bias*. Of course, even using *mean bias*, if t^2 is a consistent estimator of θ^2, it will tend to be unbiased for large samples, as in subfigure (2) of Fig. 7.2. It is partly for this reason that consistency is usually considered to be a more important property than unbiasedness.

7.2. Usual Methods of Constructing Consistent Estimators

Suppose that N observations X_i have been made of a random variable X with probability density function $f(X, \theta)$, where θ is an unknown parameter. If the purpose of the observations is to furnish information about the true value θ_0 of θ, one needs a *consistent estimator* $\hat{\theta}$ of θ_0. Thus, $\hat{\theta}$ should converge to θ_0 as N increases.

A quantity with such convergence properties can easily be constructed by making use of the *law of large numbers* (Section 3.3). This law states that

$$N^{-1} \sum_{i=1}^{N} a(X_i) \xrightarrow[N \to \infty]{} E[a(X)] = \int a(X) f(X, \theta) dX , \qquad (7.2)$$

where $a(X)$ is any function of X with finite variance $V[a(X)]$. This says that if $V[a(X)]$ is finite, the sample mean (the left-hand-side of Eq. 7.2) is a consistent estimator of $a(X)$.

The three most usual methods of estimation, each making use of the law of large numbers, are the *moments method*, the *maximum likelihood method* and the *least squares method*. In the present section we shall see how the estimators are constructed by these methods, and we shall show that they are consistent. Other properties of these methods will be discussed later in this chapter and in the next one.

7.2.1. *The moments method*

Suppose that it is possible to find such a function $a(X)$, that the right-hand side of Eq. (7.2) becomes a known function h of θ

$$E[a(X)] = \int a(X) f(X, \theta) dX = h(\theta) . \qquad (7.3)$$

If, at the true value θ_0, there exists an "inverse function"[b] h^{-1} of h, defined by

$$h^{-1}[h(\theta_0)] \equiv \theta_0 .$$

one can "solve" Eq. (7.3) for θ_0:

$$\theta_0 = h^{-1}\{E[a(X)]\} = h^{-1} \int a(X) f(X, \theta_0) dX . \qquad (7.4)$$

An intuitive way of obtaining an estimator $\hat{\theta}$ of θ_0 is to make the replacement (7.2) on the right-hand side of Eq. (7.4):

$$\hat{\theta} = h^{-1} \left[N^{-1} \sum_{i=1}^{N} a(X_i) \right] . \qquad (7.5)$$

---

[b]If there exist many inverse functions, only one of them will, in general, give a consistent estimator.

Thus the estimator $\hat{\theta}$ becomes a function of the known observations X_i rather than of the unknown p.d.f., $f(X, \theta_0)$, and it is consistent by the law of large numbers, Eq. (7.2).

In the one-dimensional case of this method, one may choose simply

$$a(X) = X . \tag{7.6}$$

From Eq. (7.3), $h(\theta_0)$ is then just the mean $\mu(\theta_0)$, and the consistent estimator $\hat{\theta}$ is given by

$$\hat{\theta} = \mu^{-1} \left(N^{-1} \sum_{i=1}^{N} X_i \right) .$$

When there are several parameters $\boldsymbol{\theta} = (\theta_1, \ldots, \theta_r)$ to be estimated, several functions $a_j(X)$ are required. Equation (7.3) then becomes

$$E[a_j(X)] = h_j(\boldsymbol{\theta}) , \quad j = 1, \ldots, r , \tag{7.7}$$

and the estimates $\hat{\theta}_j$ are the solutions of the simultaneous equations obtained by replacing

$$N^{-1} \sum_{i=1}^{N} a_j(X_i) \longrightarrow E[a_j(X)]$$

in Eq. (7.7).

In the particular case when

$$a_j(X_i) = X^j ,$$

the functions $h_j(\boldsymbol{\theta})$ become the moments $\mu'_j(\boldsymbol{\theta})$ of the distribution $f(X, \boldsymbol{\theta})$. This is the *moments method* of estimation.

7.2.2. *Implicitly defined estimators*

A more general alternative is to choose the function $a(X)$ in Eq. (7.3) to be a function of both X and θ, $a(X, \theta)$. Suppose that it is possible to find such a function $a(X, \theta)$, that $h(\theta)$ has a zero at the true value of the parameter, $\theta = \theta_0$:

$$E[a(X, \theta_0)] = \int a(X, \theta_0) f(X, \theta_0) dX = h(\theta_0) = 0 . \tag{7.8}$$

Equation (7.8) and the law of large numbers (7.2) then lead to an implicit method of finding an estimator. Let us define the experimental function ξ of the observations X_i,

$$\xi(\theta) \equiv N^{-1} \sum_{i=1}^{N} a(X_i, \theta) . \tag{7.9}$$

By the law of large numbers (7.2), $\xi(\theta)$ has asymptotically the same roots as Eq. (7.8):

$$\xi(\theta_0) \xrightarrow[N\to\infty]{} E[a(X,\theta_0)] = 0 \,.$$

An estimator $\hat{\theta}$ can now be defined implicitly as a root of the equation

$$\xi(\theta) = 0 \,. \tag{7.10}$$

The method of moments is obviously a particular case, with $a(X_i, \theta)$ in Eq. (7.9) replaced by

$$a(X_i) - E[a(X)] \,.$$

Equation (7.10) then yields the result of Eq. (7.5).

One can prove that one of the roots of Eq. (7.8) gives a consistent estimate of θ_0, provided some regularity conditions hold (in particular, that ξ is differentiable, and that the means of ξ and $\partial\xi/\partial\theta$ exist), and provided that in the limit $N \to \infty$

$$E\left\{\left[\frac{\partial\xi(\theta)}{\partial\theta}\right]_{\theta=\theta_0}\right\} \neq 0 \,. \tag{7.11}$$

Condition (7.11) ensures that the left-hand side of Eq. (7.9) has an inverse.

It is easy to sketch the proof in one dimension. In Fig. 7.3 we plot the function $\xi(\theta)$ versus θ. Let us denote the roots of Eq. (7.10) by A, $A', \ldots$, the point θ_0, $\xi(\theta_0)$ by B, and the point $(\theta_0, 0)$ by C. By the law of large numbers, B converges to C, and the slope of $\xi(\theta)$ at B converges to the constant (7.11) which, by hypothesis, is different from zero.

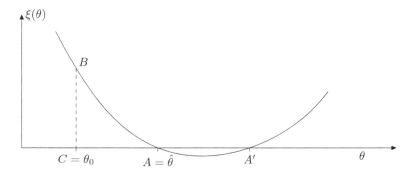

Fig. 7.3. An example of a function $\xi(\theta)$.

Therefore, at least one of the points A, $A', \ldots$ converges to C. Thus one of the roots $\hat{\theta}$ is a consistent estimate of θ [Fourgeaud, p. 214].

One obvious difficulty with such implicit methods is that Eq. (7.10) may have many solutions, and there is no way to distinguish between them. The method is certain to be good only when the function (7.8) has one unique zero.

In practice, one usually finds Eq. (7.10) by maximizing or minimizing some other function $g(X, \theta)$, related to $a(X, \theta)$ by

$$a(X_i, \theta) = \frac{\partial}{\partial \theta} g(X_i, \theta).$$

Typical examples are the well-known methods of maximum likelihood and minimum χ^2, to which we shall shortly return.

Under the conditions (7.8) and (7.11), one of the maxima or minima of the experimental quantity

$$N^{-1} \sum_{i=1}^{N} g(X_i, \theta)$$

provides a consistent estimator $\hat{\theta}$ of θ_0, and this $\hat{\theta}$ is a root of the equation

$$\xi(\theta) = \frac{1}{N} \frac{\partial}{\partial \theta} \sum_{i=1}^{N} g(X_i, \theta) = 0. \tag{7.12}$$

A choice of function $a(X_i, \theta)$, which by construction meets the condition (7.8), is

$$a(X_i, \theta_0) = \frac{\partial}{\partial \theta} \{ g(X_i, \theta) - E[g(X_i, \theta)] \}_{\theta=\theta_0},$$

provided the operators E and $\partial/\partial\theta$ commute. In this case,

$$E[a(X_i, \theta_0)] = E\left\{ \frac{\partial g(X_i, \theta)}{\partial \theta} - E\left[\frac{\partial g(X_i, \theta)}{\partial \theta} \right] \right\}_{\theta=\theta_0} \equiv 0.$$

Practically, $E(\partial g/\partial \theta)$ must not depend on θ_0; if it does, the estimation is useless since the true value θ_0 is unknown. In practice, therefore, Eq. (7.8) has to be supplemented by the further condition

$$\frac{\partial}{\partial \theta} E\left[\frac{\partial g(X_i, \theta)}{\partial \theta} \right]_{\theta=\theta_0} = 0.$$

The method of moments clearly corresponds to the particular choice

$$g(X_i, \theta) = a(X_i)\theta - \int_{\theta_0}^{\theta} \int a(X) f(X, \theta) dX \, d\theta. \tag{7.13}$$

7.2.3. *The maximum likelihood method*

An important case of the above method is the *maximum likelihood method*. In Section 5.1.1 we have defined the likelihood of a set of N independent observations $\mathbf{X} = X_1, X_2, \ldots, X_N$ as

$$L(\mathbf{X}|\theta) = \prod_{i=1}^{N} f(X_i, \theta),$$

where $f(X, \theta)$ is the p.d.f. of any observation X.

The *maximum likelihood estimate* of the parameter θ is that value $\hat{\theta}$ for which $L(\mathbf{X}|\theta)$ has its maximum, given the particular observations $\mathbf{X}$.

The maximum likelihood method[c] is a particular case of the method outlined in the previous section, with the choice

$$g(X_i, \theta) = \ln f(X_i, \theta). \tag{7.14}$$

The Eq. (7.12) becomes

$$\frac{\partial}{\partial \theta} \sum_{i=1}^{N} \ln f(X_i, \theta) = \frac{\partial}{\partial \theta} \ln L(\mathbf{X}, \theta) = 0. \tag{7.15}$$

Equation (7.15) is called the *likelihood equation*. It is a necessary condition for the existence of a maximum in $L(\mathbf{X}, \theta)$, provided that the maximum does not occur at a limit of the range of θ. The maximum likelihood estimate $\hat{\theta}$ is a root of Eqation (7.15).

It is easy to show that one of these roots is a consistent estimate, provided that f is twice differentiable, that the means of

$$\left. \frac{\partial}{\partial \theta} \ln f(X, \theta) \right|_{\theta=\theta_0} \quad \text{and} \quad \left. \frac{\partial^2}{\partial \theta^2} \ln f(X, \theta) \right|_{\theta=\theta_0}$$

exist, and that the operators

$$\int dX \quad \text{and} \quad \partial/\partial \theta$$

commute. Using the fact that

$$\int f(X, \theta_0) dX = 1$$

---

[c]Note that maximizing $\ln L$ or L gives the same result.

we have

$$\int \frac{\partial}{\partial\theta}[f(X,\theta)]_{\theta=\theta_0}dX = 0$$

or

$$\int \frac{\partial f}{\partial\theta}\frac{1}{f}fdX = 0. \tag{7.16}$$

Using the expectation operator, this can be rewritten as

$$E\left[\frac{\partial}{\partial\theta}\ln f(X,\theta)\right]_{\theta=\theta_0} = 0.$$

Differentiating Eq. (7.16) once more with respect to θ,

$$\int \frac{\partial^2 \ln f}{\partial\theta^2}fdX + \int \left(\frac{\partial \ln f}{\partial\theta}\right)^2 fdX = 0,$$

it follows that the experimental quantity

$$E\left[\frac{\partial\xi}{\partial\theta}\Big|_{\theta=\theta_0}\right] \equiv E\left[\frac{1}{N}\frac{\partial^2}{\partial\theta^2}\ln L(\mathbf{X},\theta)\right]_{\theta=\theta_0},$$

by the law of large numbers approaches a negative constant

$$E\left(\frac{\partial\xi}{\partial\theta}\right)_{\theta=\theta_0} \xrightarrow[N\to\infty]{} E\left[\frac{\partial^2}{\partial\theta^2}\ln f(X,\theta)\right]_{\theta=\theta_0}$$

$$= -E\left[\left(\frac{\partial \ln f(X,\theta)}{\partial\theta}\right)^2\right]_{\theta=\theta_0} < 0. \tag{7.17}$$

Thus, from Section 7.2.2 the root $\hat{\theta}$ of the likelihood equation (7.15) does indeed correspond to a maximum of the likelihood function.

It also follows from Eq. (7.16) that

$$V\left(\frac{\partial}{\partial\theta}\ln f(X,\theta)\right) = \int \left(\frac{\partial \ln f(X,\theta)}{\partial\theta^2}\right)^2 f(X,\theta)dX.$$

Thus the quantity

$$\frac{\partial}{\partial\theta}\ln L(\mathbf{X},\theta) = \frac{\partial}{\partial\theta}\left[\sum_{i=1}^{N}\ln f(X_i,\theta)\right]$$

is distributed with mean 0 and variance $NI(\theta)$ for all values of N (see Eqn. 5.4).

Moreover, the consistent solution will (asymptotically) correspond to the absolute maximum of the likelihood. In order to prove this, it is sufficient to show the following:

(i) By the law of large numbers, $\xi(\theta)$ converges to a constant function

$$\int \frac{\partial f(X, \theta)}{\partial \theta} \frac{f(X, \theta_0)}{f(X, \theta)} dX \,,$$

and therefore the roots of

$$\xi(\theta) = 0$$

converge to constants, $\tilde{\theta}_1, \tilde{\theta}_2, \ldots$

(ii) Since the function ln is strictly concave,

$$\int f(X, \theta_0) \ln \left[\frac{f(X, \theta)}{f(X, \theta_0)} \right] dX < \ln \int f(X, \theta_0) \frac{f(X, \theta)}{f(X, \theta_0)} dX = 0 \,,$$

for all $\theta \neq \theta_0$. Therefore

$$\int f(X, \theta_0) \ln\, f(X, \theta) dX < \int f(X, \theta_0) \ln\, f(X, \theta_0) dX$$

for all $\theta \neq \theta_0$.

Then if $\hat{\theta}'_N$ is a non-consistent sequence of solutions, and $\hat{\theta}_N$ a consistent one,

$$\lim_{N \to \infty} \frac{1}{N} \sum_{i=1}^{N} \ln\, f(X_i, \hat{\theta}'_N) = E[\ln\, f(X, \lim_{N \to \infty} \hat{\theta}'_N)] < E[\ln\, f(X, \theta_0)]$$

$$= \lim_{N \to \infty} \frac{1}{N} \sum_{i=1}^{N} \ln\, f(X_i, \hat{\theta}_N) \,.$$

Since this proof is true for the experimental likelihood only in the limit $N \to \infty$, one cannot be sure that a maximum found for N finite is the absolute maximum. As N increases, it is conceivable that the relative heights of different maxima of $\ln\, L$ change. This reflects the fundamental uncertainty about the true value θ_0, present in any *finite* sample.

7.2.4. *Least squares methods*

Consider a set of observations $\mathbf{Y} = \{Y_1, \ldots, Y_N\}$ from a distribution with expectations

$$E(\mathbf{Y}) = \mathbf{M}(\boldsymbol{\theta}) = \{m_i(\boldsymbol{\theta})\} \,, \quad i = 1, \ldots, N \,.$$

$\mathbf{M}$ is an N-dimensional function of r unknown parameters $\boldsymbol{\theta} = \{\theta_1, \ldots, \theta_r\}$.

The *least squares method* minimizes the quadratic from

$$g(\mathbf{Y}, \boldsymbol{\theta}) = [\mathbf{Y} - \mathbf{M}(\boldsymbol{\theta})]^{\mathrm{T}} \underset{\sim}{W} [\mathbf{Y} - \mathbf{M}(\boldsymbol{\theta})] \,, \tag{7.18}$$

where $\underset{\sim}{W}$ is a matrix of weights which must be independent of $\boldsymbol{\theta}$, otherwise a linear problem becomes non-linear and Eq. (7.2) is not valid.

The *least squares estimate* $\hat{\boldsymbol{\theta}}$ is again a root of Eq. (7.10), which here takes the form

$$\xi(\boldsymbol{\theta}) = \frac{\partial g}{\partial \boldsymbol{\theta}} = -2 \frac{\partial \mathbf{M}(\boldsymbol{\theta})^{\mathrm{T}}}{\partial \boldsymbol{\theta}} \underset{\sim}{W} [\mathbf{Y} - \mathbf{M}(\boldsymbol{\theta})] = 0. \tag{7.19}$$

If $\underset{\sim}{W}$ remains positive definite, and if the covariance matrix of the random variables Y_i remains bounded as N goes to infinity, one can prove that one of the roots $\hat{\boldsymbol{\theta}}$ of Eq. (7.19) is a *consistent estimator*.

In order to do so, let us assume that $\underset{\sim}{W}$ is diagonal (if it is not, it can always be diagonalized without affecting our proof). Defining

$$\mathbf{X}_i(\boldsymbol{\theta}) = \left\{ -\frac{\partial m_i(\boldsymbol{\theta})}{\partial \theta_k} W_{ii}[Y_i - m_i(\boldsymbol{\theta})] \right\}, \quad k = 1, \ldots, r,$$

the vector $\boldsymbol{\xi}(\boldsymbol{\theta})$, with r components in $\boldsymbol{\theta}$-space,

$$\xi_k(\boldsymbol{\theta}) = -\sum_{i=1}^{N} \frac{\partial m_i(\boldsymbol{\theta})}{\partial \theta_k} W_{ii}[Y_i - m_i(\boldsymbol{\theta})]$$

can be written

$$\boldsymbol{\xi}(\boldsymbol{\theta}) = \sum_{i=1}^{N} \mathbf{X}_i(\boldsymbol{\theta}). \tag{7.20}$$

By construction,

$$E[\mathbf{X}_i(\boldsymbol{\theta}_0)] = 0.$$

Now the arguments of Section 7.2.2 can be applied directly, since Eq. (7.20) is of the form (7.9). Therefore, if Eq. (7.11) is true in the limit $N \to \infty$, one of the roots of Eq. (7.19) converges to $\boldsymbol{\theta}_0$ provided $\partial \boldsymbol{\xi}/\partial \boldsymbol{\theta}$ converges to a non-zero constant. In fact

$$\frac{\partial \boldsymbol{\xi}}{\partial \boldsymbol{\theta}} = -\frac{\partial^2 \mathbf{M}^{\mathrm{T}}}{\partial \boldsymbol{\theta}^2} \underset{\sim}{W} [\mathbf{Y} - \mathbf{M}(\boldsymbol{\theta})] + \frac{\partial \mathbf{M}^{\mathrm{T}}}{\partial \boldsymbol{\theta}} \underset{\sim}{W} \frac{\partial \mathbf{M}}{\partial \boldsymbol{\theta}}. \tag{7.21}$$

By the law of large numbers, the first term on the right-hand side of Eq. (7.21) vanishes in the limit $N \to \infty$. At $\boldsymbol{\theta} = \boldsymbol{\theta}_0$ the second term is positive definite. Thus the consistent root $\hat{\boldsymbol{\theta}}$ of Eq. (7.19) corresponds (in the limit $N \to \infty$) to a minimum.

A case of particular importance is the *linear least squares method*, in which $\mathbf{M}(\boldsymbol{\theta})$ is a linear function of $\boldsymbol{\theta}$, thus

$$\mathbf{M}(\boldsymbol{\theta}) = E(\mathbf{Y}) = \underset{\sim}{A} \boldsymbol{\theta}.$$

The matrix $\mathcal{A}$ with constant elements is called the *design matrix*.

The estimator is then the *unique* root $\hat{\boldsymbol{\theta}}$ of

$$\xi(\boldsymbol{\theta}) = -2 \, \mathcal{A}^{\mathrm{T}} \underset{\sim}{W} \, (\mathbf{Y} - \mathcal{A} \, \boldsymbol{\theta}) = 0 \,. \tag{7.22}$$

This root is

$$\hat{\boldsymbol{\theta}} = (\mathcal{A}^{\mathrm{T}} \underset{\sim}{W} \mathcal{A})^{-1} \, \mathcal{A}^{\mathrm{T}} \underset{\sim}{W} \, \mathbf{Y} \,. \tag{7.23}$$

Clearly if $\mathcal{A}$ is such that $\mathcal{A}^{\mathrm{T}} \underset{\sim}{W} \mathcal{A}$ is non-singular, $\hat{\boldsymbol{\theta}}$ is also consistent.

7.3. Asymptotic Distributions of Estimates

By asymptotic we always mean in the limit as the number of observations, N, becomes infinitely large.

7.3.1. *Asymptotic Normality*

To obtain the asymptotic distribution of consistent estimates we invoke the *Central Limit theorem* (Section 3.3.2), which goes beyond the statement (7.2). It says that

$$N^{-1} \sum_{i=1}^{N} a(X_i) \tag{7.24}$$

is asymptotically Normally distributed, with mean $E[a(X)]$ and variance $N^{-1} V[a(X)]$, provided $V[a(X)]$ is finite.

We can express $\hat{\boldsymbol{\theta}}$ in Eq. (7.5) by a series expansion around $E[a(X)]$.

$$\hat{\boldsymbol{\theta}} = h^{-1} \left[N^{-1} \sum_{i=1}^{N} a(X_i) \right]$$

$$= h^{-1} \{ E[a(X)] \} + \frac{\partial h^{-1}}{\partial a} \{ E[a(X)] \} \left\{ N^{-1} \sum_{i=1}^{N} a(X_i) - E[a(X)] \right\} + O(N^{-1}) \,. \tag{7.25}$$

To this order, $\hat{\theta}$ is obviously asymptotically Normally distributed, since the expression (7.24) is, and all other terms entering the right-hand side of Eq. (7.25) are constants.

The Normality of estimators of the form (7.5) is, however, very much an asymptotic property. This is clear from the fact that the random variable (7.24) is Normal for a finite number of observations only if the $a(X_i)$ are Normal. Even if this were the case, the function h will, in general, destroy the Normality for a finite number of observations.

The variance of the random variable (7.24) is proportional to N^{-1}. Therefore, to the order of approximation given in Eq. (7.25), the variance of $\hat{\theta}$ is proportional to N^{-1}. Asymptotically, the variance of $\hat{\theta}$ is easily seen to be

$$V(\hat{\theta}) = E(\hat{\theta} - \theta_0)^2 = \frac{1}{N}\left(\frac{\partial h^{-1}}{\partial a}\right)^2 V(a),$$

where $V(a)$ is the variance of $a(X)$.

When there are many parameters, $\boldsymbol{\theta} = (\theta_1, \theta_2, \ldots, \theta_r)$, the functions h^{-1} and a become vector functions $\mathbf{h}^{-1}$ and $\mathbf{a}$. The covariance matrix of the asymptotic distribution of $\widehat{\boldsymbol{\theta}}$ then becomes

$$\underset{\sim}{V}(\widehat{\boldsymbol{\theta}}) = E[(\widehat{\boldsymbol{\theta}} - \boldsymbol{\theta}_0)(\widehat{\boldsymbol{\theta}} - \boldsymbol{\theta}_0)^{\mathrm{T}}].$$

$$= \frac{1}{N}\left[\frac{\partial \mathbf{h}^{-1}}{\partial \mathbf{a}}\right] \underset{\sim}{V}(\mathbf{a}) \left[\frac{\partial \mathbf{h}^{-1}}{\partial \mathbf{a}}\right]^{\mathrm{T}},$$

where $\underset{\sim}{V}(\mathbf{a})$ is the covariance matrix of $\mathbf{a}$, $\partial \mathbf{h}^{-1}/\partial \mathbf{a}$ is a matrix with elements $\partial h_i^{-1}/\partial a_j$, and $h_i^{-1}(\mathbf{a}) = \theta_{0i}$.

The bias comes from terms of order N^{-1} which have been neglected on the right-hand side of Eq. (7.25). It follows that the bias is at most of order N^{-1}.

Analogous results are obtained when one uses implicitly defined estimators. Normality, for instance, follows from the fact that B in Fig. 7.3 is Normally distributed around C. That is, applying the *Central Limit theorem*, $\boldsymbol{\xi}(\boldsymbol{\theta}_0)$, in Eq. (7.9), is asymptotically Normally distributed about zero. Then in the limit, as AC becomes proportional to BC, $(\boldsymbol{\theta} - \boldsymbol{\theta}_0)$ is also Normally distributed. The variance matrix of $\boldsymbol{\theta}$ is given by

$$\underset{\sim}{V}(\widehat{\boldsymbol{\theta}}) = \frac{1}{N}\left[E\left(\frac{\partial \boldsymbol{\xi}(\boldsymbol{\theta})}{\partial \boldsymbol{\theta}}\right)_{\boldsymbol{\theta}=\boldsymbol{\theta}_0}\right]^{-1} \underset{\sim}{V}[\mathbf{a}(\mathbf{X}, \boldsymbol{\theta}_0)] \left[E\left(\frac{\partial \boldsymbol{\xi}(\boldsymbol{\theta})}{\partial \boldsymbol{\theta}}\right)_{\boldsymbol{\theta}=\boldsymbol{\theta}_0}\right]^{-1} \quad (7.26)$$

since

$$E\left[\left(\frac{\partial \boldsymbol{\xi}(\boldsymbol{\theta})}{\partial \boldsymbol{\theta}}\right)\right]_{\boldsymbol{\theta}=\boldsymbol{\theta}_0}$$

is the asymptotic slope of BA, and $N^{-1}\underset{\sim}{V}[\mathbf{a}(\mathbf{X}, \boldsymbol{\theta}_0)]$ is the asymptotic variance of BC.

As an example, let us calculate the asymptotic $\underset{\sim}{V}(\widehat{\boldsymbol{\theta}})$ for the maximum likelihood method, where

$$\mathbf{a}(\mathbf{X}, \boldsymbol{\theta}) = \frac{\partial \ln \mathbf{f}(\mathbf{X}, \boldsymbol{\theta})}{\partial \boldsymbol{\theta}}.$$

Then we have

$$\underset{\sim}{V} [\mathbf{a}(\mathbf{X}, \boldsymbol{\theta_0})] = E\left[\left(\frac{\partial \ln f(\mathbf{X}, \boldsymbol{\theta})}{\partial \boldsymbol{\theta}}\right)^2\right]_{\boldsymbol{\theta}=\boldsymbol{\theta}_0}$$

$$= -E\left(\frac{\partial^2 \ln f(\mathbf{X}, \boldsymbol{\theta})}{\partial \boldsymbol{\theta}^2}\right)_{\boldsymbol{\theta}=\boldsymbol{\theta}_0}$$

from Eq. (7.17), and

$$E\left[\left(\frac{\partial \boldsymbol{\xi}(\boldsymbol{\theta})}{\partial \boldsymbol{\theta}}\right)_{\boldsymbol{\theta}=\boldsymbol{\theta}_0}\right] = E\left(\frac{\partial^2 \ln f(\mathbf{X}, \boldsymbol{\theta})}{\partial \boldsymbol{\theta}^2}\right)_{\boldsymbol{\theta}=\boldsymbol{\theta}_0}$$

by the law of large numbers. It follows from Eq. (7.26) that, asymptotically

$$\underset{\sim}{V}(\hat{\boldsymbol{\theta}}) = \frac{1}{N}\left[E\left(\frac{\partial \ln f(\mathbf{X}, \boldsymbol{\theta})}{\partial \boldsymbol{\theta}}\right)^2\right]_{\boldsymbol{\theta}=\boldsymbol{\theta}_0}^{-1} = -\frac{1}{N}\left[E\left(\frac{\partial^2 \ln f(\mathbf{X}, \boldsymbol{\theta})}{\partial \boldsymbol{\theta}^2}\right)^2\right]_{\boldsymbol{\theta}=\boldsymbol{\theta}_0}^{-1}$$

$$= [NI(\boldsymbol{\theta}_0)]^{-1}, \tag{7.27}$$

the inverse of the information matrix.

Thus, for the maximum likelihood method, the quantity $\sqrt{N}(\hat{\boldsymbol{\theta}} - \boldsymbol{\theta}_0)$ has asymptotically the Normal distribution $N[0, I_1^{-1}(\boldsymbol{\theta}_0)]$. Also the quantity $\partial \ln L(\mathbf{X}, \boldsymbol{\theta})/\partial \boldsymbol{\theta}$ is asymptotically Normally distributed with mean 0 and variance matrix $I_N(\boldsymbol{\theta}) = NI(\boldsymbol{\theta})$.

7.3.2. *Asymptotic expansion of moments of estimates*

In this section we shall demonstrate how the *moments of an estimator* can be computed, using a series expansion in N^{-1}. Of particular interest are the moments

$$E(\hat{\theta} - \theta_0), \quad V(\hat{\theta})$$

the *bias* and the *variance* of $\hat{\theta}$, respectively.

Recall that

$$E(\hat{\theta} - \theta_0) = E(\hat{\theta}) - \theta_0, \quad V(\hat{\theta} - \theta_0) = V(\hat{\theta}),$$

$$E[(\hat{\theta} - \theta_0)^2] = V(\hat{\theta}) + [E(\hat{\theta} - \theta_0)]^2.$$

Suppose that the estimator $\hat{\theta}$ is the root of Eq. (7.12), with

$$E[\xi(\theta_0)] = 0$$

valid, and Eq. (7.11) valid. Let us expand $\xi(\hat{\theta})$ around the true, but unknown value θ_0, making use of Eq. (7.12):

$$\xi(\hat{\theta}) = \xi(\theta_0) + (\hat{\theta} - \theta_0)\frac{\partial \xi}{\partial \theta}\bigg|_{\theta=\theta_0} + \frac{1}{2}(\hat{\theta} - \theta_0)^2\frac{\partial^2 \xi}{\partial \theta^2}\bigg|_{\theta=\theta_0} + \cdots = 0$$

by construction. Then

$$\xi(\theta_0) = -(\hat{\theta} - \theta_0)\frac{\partial \xi}{\partial \theta}\bigg|_{\theta=\theta_0} - \frac{1}{2}(\hat{\theta} - \theta_0)^2\frac{\partial^2 \xi}{\partial \theta^2}\bigg|_{\theta=\theta_0} - \cdots . \tag{7.28}$$

In order to express $(\hat{\theta} - \theta_0)$ in terms of $\xi(\theta_0)$, we use Lagrange's theorem of inversion of a series, which states that if

$$u = \sum_{i=1}^{\infty} a_i v^i \tag{7.29}$$

then

$$v = \sum_{j=1}^{\infty} \frac{u^j}{j!}\left[\frac{d^{j-1}}{dx^{j-1}}\left(\sum_{i=1}^{\infty} a_i x^{i-1}\right)^{-j}\right]_{x=0}$$

provided that $a_1 \neq 0$, and that the series (7.29) converges. Thus the inversion of the series (7.28) becomes

$$\hat{\theta} - \theta_0 = \left[\frac{\xi(\theta_0)}{-\partial \xi/\partial \theta} + \frac{1}{2}\frac{\xi^2(\theta_0)\partial^2 \xi/\partial \theta^2}{(-\partial \xi/\partial \theta)^3} + \cdots\right]_{\theta=\theta_0}. \tag{7.30}$$

By the law of large numbers, all the derivatives

$$\frac{\partial^n \xi(\theta)}{\partial \theta^n}\bigg|_{\theta=\theta_0}$$

converge to fixed values for $N \to \infty$, in particular

$$\frac{\partial \xi}{\partial \theta}\bigg|_{\theta=\theta_0} \xrightarrow[N\to\infty]{} E\left[\left(\frac{\partial \theta}{\partial \theta}\right)_{\theta=\theta_0}\right] \neq 0$$

by hypothesis, Eq. (7.11). The use of Lagrange's theorem is, therefore, justified for N large enough. From the law of large numbers, also $\xi(\theta_0)$ converges to

$$E[\xi(\theta_0)] = 0.$$

Now let us define new random variables X, Y, Z through

$$\xi(\theta_0) = E[\xi(\theta_0)] + X = X$$

$$\frac{\partial \xi}{\partial \theta}\bigg|_{\theta=\theta_0} = E\left[\left(\frac{\partial \xi}{\partial \theta}\right)_{\theta=\theta_0}\right] + Y \equiv -a + Y$$

$$\frac{\partial^2 \xi}{\partial \theta^2}\bigg|_{\theta=\theta_0} = E\left[\left(\frac{\partial^2 \xi}{\partial \theta^2}\right)_{\theta=\theta_0}\right] + Z \equiv b + Z,$$

where a and b have been defined for convenience of notation.

By construction

$$E(X) = E(Y) = E(Z) = 0.$$

Then one can easily show[d] that the moments have an N-dependence give by

$$E(X^\alpha Y^\beta Z^\gamma) \propto N^{1-(\alpha+\beta+\gamma)}.$$

Rewriting Eq. (7.30) in terms of the new variables, and retaining terms of order N^{-1}, one gets

$$\hat{\theta} - \theta_0 = \frac{X}{a} + \frac{1}{a^2}\left(XY + \frac{b}{2a}X^2\right) + O\left(\frac{1}{N^2}\right). \tag{7.31}$$

Taking expectations and variances, respectively, one finally obtains

$$E(\hat{\theta} - \theta_0) = \frac{1}{a^2}\left[E(XY) + \frac{b}{2a}E(X^2)\right] + O\left(\frac{1}{N^2}\right),$$

$$V(\hat{\theta} - \theta_0) = \frac{1}{a^2}E(X^2) + O\left(\frac{1}{N^2}\right).$$

---

[d]One uses the independence of the terms of the sum

$$X = \frac{1}{N}\sum_{i=1}^{N} a(X_i, \theta_0),$$

and of the similar sums for Y and Z. The statement holds when at least two of the three exponents α, β, and γ are ≤ 1.

Going back to the functions g (which were to be minimized or maximized), we can express the bias and the variance to order $1/N$ as

$$b_N(\hat{\theta}) = E(\hat{\theta} - \theta_0)$$

$$= N^{-1} \left[E\left(\frac{\partial^2 g}{\partial \theta^2}\right) \right]_{\theta=\theta_0} \left\{ \text{cov}\left(\frac{\partial g}{\partial \theta}, \frac{\partial^2 g}{\partial \theta^2}\right) \right.$$

$$\left. - \frac{1}{2} \frac{E(\partial^3 g/\partial \theta^3)}{E(\partial^2 g/\partial \theta^2)} E\left[\left(\frac{\partial g}{\partial \theta}\right)^2\right] \right\}_{\theta=\theta_0} + O\left(\frac{1}{N^2}\right) \quad (7.32)$$

$$V(\hat{\theta} - \theta_0) = \frac{1}{N} \left\{ \frac{E[(\partial g/\partial \theta)^2]}{[E(\partial^2 g/\partial \theta^2)]^2} \right\}_{\theta=\theta_0} + O\left(\frac{1}{N^2}\right),$$

where

$$\text{cov}\left(\frac{\partial g}{\partial \theta}, \frac{\partial^2 g}{\partial \theta^2}\right) = E\left(\frac{\partial g}{\partial \theta}\frac{\partial^2 g}{\partial \theta^2}\right) - E\left(\frac{\partial g}{\partial \theta}\right) E\left(\frac{\partial^2 g}{\partial \theta^2}\right).$$

One still has the liberty of choice of different functions $g(X, \theta)$, and this may be used to minimize the bias or/and the variance. For instance, one could compute a new estimate

$$\hat{\theta}' = \hat{\theta} - E(\hat{\theta} - \theta_0),$$

which would be unbiased. But usually, $E(\hat{\theta} - \theta_0)$ depends on θ_0, and the best one can do is just to use the leading term in N of $E(\hat{\theta} - \theta_0)$ thereby reducing the bias to still higher orders of N^{-1}. However, the price for this reduction in bias is an increase in the variance. In most cases

$$V(\hat{\theta}' - \theta_0) > V(\hat{\theta} - \theta_0).$$

7.3.3. *Asymptotic bias and variance of the usual estimators*

In this section we shall explicitly calculate the bias and the variance from Eq. (7.32) for the three usual estimators. We specialize to the case of only one parameter θ.

(i) The *moments method*
   From Eqs. (7.6) and (7.13) the function g is

$$g(X_i, \theta) = X_i \theta - \int_{\theta_0}^{\theta} \int X f(X, \theta) dX d\theta.$$

It follows that

$$E[\xi(\theta_0)] = E\left(\frac{\partial g}{\partial \theta}\right)_{\theta=\theta_0} = E(X) - \mu_1(\theta_0) = 0\,,$$

$$E\left(\frac{\partial^2 g}{\partial \theta^2}\right) = -\frac{\partial \mu_1'(\theta)}{\partial \theta}$$

$$E\left(\frac{\partial g}{\partial \theta}\frac{\partial^2 g}{\partial \theta^2}\right) = 0$$

$$V\left(\frac{\partial g}{\partial \theta}\right) = E\left[\left(\frac{\partial g}{\partial \theta}\right)^2\right] = \mu_2(\theta_0)$$

$$E\left(\frac{\partial^3 g}{\partial \theta^3}\right) = -\frac{\partial^2 \mu_1'(\theta_0)}{\partial \theta^2}\,,$$

where the notation of moments is as defined in Section 2.4.6. Using Eq. (7.32) we obtain the bias and the variance of $\hat\theta - \theta_0$,

$$b_N(\hat\theta) = E(\hat\theta - \theta_0) = -\frac{1}{2N}\mu_2 \frac{(\partial^2 \mu_1'/\partial\theta^2)}{(\partial \mu_1'/\partial\theta)^3}\bigg|_{\theta=\theta_0}\,,$$

$$V(\hat\theta - \theta_0) = \frac{1}{N}\frac{\mu_2}{(\partial \mu_1'/\partial\theta)^2}\bigg|_{\theta=\theta_0}\,.$$

(ii) The *linear least squares method*

Here g is given by Eq. (7.18) and ξ by Eq. (7.22), with

$$\underset{\sim}{A} = a\,, \quad \underset{\sim}{W} = \frac{1}{\sigma^2}$$

for the one-dimensional case. Thus

$$E[\xi(\theta_0)] = E\left(\frac{\partial g}{\partial \theta}\right)_{\theta=\theta_0} = -2E\left[\frac{a(Y-a\theta)}{\sigma^2}\right] = 0\,.$$

Furthermore,

$$E\left(\frac{\partial^2 g}{\partial \theta^2}\right) = 2\left(\frac{a}{\sigma}\right)^2$$

$$E\left(\frac{\partial g}{\partial \theta}\frac{\partial^2 g}{\partial \theta^2}\right) = 0$$

$$\frac{\partial^n g}{\partial \theta^n} = 0 \quad \text{for} \quad n \geq 3\,.$$

Thus one verifies that the linear least squares estimator (7.23) is unbiased to all orders of N [the terms grouped into $O(N^{-2})$ in Eq. (7.32) are proportional to higher-order derivatives, which all vanish].

(iii) The *maximum likelihood method*

Here g is given by Eq. (7.14), and we have already seen in Eq. (7.16) that $E[\xi(\theta_0)] = 0$. Using the relation (7.27) between the second derivatives of g and the information I (per event, on θ_0), and the relation (7.17), the bias and the variance take the form

$$b_N(\hat{\theta}) = E(\hat{\theta} - \theta_0) = \frac{K + 2J}{2NI^2}$$

$$\tag{7.33}$$

$$V(\hat{\theta} - \theta_0) = \frac{1}{NI},$$

where

$$K \equiv E\left(\frac{\partial^3 \ln f}{\partial \theta^3}\right)_{\theta=\theta_0}$$

$$J \equiv E\left(\frac{\partial^2 \ln f}{\partial \theta^2}\frac{\partial \ln f}{\partial \theta}\right)_{\theta=\theta_0}.$$

In this notation Eq. (7.31) becomes

$$\hat{\theta} - \theta_0 = \frac{X}{I} + \frac{XY}{NI^2} + \frac{1}{2}\frac{X^2K}{NI^3} + O\left(\frac{1}{N^2}\right).$$

When I, J, and K are estimated using the derivatives of the experimental likelihood function at its maximum, random terms of order $N^{-\frac{1}{2}}$ get introduced. Therefore, the expressions (7.33) for the bias and the variance are correct excluding terms of order $N^{-3/2}$.

We shall come back to the question of reduction of bias in Section 8.6. Note that all these expressions are not correct as they stand for the many-dimensional case. However, the same method may be used.

7.4. Information and the Precision of an Estimator

Recall that an estimate (the numerical result of applying an estimator) is a random variable, and it therefore has a probability distribution called the *sampling distribution* of the estimator. In Section 7.1 we defined consistency and unbiasedness in such a way that the sampling distribution of a *consistent estimator* is centred arbitrarily close to the true value θ_0, for a sufficiently large number of observations, and the sampling distribution of an unbiased estimator is always centred on θ_0.

In this section we shall describe the relation between the information of an estimator and the variance of its sampling distribution.

7.4.1. *Lower bounds for the variance — Cramér–Rao inequality*

Let **X** be observations from a distribution with density function $f(\mathbf{X}|\theta)$, and let the estimator $\hat{\theta}$ have the sampling distribution $q(\hat{\theta}|\theta)$. Denote the likelihood function of the observations $L_{\mathbf{X}} = L(\mathbf{X}|\theta)$, the likelihood of the estimator $L_{\hat{\theta}} = L(\mathbf{X}|\hat{\theta})$, and the corresponding informations $I_{\mathbf{X}}$ and $I_{\hat{\theta}}$.

From Eq. (7.1), the bias is a function of the true value θ_0

$$b = E(\hat{\theta}) - \theta_0 = \int \hat{\theta}(\mathbf{X}) f(\mathbf{X}|\theta_0) d\mathbf{X} - \theta_0 \,.$$

Let us, for the time being, drop the subscript on θ. The variance of the sampling distribution,

$$V(\hat{\theta}) = \int [\hat{\theta} - E(\hat{\theta})]^2 q(\hat{\theta}|\theta) d\hat{\theta} \,, \qquad (7.34)$$

is related to the information by the *Cramér–Rao inequality*.

$$V(\hat{\theta}) \geq \frac{[1 + (db/d\theta)]^2}{I_{\hat{\theta}}} \geq \frac{[1 + (db/d\theta)]^2}{I_{\mathbf{X}}} \,. \qquad (7.35)$$

The first part of this important inequality states that the variance of an unbiased estimate is bounded below by the inverse of the information it carries.

The second part of the inequality follows directly from Eq. (5.10), changing t to $\hat{\theta}$, and it states that the variance of any unbiased estimate is bounded below by the inverse of the information contained in the observations. Replacing $I_{\mathbf{X}}$ by its definition (5.2), we have for any estimate

$$V(\hat{\theta}) \geq \frac{\left(\dfrac{d\tau(\theta)}{d\theta}\right)^2}{E\left[\left(\dfrac{\partial \ln L_{\mathbf{X}}}{\partial \theta}\right)^2\right]} \,, \qquad (7.36)$$

where

$$\tau(\theta) \equiv E(\hat{\theta}) = \theta + b(\theta) \,. \qquad (7.37)$$

The conditions under which the inequalities (7.35) and (7.36) hold are that

(i) the range of observables **X** should not depend on θ;
(ii) L_X must be sufficiently regular that differentiation with respect to θ and integration over **X** commute.

The proof of the inequality is given below.

Let us note first that, by definition,

$$\int L_{\hat{\theta}}d\hat{\theta} = \int q(\hat{\theta}|\theta)d\hat{\theta} = 1 .$$

It is, therefore, immaterial whether we call the sampling distribution $q(\hat{\theta}|\theta)$ or $L_{\hat{\theta}}$. Differentiating with respect to θ, gives

$$\int \frac{\partial q}{\partial \theta}d\hat{\theta} = \int \frac{\partial(\ln q)}{\partial \theta}q d\hat{\theta} = 0 . \qquad (7.38)$$

Differentiating Eq. (7.37) with respect to θ gives

$$\frac{\partial}{\partial \theta}\int \hat{\theta}q(\hat{\theta}|\theta)d\hat{\theta} = \int \hat{\theta}\frac{\partial(\ln q)}{\partial \theta}q d\hat{\theta} = 1 + \frac{db}{d\theta} . \qquad (7.39)$$

Noting that $\theta + b(\theta)$ is a constant, we can multiply it into Eq. (7.38), to obtain zero, and subtract it from Eq. (7.39):

$$\int [\hat{\theta} - \theta - b(\theta)]\frac{(\ln q)}{\partial \theta}q d\hat{\theta} = 1 + \frac{db}{d\theta} .$$

Application of Schwarz' inequality now yields

$$\int [\hat{\theta} - \theta - b(\theta)]^2 q(\hat{\theta}|\theta)d\hat{\theta} \int \left[\frac{\partial(\ln q)}{\partial \theta}\right]^2 q(\hat{\theta}|\theta)d\hat{\theta} \geq \left(1 + \frac{db}{d\theta}\right)^2 . \qquad (7.40)$$

Since the first integral is an expression for $V(\hat{\theta})$, as defined by Eqs. (7.34) and (7.37), the inequality (7.40) becomes

$$V(\hat{\theta}) \geq \frac{\left(1 + \dfrac{db}{d\theta}\right)^2}{E\left[\left(\dfrac{\partial \ln q}{\partial \theta}\right)^2\right]} = \frac{\left(1 + \dfrac{db}{d\theta}\right)^2}{E\left[\left(\dfrac{\partial \ln L_{\hat{\theta}}}{\partial \theta}\right)^2\right]} = \frac{\left(1 + \dfrac{db}{d\theta}\right)^2}{I_{\hat{\theta}}} ,$$

which proves the Cramér–Rao inequality (7.35).

7.4.2. *Efficiency and minimum variance*

Consider first the conditions under which the inequality (7.40), and the first inequality in (7.35), become equalities. This occurs when

$$\frac{\partial \ln L_{\hat{\theta}}}{\partial \theta} = A(\theta)[\hat{\theta} - \theta - b(\theta)] . \tag{7.41}$$

It follows from Eq. (7.41) that *minimum variance* for the statistic $\hat{\theta}$,

$$V(\hat{\theta}) = (I_{\hat{\theta}})^{-1}\left(1 + \frac{db}{d\theta}\right)^2 , \tag{7.42}$$

is attained when the sampling distribution of $\hat{\theta}$ is of the exponential form (5.8)

$$L_{\hat{\theta}} = q(\hat{\theta}|\theta) = \exp[a(\theta)\hat{\theta} + \beta(\hat{\theta}) + c(\theta)] . \tag{7.43}$$

Here $a(\theta)$ is the integral of the arbitrary function $A(\theta)$ in Eq. (7.41), $\beta(\hat{\theta})$ is an integration constant and $c(\theta)$ is an arbitrary function.

When also the second inequality in Eq. (7.35) holds as equality,

$$I_{\hat{\theta}} = I_{\mathbf{X}} , \tag{7.44}$$

the variance $V(\hat{\theta})$ attains its *minimum variance bound*, and the statistic $\hat{\theta}$ is said to be an *efficient estimator*. It is important to note the distinction between

(i) *minimum variance* when the particular estimator $\hat{\theta}$ has the lowest variance of the family of estimators considered.
(ii) *minimum variance bound* for the set of data $\mathbf{X}$, attained when both Eqs. (7.42) and (7.44) hold. (The estimator $\hat{\theta}$ is *efficient*.)

An important case is when the sampling distribution $q(\hat{\theta}|\theta)$ is Normal. It is then completely equivalent to speak of the estimate $\hat{\theta}$ reaching the minimum variance bound, and maximum information.

Equation (7.44) is a necessary and sufficient condition for $\hat{\theta}$ to be a *sufficient statistic* for θ.

When a sufficient statistic t exists, the probability density $f(\mathbf{X}|\theta)$ is of the exponential form (5.8), by Darmois' theorem (Section 5.3.4). Inversely, when $f(\mathbf{X}, \theta)$ is of the exponential form (5.8), then

$$t = N^{-1}\sum_{i=1}^{N} \alpha(X_i) \tag{7.45}$$

is a sufficient statistic, from a sampling distribution of exponential form,

$$q(t|\theta) = \exp[a(\theta)t + \beta_1(t) + c(\theta)],$$

where β_1 is deduced from β by Eq. (7.45). It can be shown that

$$E(t) = -(dc/d\theta)/(da/d\theta) \equiv r(\theta).$$

The statistic (7.45) is therefore an unbiased estimator[e] of $r(\theta)$ and since $q(t|\theta)$ is of the exponential form, it follows that

$$V(t) = I_t^{-1}[r(\theta)] = I_{\mathbf{X}}^{-1}[r(\theta)].$$

In conclusion, a minimum variance bound unbiased estimator t exists if, and only if, the distribution $f(\mathbf{X}, \theta)$ admits sufficient statistics. In this case, the function $r(\theta)$ estimated by t is unique, to within a linear transformation [Fourgeaud, p. 206].

Examples

To illustrate these results, consider the examples of Darmois theorem in Section 5.3.4.

In case (i), with μ unknown and σ^2 known,

$$r(\mu) = -\frac{\mu}{\sigma^2},$$

and μ can be estimated efficiently.

In case (ii), with μ known and σ^2 unknown

$$r(\sigma) = \frac{\mu^2}{2} - \frac{\sigma^2}{2\sqrt{2\pi}},$$

and only the function σ^2 can be estimated efficiently without bias.

Finally, in case (iii), when both μ and σ^2 are unknown, we have

$$r_1(\mu) = \mu$$

and

$$r_2(\sigma) = \frac{\mu^2}{2} - \frac{\sigma^2}{2\sqrt{2\pi}}.$$

Thus μ and σ^2 can be estimated efficiently without bias.

---

[e]It is also the maximum likelihood estimate.

This example shows why one usually estimates σ^2 rather than σ: σ^2 is the unique function of σ which can be estimated efficiently without bias.

On the other hand, it is possible to construct an unbiased estimate of σ, which then does not attain the minimum variance bound. In other words, it does not contain the maximum information about σ, it is less efficient. Such an estimator is

$$T = \sqrt{\frac{N}{2}} \frac{\Gamma\left(\dfrac{N}{2}\right)}{\Gamma\left(\dfrac{N+1}{2}\right)} \sqrt{\frac{1}{N}\sum_{i=1}^{N}(X_i - \mu)^2}\,, \qquad (7.46)$$

[Fourgeaud, p. 204]. Among all unbiased estimators, T is of minimum variance. If one insisted upon an estimator of even smaller variance, one would have to give up unbiasedness. This brings us to an important result, of greater generality than the above example.

If a sufficient statistic t exists, a necessary and sufficient condition, for an unbiased estimate T to have minimum variance, is that it is a function of t only

$$T = T(t)\,. \qquad (7.47)$$

In the example of Eq. (7.46), T is a function of the sufficient statistic

$$t = \sqrt{\frac{1}{N}\sum_{i=1}^{N}(X_i - \mu)^2}\,.$$

The statistic T is not, in general, efficient, that is, it does not necessarily attain the minimum variance bound, since there is only one function of the parameter which can be estimated in an unbiased and efficient way.

7.4.3. *Cramér–Rao inequality for several parameters*

The above results generalize in a straightforward way to the case of several parameters,

$$\boldsymbol{\theta} = (\theta_1, \ldots, \theta_k)\,.$$

As shown in Section 5.2.1, the information becomes a matrix $\mathcal{I}_{\mathbf{X}}$, with elements given by Eq. (5.3). The *Cramér–Rao inequality* becomes, for unbiased estimators $\widehat{\boldsymbol{\theta}}$,

$$V(\hat{\theta}_i) \geq \left[\,\mathcal{I}_{\hat{\theta}}^{-1}(\boldsymbol{\theta})\right]_{ii} \geq \left[\,\mathcal{I}_{\mathbf{X}}^{-1}(\boldsymbol{\theta})\right]_{ii}\,, \quad i = 1, \ldots, k\,, \qquad (7.48)$$

where $[\mathcal{I}^{-1}]_{ii}$ denotes the ith diagonal element of the inverse of the information matrix.

More generally, let us define ellipsoids in $\boldsymbol{\theta}$-space by

$$(\mathbf{u} - \boldsymbol{\theta})^{\text{T}} \underset{\sim}{M} (\mathbf{u} - \boldsymbol{\theta}) = \text{const.},$$

where $\mathbf{u}$ is any vector variable, and $\underset{\sim}{M}$ is a positive-definite matrix. Then the Cramér–Rao inequality states that the ellipsoids

$$(\mathbf{u} - \boldsymbol{\theta})^{\text{T}} \mathcal{I}_{\hat{\theta}} (\mathbf{u} - \boldsymbol{\theta}) = c,$$

lie everywhere inside the ellipsoid

$$(\mathbf{u} - \boldsymbol{\theta})^{\text{T}} \underset{\sim}{V} (\widehat{\boldsymbol{\theta}})(\mathbf{u} - \boldsymbol{\theta}) = c,$$

where $\underset{\sim}{V} (\hat{\theta})$ denotes the covariance matrix of the estimates $\widehat{\boldsymbol{\theta}}$, with elements

$$v_{ij} = E(\hat{\theta}_i \hat{\theta}_j) - E(\hat{\theta}_i)E(\hat{\theta}_j).$$

The equality signs in Eq. (7.48) hold when the estimates $\widehat{\boldsymbol{\theta}}$ are jointly sufficient for $\boldsymbol{\theta}$ [Fourgeaud, p. 198].

7.4.4. *The Gauss–Markov theorem*

In the case where sufficient statistics do not exist, we must reduce our ambition. Let us consider the class of *unbiased* estimators which are *linear functions* of the observations.

Given a set of N observations $Y_1, \ldots, Y_N$ with probability density function $f(\mathbf{Y}|\boldsymbol{\theta})$, dependent on k unknown parameters $\boldsymbol{\theta}$, and with covariance matrix $\sigma^2 \underset{\sim}{V}$, a necessary condition for the existence of unbiased linear estimators $\hat{\boldsymbol{\theta}}$, is

$$E(\mathbf{Y}) = \underset{\sim}{A} \boldsymbol{\theta},$$

where $\underset{\sim}{A}$ is an $N \times k$ matrix of constants. The estimators are then given by Eq. (7.23), with $\underset{\sim}{W} = \underset{\sim}{V}^{-1}$

$$\hat{\boldsymbol{\theta}} = (\underset{\sim}{A}^{\text{T}} \underset{\sim}{V}^{-1} \underset{\sim}{A})^{-1} \underset{\sim}{A}^{\text{T}} \underset{\sim}{V}^{-1} \mathbf{Y}. \tag{7.49}$$

The Gauss–Markov theorem states that among the estimators of the class considered, the linear least squares estimators (7.49) have minimum variance.

Suppose that $\mathbf{t} = \underset{\sim}{L} \mathbf{Y}$ is an unbiased estimator of θ. We shall show that $V(t_j)$, $j = 1, \ldots, k$, is minimal when

$$\underset{\sim}{L} = (\underset{\sim}{A}^{\text{T}} \underset{\sim}{V}^{-1} \underset{\sim}{A})^{-1} \underset{\sim}{A}^{\text{T}} \underset{\sim}{V}^{-1}. \tag{7.50}$$

The covariance matrix for **t** is

$$\text{cov}(\mathbf{t}) = E[(\underset{\sim}{L}\,\mathbf{Y} - \boldsymbol{\theta})(\underset{\sim}{L}\,\mathbf{Y} - \boldsymbol{\theta})^{\mathrm{T}}]$$
$$= E[\underset{\sim}{L}\,(\mathbf{Y} - \underset{\sim}{A}\,\boldsymbol{\theta})(\mathbf{Y} - \underset{\sim}{A}\,\boldsymbol{\theta})^{\mathrm{T}}\,\underset{\sim}{L}^{\mathrm{T}}]$$
$$= \sigma^2\,\underset{\sim}{L}\underset{\sim}{V}\underset{\sim}{L}^{\mathrm{T}}\,. \tag{7.51}$$

Since **t** is an unbiased estimate of $\boldsymbol{\theta}$, it follows that

$$E(\mathbf{t}) = E(\underset{\sim}{L}\,\mathbf{Y}) = \underset{\sim}{L}\underset{\sim}{A}\,\boldsymbol{\theta} = \boldsymbol{\theta}$$

for all $\boldsymbol{\theta}$. Then Eq. (7.51) can be written

$$\sigma^2\,\underset{\sim}{L}\underset{\sim}{V}\underset{\sim}{L}^{\mathrm{T}} = \sigma^2\{[(\underset{\sim}{A}^{\mathrm{T}}\underset{\sim}{V}^{-1}\underset{\sim}{A})^{-1}\,\underset{\sim}{A}^{\mathrm{T}}\underset{\sim}{V}^{-1}]\,\underset{\sim}{V}\,[(\underset{\sim}{A}^{\mathrm{T}}\underset{\sim}{V}^{-1}\underset{\sim}{A})^{-1}\,\underset{\sim}{A}^{\mathrm{T}}\underset{\sim}{V}^{-1}]^{\mathrm{T}}$$
$$+\,[\underset{\sim}{L} - (\underset{\sim}{A}^{\mathrm{T}}\underset{\sim}{V}^{-1}\underset{\sim}{A})^{-1}\,\underset{\sim}{A}^{\mathrm{T}}\underset{\sim}{V}^{-1}]\,\underset{\sim}{V}\,[\underset{\sim}{L} - (\underset{\sim}{A}^{\mathrm{T}}\underset{\sim}{V}^{-1}\underset{\sim}{A})^{-1}\,\underset{\sim}{A}^{\mathrm{T}}\underset{\sim}{V}^{-1}]^{\mathrm{T}}\},$$

which is easily verified by evaluating both terms of the right-hand side. Each of these terms is of the form $\underset{\sim}{C}\underset{\sim}{C}^{\mathrm{T}}$, and so has non-negative diagonal elements. Since only the second terms is a function of $\underset{\sim}{L}$, the diagonal elements of $\underset{\sim}{L}\underset{\sim}{V}\underset{\sim}{L}^{\mathrm{T}}$ are strictly minimal when the second term is zero, i.e. when Eq. (7.50) holds. This proves the theorem.

It is important to note that no assumptions have been made about the distribution of the observations **Y**, other than that their mean value is a linear function of the parameters. The optimal properties of the least squares estimates, namely minimum variance and unbiasedness, follow directly from the linearity of the problem. Further, these properties hold for any number of observations.

7.4.5. *Asymptotic efficiency*

Let us show in this section that the maximum likelihood estimate $\hat{\theta}$ is asymptotically efficient.

Since $\hat{\theta}$ is a consistent estimate of θ (cf. Section 7.2.3), we can expand the p.d.f. of the observations **X** around $\hat{\theta}$

$$f(\mathbf{X}, \theta) = \exp\left[\ln\,f(\mathbf{X}, \hat{\theta}) + \frac{1}{2}(\hat{\theta} - \theta)^2 \frac{\partial^2\,\ln\,f(\mathbf{X}, \theta)}{\partial\theta^2}\bigg|_{\theta=\hat{\theta}} + O(N^{-3})\right]. \tag{7.52}$$

Note that the second derivative converges to a constant

$$\frac{\partial^2\,\ln\,f}{\partial\theta^2}\bigg|_{\theta=\hat{\theta}} \xrightarrow[N\to\infty]{} E\left(\frac{\partial^2\,\ln\,f}{\partial\theta^2}\right) = -I_{\mathbf{X}}\,.$$

The p.d.f. (7.52) has the exponential form (5.8), and it then follows that $\hat{\theta}$ asymptotically is a sufficient statistic for θ. From the results of Section 7.4.2, $\hat{\theta}$ is efficient, and from Eq. (7.52), $\hat{\theta}$ is Normally distributed around θ [Fourgeaud, p. 212].

It now follows from the uniqueness of efficient unbiased estimates[f] that all asymptotically efficient estimates are perfectly correlated and, in particular, that they are all functions of the maximum likelihood estimate.

Note that this property of maximum likelihood estimates is again very much an asymptotic property.

7.5. Bayesian Inference

7.5.1. *Choice of prior density*

In Chapter 6, on decision theory, we have discussed the interpretation of subjective prior densities as the expression of one's current belief about the unknown parameters. In order to represent a clear expression of current knowledge, a reasonable approach is first to select a family of distributions with a range of possible shapes, and then choose whichever member of the family best represents one's beliefs about the unknown parameter.

For example, if one is concerned with a parameter with possible values in the range (0, 1), then the beta distributions (4.36), give a wide range of possible shapes, see Fig. 4.3. For $a = b = 0$, this is the uniform distribution, for $a = b = 10$, a distribution sharply peaked around $\theta = \frac{1}{2}$, while for $a = b = -\frac{1}{2}$, it is a distribution with high density near $\theta = 0$ and $\theta = 1$, reflecting the belief that the parameter could be either 0 or 1 but is not likely to be $\frac{1}{2}$. When the range of possible values is from zero to infinity, a convenient family is the gamma distributions (4.38).

When the range of values covers the whole real line, then the Normal distributions $N(\mu, \tau^2)$ may give a suitable set of prior densities. The most likely value of the parameter θ is μ, and τ^{-2} reflects the precision of one's knowledge (belief).

The question now arises of how to represent total lack of knowledge. In the spirit of the paragraph above, one could take a Gaussian prior and let the variance τ^2 go to infinity. Then the Gaussian tends to a *uniform prior* in θ. However, if one represents lack of knowledge by a uniform distribution in θ over some range of values, the criticism can be made that another physicist, who also

---

[f]Since the estimators we consider are consistent, they behave at infinity as unbiased estimators. The uniqueness property may be found in [Fourgeaud, p. 207].

knows nothing except the range, but who is interested in some function $g(\theta)$, would take as his prior a uniform distribution in $g(\theta)$. These two distributions are inconsistent, and lead to different posterior densities for θ, for the same data. It is sometimes suggested to use the "natural" physical variable, but who is to say whether an angle is more natural than its cosine, whether mass is more natural than mass squared, or lifetime more natural than decay rate.

In practice, the *uniform prior* is probably the most popular, but this may be largely because it is the most convenient numerically or because it gives the illusion of being no prior at all. It is sometimes justified by citing Laplace's *principle of insufficient reason*, which states that in the absence of any other knowledge, all possible hypotheses are equally probable. We have already seen that this is not so obvious when applied to continuous variables, since a p.d.f. uniform in one variable is not uniform when expressed in terms of some function of that variable. When the range is infinite or semi-infinite, an additional undesirable effect of uniform priors is that all the prior probability ends up at infinity, in the sense that the total integrated prior probability between any two finite values is zero. This cannot correspond to anyone's real prior belief about the value of that parameter. This problem is aggravated considerably when multiple parameters are being estimated simultaneously, which requires a multidimensional prior density.

It may seem that the above problems arise only because the hypotheses are continuous, and that Laplace's principle should be straightforward to apply when hypotheses are discrete and finite in number. Consider however the case where we wish to determine whether an object is red or white, where it is known that it must be one of the two. Clearly the prior expressing lack of knowledge gives equal probability to each colour. Suppose however that a hundred different shades of red are possible, but only one shade of white. We are still interested only in whether the object is red or white, but now has the probability of white become $1/101$, or is it still one-half?

The physicist Harold Jeffreys has done the most important work [Jeffreys] in the search for priors expressing lack of knowledge, which he called *objective priors*. These are now known as *Jeffreys priors*. Applying the principle of scale invariance, he postulated the prior $\pi(\theta) = 1/\theta$ for the case $0 \leq \theta \leq \infty$. Applying the principle of minimum *Fisher information*, he derived a family of priors which are in fact just the square root of the Fisher information:

$$\pi(\theta) = -\sqrt{E_\theta \left(\frac{\partial^2 \ln \left(f(x|\theta) \right)}{\partial \theta^2} \right)}.$$

According to this latter principle, the Jeffreys prior for an unbounded parameter θ turns out to be just the uniform prior, but the prior for $0 \leq \theta \leq \infty$ is $\pi(\theta) = 1/\sqrt{\theta}$. We shall consider the behaviour of these *Jeffreys priors* for the important case of the estimation of the parameter of a Poisson distribution before going on with the equally important case of Normally distributed data. For a more extensive treatment of standard priors, see [Kass].

7.5.2. *Bayesian inference about the Poisson parameter*

For Poisson-distributed data

$$P(n|\mu) = \frac{e^{-\mu}\mu^n}{n!}$$

we recall that the *expectation* of n is just $E(n) = \mu$. Now in Bayesian estimation it is possible, for any given prior $\pi(\mu)$ to calculate also the expectation $E(\mu)$ for a given observation n. For the uniform prior, this expectation is $E(\mu) = n + 1$, which does not seem right since $E(n) = \mu$. This problem is indeed fixed by using the Jeffreys scale-invariant prior $\pi(\mu) = 1/\mu$, for which $E(\mu) = n$ as expected.

Unfortunately, the prior $\pi(\mu) = 1/\mu$ also has problems. One problem appears when the observed value of n is $n = 0$. In this case, the posterior density $\pi(\mu \,|\, n = 0)$ is a delta-function at $\mu = 0$ which means there is no probability that μ can be anything but zero. The uniform prior for this case happens to give the same upper limit as the classical method which is widely used and which is much more reasonable, and that is one reason physicists often prefer the uniform prior.

Another problem with the Jeffreys priors appears when there is background of expectation b in addition to the signal μ, which is nearly always true in real experiments:

$$P(n|\mu, b) = \frac{e^{-\mu+b}(\mu+b)^n}{n!}.$$

In this case, the uniform prior gives a reasonable posterior, whereas both the Jeffreys priors produce posterior densities that cannot be used to calculate credible intervals (see Chapter 9) because the relevant integrals do not converge. This can in turn be fixed by using an extended Jeffreys prior $\pi(\mu) = 1/\sqrt{\mu + b}$ which gives a reasonable posterior but implies that one's prior belief about the signal depends on the background, which cannot be true.

Probably the reason why Jeffreys was unable to find a suitable prior representing lack of knowledge, is because prior densities don't really represent knowledge, they represent belief. We should not be surprised that there is no function representing at the same time belief and lack of belief. The "lack of knowledge" is represented by the likelihood function for "no data", which is indeed uniform. But, unlike the uniform prior *p.d.f.*, the uniform *likelihood function* is invariant under change of variables. Once again, the transformation properties of functions give an important clue about how they should be used.

7.5.3. *Priors closed under sampling*

When a prior density and the resulting posterior density belong to the same family, the prior density is said to be *closed under sampling*. Suppose one has observations, $\mathbf{X}$ with p.d.f. $f(\mathbf{X}|\theta)$. Let $\xi(\theta; \alpha)$ be the prior density of θ, where α indicates a particular member of the family ξ. Then the density is closed under sampling if the posterior density can be written as $\xi(\theta; \beta)$, where $\beta \equiv \beta(\alpha, X)$; that is

$$\xi(\theta; \beta) = \frac{\left[\prod_{i=1}^{N} f(X_i|\theta)\right]\xi(\theta, \alpha)}{\int \left[\prod_{i=1}^{N} f(X_i|\theta)\right]\xi(\theta, \alpha)d\theta}.$$

In this case, the change in knowledge resulting from the observations $\mathbf{X}$, is summarized in the change in the index from α to $\beta(\alpha, \mathbf{X})$.

Any discrete prior distribution is closed under sampling. For continuous distributions, it can be shown that a closed family only exists if the p.d.f. $f(X|\theta)$ is of the exponential form. The family of distributions ξ then has the form

$$\xi(\theta) \propto G(\theta)^{\alpha} \exp\{\beta A(\theta)\}.$$

where α, β are independent of θ, and A, G are arbitrary functions of θ.

We now illustrate the application of the Bayesian approach to inference about the parameters of a Normal distribution, given a set of observations $X_1, \ldots, X_N$ from a Normal distribution $N(\mu, \sigma^2)$.

7.5.4. *Bayesian inference about the mean, when the variance is known*

Suppose that the observations $\mathbf{X}$ come from a distribution $N(\theta, \sigma^2)$, where θ is unknown and σ^2 is known.

The posterior density of θ, given the observations $\mathbf{X}$ and the prior density $\pi(\theta)$, is

$$\pi(\theta|\mathbf{X}) \propto \frac{1}{(2\pi\sigma^2)^{N/2}} \exp\left[-\sum_{i=1}^{N} \frac{(X_i - \theta)^2}{2\sigma^2}\right] \pi(\theta). \tag{7.53}$$

Choosing $\pi(\theta)$ as $N(\mu_0, \sigma_0^2)$, it is easily seen that $\pi(\theta|\mathbf{X})$ is also Normal. Ignoring terms not involving θ, Eq. (7.53) becomes

$$\pi(\theta|\mathbf{X}) \propto \exp\left[-\frac{1}{2}\frac{(\overline{X} - \theta)^2 N}{\sigma^2} - \frac{1}{2}\frac{(\theta - \mu_0)^2}{\sigma_0^2}\right]$$

$$\propto \exp\left[-\frac{1}{2}\theta^2\left(\frac{N}{\sigma^2} + \frac{1}{\sigma_0^2}\right) + \theta\left(\frac{N\overline{X}}{\sigma^2} + \frac{\mu_0}{\sigma_0^2}\right)\right]. \tag{7.54}$$

Therefore the posterior density of θ is Normal, $N(\nu, \tau^2)$, with mean

$$\nu = \frac{\dfrac{N\overline{X}}{\sigma^2} + \dfrac{\mu_0}{\sigma_0^2}}{\dfrac{N}{\sigma^2} + \dfrac{1}{\sigma_0^2}}, \tag{7.55}$$

and variance

$$\tau^2 = \frac{1}{\dfrac{N}{\sigma^2} + \dfrac{1}{\sigma_0^2}}.$$

It is in fact a remarkable property of the Normal distribution that the product of two Normal variables is also a Normal variable, even when the two input means are very far apart with respect to their standard deviations. In this case the product density has almost all its probability in a region where integrated prior probability is arbitrarily close to zero and also arbitrarily many standard deviations from the mean of the data.

The mean of the posterior density of θ is thus a weighted average of the prior mean, μ_0, and the observed mean $\overline{X}$, the weights being proportional to the reciprocal of the variances (precision). The precision τ^{-2} of the posterior density is the sum of the precisions of the prior density and of the observed mean.

The posterior density $\pi(\theta|\mathbf{X})$ summarizes all knowledge about θ, given the observations $\mathbf{X}$. However, one must remember that the Bayesian interpretation of ν and τ^2 is different from the classical one.

From the decision theory point of view, we have seen in Chapter 6 that, given a parabolic loss function, the optimum value to choose for θ is the mean of the posterior density (7.55)

$$\theta_{\text{opt}} = \frac{\dfrac{N\overline{X}}{\sigma^2} + \dfrac{\mu_0}{\sigma_0^2}}{\dfrac{N}{\sigma^2} + \dfrac{1}{\sigma_0^2}}.$$

We may represent the case of no prior knowledge by letting $\sigma_0 \to \infty$. Then θ_{opt} becomes $\overline{X}$, as one might expect.

7.5.5. *Bayesian inference about the variance, when the mean is known*

Suppose that the observations $\mathbf{X}$ come from a distribution $N(\mu, \theta)$, where only μ is known. Writing

$$\hat{\sigma}^2 = \frac{1}{N} \sum_{i=1}^{N} (X_i - \mu)^2,$$

the likelihood for the observations can be written

$$L(\mathbf{X}, \theta) = (2\pi\theta)^{-N/2} \exp\left[-\frac{N\hat{\sigma}^2}{2\theta}\right]. \tag{7.56}$$

In choosing a prior density, one looks for a family of the same form as Eq. (7.56), such as the particular parametrization

$$\pi(\theta) \propto \theta^{-\nu_0^{-1}/2} \exp\left[-\frac{\nu_0\sigma_0^2}{\theta}\right]. \tag{7.57}$$

This density has a mean

$$E(\theta) = \frac{\sigma_0^2\nu_0}{\nu_0 - 2}$$

and a most probable value $\sigma_0^2\nu_0/(\nu_0 + 2)$. The variance of the density is

$$V(\theta) = \frac{2\sigma_0^4\nu_0^2}{(\nu_0 - 2)^2(\nu_0 - 4)}, \quad \text{for } \nu_0 > 4.$$

One can locate the distribution by varying σ_0^2, and reduce the variance by increasing ν_0, thus providing a range of possible shapes for the prior density.

The density (7.57) has the other interesting property that

$$\phi = \frac{\nu_0\sigma_0^2}{\theta}$$

is distributed as a $\chi^2(\nu_0)$.

Imprecise knowledge corresponds to letting $\nu_0 \rightarrow 0$. Equation (7.57) can then be shown to correspond to a uniform distribution in $\ln \theta$ over the whole real line.

Consider now the posterior density

$$\pi(\theta|\mathbf{X}) \propto L(\mathbf{X}, \theta)\pi(\theta) \propto \theta^{-\frac{1}{2}(\nu_0+N)-1} \exp\left[-\frac{\nu_0\sigma_0^2 + N\hat{\sigma}^2}{\theta}\right]. \qquad (7.58)$$

Equation (7.58) is obviously of the same form as the prior density (7.57), with σ_0^2 replaced by

$$\sigma_1^2 = \frac{\nu_0\sigma_0^2 + N\hat{\sigma}^2}{\nu_0 + N}$$

a weighted average of the prior and sample variances, and with ν_0 replaced by

$$\nu_1 = \nu_0 + N.$$

The variable

$$\frac{\nu_1\sigma_1^2}{\theta}$$

is now distributed as $\chi^2(\nu_0 + N)$. The posterior mean, giving the optimum choice of value for θ, for a quadratic loss function is

$$E(\theta|\mathbf{X}) = \frac{\nu_0\sigma_0^2 + N\hat{\sigma}^2}{\nu_0 + N - 2}$$

and the posterior variance is

$$V(\theta|\mathbf{X}) = \frac{2(\nu_0\sigma_0^2 + N\hat{\sigma}^2)^2}{(\nu_0 + N - 2)^2(\nu_0 + N - 4)}.$$

The coefficient of variation, defined as $\sqrt{\text{Variance}}/\text{Mean}$, is given by $\sqrt{2/(\nu_0 + N - 4)}$, illustrating that as the number of observations increases, the standard error becomes small relative to the mean. As N increases, the posterior density becomes independent of the prior density. Similarly, in the case where prior knowledge is represented by a uniform distribution in $(\ln \theta)$, the posterior distribution is such that $N\hat{\sigma}^2/\theta$ is distributed as $\chi^2(N)$.

It should be noted that, although this last statement looks very similar to the result obtained classically, here the variable is $1/\theta$, whereas classically $\hat{\sigma}^2$ is the random variable.

7.5.6. *Bayesian inference about the mean and the variance*

Consider the case when the observations $\mathbf{X}$ are from a distribution $N(\theta_1, \theta_2)$, where θ_1 and θ_2 are both unknown. It is now necessary to defined a joint prior distribution for θ_1 and θ_2. Assume that θ_1 and θ_2 are independent, so that given some value of θ_1, one's beliefs about θ_2 would not change. In the case of vague prior knowledge for both parameters, we take θ_1 to be uniformly distributed over $(-\infty, \infty)$ and $\ln \theta_2$ uniformly distributed over $(0, \infty)$. As discussed above, this distribution of θ_2 is a reasonable representation of ignorance about the parameter.

Then the likelihood of the observations is

$$L(\mathbf{X}|\theta_1, \theta_2) = (2\pi\theta_2)^{-N/2} \exp\left[-\sum_{i=1}^{N}(X_i - \theta_1)^2/2\theta_2\right].$$

and the joint posterior density of θ_1 and θ_2 becomes

$$\pi(\theta_1, \theta_2|\mathbf{X}) \propto L(\mathbf{X}|\theta_1, \theta_2)\pi(\theta_1, \theta_2) \propto \theta_2^{-(N+2)/2} \exp\left[-\sum_{i=1}^{N}(X_i - \theta_1)^2/2\theta_2\right].$$

We can write

$$\sum_{i=1}^{N}(X_i - \theta_1)^2 = \sum_{i=1}^{N}(X_i - \overline{X})^2 + N(\overline{X} - \theta_1)^2$$

$$= (N-1)s^2 + N(\overline{X} - \theta_1)^2.$$

Therefore the posterior density can be written

$$\pi(\theta_1, \theta_2|\mathbf{X}) \propto \theta_2^{-(N+2)/2} \exp\{-[(N-1)s^2 + N(\mathbf{X} - \theta_1)^2]/\theta_2\}. \qquad (7.59)$$

Integrating Eq. (7.59) with respect to θ_2, we obtain the posterior density of θ_1,

$$\pi(\theta_1|\mathbf{X}) \propto \left\{1 + \frac{N(\overline{X} - \theta)^2}{(N-1)s^2}\right\}^{-N/2}.$$

There $t = \sqrt{N}(\overline{X} - \theta_1)/s$ has a t distribution with $(N-1)$ degrees of freedom.[g] Similarly, we obtain the posterior density of θ_2 as

$$\pi_2(\theta_2|\mathbf{X}) \propto \theta_2^{-(N-1)/2-1} \exp\left[-\frac{(N-1)s^2}{2\theta_2}\right].$$

---

[g]See Section 4.2.4 for Student's t-distribution.

Therefore, the quantity $[(N-1)s^2]/\theta_2$ has a χ^2 distribution with $(N-1)$ degrees of freedom.

7.5.7. *Summary of Bayesian inference for Normal parameters*

Thus if the variance of the distribution is known, then:

$$\sqrt{N}(\theta_1 - \overline{X})/\sigma \quad \text{is} \quad N(0,1).$$

If the variance is not known, $\sqrt{N}(\theta_1 - \overline{X})/s$ has a t distribution with $(N-1)$ degrees of freedom. The greater width of the t distribution relative to the Normal reflects the difference in knowledge prior to obtaining the observations. Similarly, if the mean is known, the posterior density of the variance, θ_2, is such that

$$\sum_{i=1}^{N} \frac{(X_i - \mu)^2}{\theta_2}$$

is a χ^2 with N degrees of freedom. If the mean is not known this quantity is a χ^2 with $(N-1)$ degrees of freedom. Here the reduction in prior knowledge is reflected in the reduction in the number of degrees of freedom from N to $(N-1)$.

Consider again the joint posterior density (7.59). The distribution of θ_1, conditional on θ_2, has a density

$$p(\theta_1|\theta_2, \mathbf{X}) \propto \theta_2^{-1/2} \exp\left[-\frac{N(\overline{X} - \theta_1)^2}{2\theta_2}\right],$$

that is $N[\overline{X}, (\theta_2/N)]$. The parameters θ_1, θ_2 are therefore not independent, contrary to the prior assumption. The larger is θ_2, the greater is the spread of θ_1, and the less precise our knowledge of θ_1. Each observation X_i has variance θ_2, and has more scatter about μ the larger is θ_2, and thus provides less information about μ.

Chapter 8

POINT ESTIMATION IN PRACTICE

In Section 8.1 of this Chapter we shall discuss the various requirements one may put on a good *estimator*. Many of them have been introduced in Chapter 7; others are "practical" requirements, and met here for the first time. In Secs. 8.2 to 8.4 we return to the usual estimators for a detailed discussion. Thus some repetition of Chapter 7 is unavoidable. On the other hand, we shall refer without proof to the properties of consistency, unbiasedness and efficiency of the usual estimators (although the reader may have skipped those parts of Chap 7).

The remaining sections deal with several aspects of point estimation in real life.

8.1. Choice of Estimator

When choosing a good estimator, many requirements present themselves, and they are often conflicting. In order to find an optimal choice, one must establish an order of importance between these requirements. On the basis of statistical merit, we shall propose an order of importance in Section 8.1.1. Obviously, the conditions may be such that an order must be established on some other merit, such as cost, for instance. In the subsequent sections we shall discuss how and when to compromise between different requirements.

8.1.1. *Desirable properties of estimators*

(i) *Consistency.* The primary requirement on any estimator is that it should converge to the true value of the parameter, as the number of observations increases. There is almost no situation in which one would knowingly choose an inconsistent estimator. For instance, if an estimator were found to be asymptotically biased, one should decide for a procedure to remove the bias. Should one decide, however, to use an inconsistent estimator, a minimum requirement would be to put an upper bound on the degree of inconsistency.

(ii) *Minimum loss of information.* When an estimate summarizes the results of an experiment in a single number, it is of vital interest to any subsequent user of the estimate that no other single number could contain more information about the parameter of interest.

(iii) *Minimum variance (efficient estimator).* In classical statistics, the smaller the variance of the estimator, the more certain can one be that the estimate is near the true value of the parameter (neglecting bias).

(iv) *Minimum loss* (minimum *cost*). In the alternative approach of decision theory, that value of the parameter should be chosen which leads to the minimum average loss (cost), resulting from making the wrong decision of value. Thus, this requirement is of the same relative importance as minimum variance, only it arises in an entirely different approach.

All the above properties depend on knowledge about the parent distribution of the data. When such knowledge is lacking, or founded on unsafe assumptions, a very desirable property is:

(v) *Robustness.* The estimate should be independent of the distribution, or insensitive to departures from the assumed distribution. We shall discuss in Section 8.7 some particular forms of robust estimates. In general, the information content of such estimates is reduced, since one chooses to ignore the distributional form.

(vi) *Simplicity.* When a physicist reads the published value of an estimated parameter, he usually presumes that the estimate is unbiased, Normally distributed, and uncorrelated with other estimates. It is therefore desirable that estimators should have these simple properties.

(vii) *Minimum computer time.* Although not fundamental, this requirement may lead to the preference of an inefficient estimator over an efficient one, for reasons of time limitations or economy.

(viii) *Minimum loss of physicist's time.* In practice, this requirement is often treated as the most important one. It is one of the purposes of this book to show that it may be worthwhile to "lose" a little bit more of physicists' time on the choice of a good estimator.

8.1.2. *Compromise between statistical merits*

The order of desirable properties above reflects our opinion that statistical merits, (i) to (iv), are more important than other merits. Among statistical merits, we consider minimum loss of information as more important than minimum variance. That there may be a choice between (ii) and (iii) follows from the fact that, although the variance is bounded below by a quantity related to the inverse of the information, this lower bound is not always attainable. Therefore, it is possible to have two estimates of θ, t_1 and t_2, with

$$I_{t_1}(\theta) < I_{t_2}(\theta)$$

but with

$$V_{t_1} < V_{t_2} \,.$$

Our recommendation to choose t_2 reflects the hope that the larger amount of information, carried by t_2, would later be useful when a decision about θ has to be made. Such information comes from our knowledge of the distribution of t_2.

In the *decision theory* approach, that value would be chosen for which the *loss function* $L(t, \theta)$ would average to a minimum. If $L(t, \theta)$ is a quadratic in t with a minimum at $t = \theta$, a reasonable choice of t for fixed θ is given by minimizing

$$E[(t - \theta)^2] = V(t) - [E(t - \theta)]^2 \,.$$

This is equivalent to minimizing the variance only in the case of unbiased estimates.

We have shown that the requirement of minimum loss of information is easily satisfied in only two cases:

(i) when *sufficient statistics* exist, one should use them as explained in Section 8.3.2;

(ii) in the *asymptotic limit*, the *maximum likelihood estimator* is optimal.

In situations other than these, we know of no general methods of finding an optimal point estimator. It may then be necessary to give up the aim of summarizing the experiment in one number, and report the *likelihood function*, at least in the neighbourhood of its maximum (maxima). Another possibility is to go directly to interval estimation (see the following chapter), which will anyway provide more information than point estimation.

8.1.3. *Cures to obtain simplicity*

It may often be worth the sacrifice of some information to obtain simplicity (unbiasedness, uncorrelatedness, Normality).

Estimates of several parameters can be made uncorrelated by diagonalizing the covariance matrix and finding the linear combinations of the parameters. But the new parameters often lack physical meaning.

Estimates can be made unbiased by several methods. We shall meet one method in Example (iii) below, but we defer the general discussion to Section 8.6.

Asymptotically, most usual estimators are unbiased and Normally distributed. The question arises as to how good the asymptotic approximation is in any specific experiment. Without finding the exact finite sample distribution of the estimates, this question cannot be answered. However, some hints can be found by one of the following methods:

(A) Check that the log-likelihood function is parabolic.
(B) If one has two asymptotically efficient estimators, check that the results are not statistically different. For instance, the two estimators may be the least squares estimator applied to two different binnings of the same data.
(C) Compute the first few moments of the sampling distribution to low orders in $1/N$, and check that the bias, skewness and kurtosis are small.
(D) Simulate the behaviour of the estimator by *Monte Carlo* techniques. In the most difficult cases, this may be necessary.

An estimator can sometimes be made simpler, or even optimal, by a change of variables. Let us distinguish between: (a) change of estimator, which is a change of variable dependent on the observations,

$$\hat{\theta} = g(\hat{\theta}_1, X)$$

(b) change of variable (of a theoretical parameter), independent of observations,

$$\theta_2 = g(\theta_1).$$

Change (a) modifies the variance, and there is no general method of constructing a Normally distributed estimate.

Under change (b) one may keep the chosen estimator (e.g. maximum like-lihood) and get a Normal estimate of the new parameters (in terms of the changed variables). However, in general it is impossible to eliminate both the *bias* and the non-Normality under such a change. This is illustrated in Example (iii) below [Kendall II, p. 46].

The best method may then be to use the first chosen estimator to find the estimate $\hat{\theta}$ of the parameter θ, then to estimate the first four moments of $\hat{\theta}$, and finally to use a general family of distributions (e.g. the Johnson family, Section 4.3.2) to give a better approximation to the true distribution of $\hat{\theta}$ than the asymptotic Normal distribution.

Examples

(i) Re z and Im z are better estimators of a complex number z than are $\rho = |z|$ and $\phi = \arg z$, since ϕ becomes badly determined when ρ is small. In fact, the information matrix is singular when $\rho = 0$.

(ii) If one has two *correlated estimates* X and Y, with correlation coefficient ρ and variances σ_X^2 and σ_Y^2, two uncorrelated estimates are

$$u = X + \alpha Y$$

$$v = X - \alpha Y,$$

with $\alpha = \sigma_X/\sigma_Y$. The covariance obviously vanishes:

$$\mathrm{cov}(u,v) = \sigma_X^2 - \alpha^2 \sigma_Y^2 + (\alpha - \alpha)\rho \sigma_X \sigma_Y = 0.$$

(iii) *Maximum likelihood fit of a histogram*

Suppose one has N events in a histogram with k bins, n_i in the ith bin, and a probability content $p_i(\theta)$ for the ith bin, where θ is an unknown parameter. The likelihood function is

$$L(\mathbf{n}|\theta) = N! \prod_{i=1}^{k} \left(\frac{p_i^{n_i}}{n_i!} \right).$$

Then Haldane and Smith have shown [Haldane] that the bias, variance, and skewness of the maximum likelihood estimate $\hat{\theta}$ are, respectively

$$b(\hat{\theta}) = -\frac{1}{2}N^{-1}A_1^{-2}B_1 + O(N^{-2}),$$

$$V(\hat{\theta}) = N^{-1}A_1^{-1} + O(N^{-2}),$$

$$\gamma_1(\hat{\theta}) = N^{-1/2}A^{-3/2}(A_2 - 3B_1) + O(N^{-2}),$$

where

$$A_1 = \sum_{i=1}^{k} \frac{1}{p_i} \left(\frac{\partial p_i}{\partial \theta} \right)^2 ,$$

$$A_2 = \sum_{i=1}^{k} \frac{1}{p_i^2} \left(\frac{\partial p_i}{\partial \theta} \right)^3$$

$$B_1 = \sum_{i=1}^{k} \frac{1}{p_i} \left(\frac{\partial p_i}{\partial \theta} \right) \left(\frac{\partial^2 p_i}{\partial \theta^2} \right) ,$$

Note that A_1 is the *information* defined in Chapter 5.

To suppress the *bias*, one looks for a transformation $\tau = \tau(\theta)$ such that $B_1(\tau) = 0$. Thus

$$\frac{\partial p_i}{\partial \tau} = \frac{\partial p_i}{\partial \theta} \frac{\partial \theta}{\partial \tau}$$

$$\frac{\partial^2 p_i}{\partial \tau^2} = \frac{\partial^2 p_i}{\partial \theta^2} \left(\frac{\partial \theta}{\partial \tau} \right)^2 + \left(\frac{\partial p_i}{\partial \theta} \right) \left(\frac{\partial^2 \theta}{\partial \tau^2} \right) ,$$

giving

$$B_1(\tau) = B_1 \left(\frac{\partial \theta}{\partial \tau} \right)^3 + A_1 \frac{\partial \theta}{\partial \tau} \frac{\partial^2 \theta}{\partial \tau^2} = 0 .$$

Here B_1 and A_1 refer to $B_1(\tau = \theta)$, $A_1(\tau = \theta)$, in which case $\partial \theta / \partial \tau = 1$, $\partial^2 \theta / \partial \tau^2 = 0$. This differential equation in $\partial \tau / \partial \theta$ may be solved for the parameter τ, which is estimated unbiasedly by the maximum likelihood method. However, once the bias is suppressed, the skewness is

$$\gamma_1(\hat{\tau}) = N^{-1/2} A_1^{-3/2} A_2 ,$$

and A_2 is, in general, not equal to zero. Consequently, the distribution of $\hat{\tau}$ cannot be made Normal for finite sets of data. Note that

$$A_1(\tau)^{-3/2} A_2(\tau) = A_1^{-3/2} A_2 .$$

8.1.4. *Economic considerations*

Economy usually implies fast computing. Optimal estimation is frequently iterative, requiring much computer time. Fast estimation can, therefore, usually be bought only at the cost of decreased statistical efficiency. Let us consider three methods to compromise between efficiency and cost.

(A) *Linear methods.* The fastest computing is offered by linear methods, since they do not require iteration. These can naturally be used when the expected values of the observations are linear functions of the parameters. Among *linear unbiased estimators*, the *least squares method* is the best (*Gauss–Markov theorem*, Section 7.4.4).

When one has the freedom of choice of distribution, as for instance in the case of empirical fits, it should be chosen from the exponential family (5.8) if possible. This facilitates the computing and has optimal properties (sections 8.3.1 and 7.4.2).

(B) *Two-step methods.* Some computer time may be saved by breaking up the estimation into two step:

(i) estimate the parameters by a simple, fast, inefficient method;

(ii) use these estimates as starting values for an optimal estimation.

Although more physicist's time may be required to evaluate the results of the first step, a fuller understanding of the situation may be gained (in addition to saving computer time).

(C) *Three-step method.* The previous methods may be improved by adding a third step as follows.

(i) Extract from the data a certain number of statistics, which summarize the observations compactly, and if possible in a physically meaningful way. One example of such a summary is a *histogram*, which reduces the number of observations to the number of bins in the histogram. Another example is the summary of an angular distribution by the coefficients of the expansion of the distribution in *Legendre polynomials*. These coefficients may be estimated very rapidly by the *moments method* (Section 8.2.1) and their physical interpretation is clear.

(ii) Estimate the parameters of interest on the basis of these summary data. If they are chosen so as to have an intuitive physical meaning, this estimation problem may be greatly simplified.

(iii) Use the preliminary estimates from the second step as starting values for an optimal estimation directly from the original data. This step is usually "forgotten" because of lack of time, and because one may not realize that in the summarizing process of the first step, some information has been lost.

When there is not much information in the original data ("small statistics"), the final step is especially important. Also, when the third step is done, the second step estimation may be simplified and approximate.

8.2. The Method of Moments

Given a set of observations $X_1, \ldots, X_N$ from a parent distribution $f(X, \boldsymbol{\theta})$, one method of estimating the parameters $\boldsymbol{\theta}$ is via the *sample moments*

$$m'_j = \frac{1}{N} \sum_{i=1}^{N} X_i^j \,. \tag{8.1}$$

Recalling the definition of moments, Section 2.4.6,

$$\mu'_j(\boldsymbol{\theta}) = \int X^j(X, \boldsymbol{\theta}) dX \,, \tag{8.2}$$

one may take the experimental (sample) moments (8.1) as estimates of the theoretical (parent) moments (8.2),

$$m'_j = \widehat{\mu'_j(\boldsymbol{\theta})} \,. \tag{8.3}$$

Solving this system of equations for the unknown parameters $\boldsymbol{\theta}$, gives estimators of $\boldsymbol{\theta}$. This is the usual *method of moments*, and it leads to estimators which are consistent (shown in Section 7.2.1) and asymptotically unbiased (shown in Section 7.3.3). It is easy to show that they have variance $V(m'_j) = (\mu'_{2j} - \mu'^2_j)/N$ and covariance $\mathrm{cov}(m'_j, m'_j) = (\mu'_{i+j} - \mu'_i \mu'_j)/N$ [Kendall I, p. 229].

In some cases this method may be straightforward from a computational point of view, but, in general, the estimates are not as efficient as the maximum likelihood estimates, since the sample moments are not in general sufficient statistics for the parent moments. However, if more moments than parameters are used, then some of the lost information may be regained.

8.2.1. *Orthogonal functions*

A special case of the moments method arises when the p.d.f. of the observations $\mathbf{X}$ can be expressed in terms of *orthogonal functions* $g_k(X)$,

$$f(X, \boldsymbol{\theta}) = \alpha + \sum_{k=1}^{r} \theta_k g_k(X) \,, \tag{8.4}$$

where α is a normalization constant.

The orthogonality is expressed by the conditions

$$\int g_j(X) g_k(X) dX = \delta_{jk}$$

$$\int g_j(X) dX = 0 \,.$$

The expectations of the functions $g_k(X)$ are

$$E[g_k(X)] = \int g_k(X)f(X,\boldsymbol{\theta})dX$$

$$= \int g_k(X)dX + \sum_{j=1}^{r} \theta_j \int g_j(X)g_k(X)dX$$

$$= \theta_k \, .$$

It follows from the results of Section 7.2.1, that a consistent and unbiased estimator of θ_k is

$$\hat{\theta}_k = N^{-1} \sum_{i=1}^{N} g_k(X_i) \, .$$

By the *Central Limit theorem* (see Secs. 3.3.2 and 7.3.1) $\hat{\theta}_k$ is asymptotically Normally distributed about θ_k, with a variance which can be estimated by

$$S_k^2 = \frac{1}{N(N-1)} \sum_{i=1}^{N} [g_k(X_i) - \hat{\theta}_k]^2 \, .$$

Approximate *confidence regions* can be obtained from the fact that the quantity

$$\frac{\hat{\theta}_k - \theta_k}{S_k} = t_{N-1}$$

has (asymptotically) a Student's t distribution with $(N-1)$ degrees of freedom.

Example

In fitting *polarization* distributions, the form (8.4) often arises. Thus, in the decay of a Λ hyperon, the cosine of the angle of the decay pion in the c.m.s. of the Λ has a density function given by

$$f(\cos \theta_\pi, P) = \frac{1}{2}(1 + \alpha P \cos \theta_\pi),$$

where α is the asymmetry parameter, assumed known. Now,

$$\int_{-1}^{1} \cos \theta d(\cos \theta) = 0$$

and

$$\int_{-1}^{1} \frac{1}{2} \cos^2 \theta d(\cos \theta) = \frac{1}{3} \, .$$

An estimate of αP is then $\alpha \hat{P} = (3/N)\Sigma_i \cos \theta_i$.

The variance of this estimate can be evaluated as

$$V(\alpha\hat{P}) = \frac{9}{N}\left[\frac{1}{3} - \frac{(\alpha\hat{P})^2}{9}\right] = \frac{1}{N}[3 - (\alpha P)^2]. \tag{8.5}$$

8.2.2. *Comparison of likelihood and moments methods*

In this section we show explicitly that the *moments method* is less efficient than the *maximum likelihood method* (m.l.).

In the case of Eq. (8.4) the log-likelihood is given by

$$\ln L(\mathbf{X}|\boldsymbol{\theta}) = \sum_{j=1}^{N} \ln\left[\alpha + \sum_{i=1}^{r} \theta_i g_i(X_j)\right].$$

The equations for the m.l. estimates are therefore found to be

$$\sum_{j=1}^{N} \frac{g_i(X_j)}{\left[\alpha + \sum_{k=1}^{r} \theta_k g_k(X_j)\right]} = 0. \tag{8.6}$$

It can be seen at once that if $\Sigma_k \theta_k g_k$ is small compared to α [small polarizations], then the likelihood method and the moments method are equivalent, since we can write Eq. (8.6) as

$$\sum_{j=1}^{N} g_i(X_j)\left[\alpha - \sum_{k=1}^{r} \hat{\theta}_k g_k(X_j) + \cdots\right] = 0, \quad i = 1, \ldots, r.$$

Thus

$$\alpha \sum_{j=1}^{N} g_i(X_j) \simeq \sum_{k=1}^{r} \hat{\theta}_k \sum_{j=1}^{N} g_i(X_j)g_k(X_j) \simeq N\hat{\theta}_i.$$

If $\Sigma_k \theta_k g_k$ is not negligible, then the moments method is less efficient.

Consider the example in Section 8.2.1 and take $\alpha P = \theta$. Then the density function is

$$f(X, \theta) = \frac{1}{2}(1 + \theta X), \quad -1 \leq X \leq 1.$$

The variance of the moment estimate of θ from Eq. (8.5) is

$$\frac{1}{N}(3 - \theta^2). \tag{8.7}$$

The asymptotic variance of the m.l. estimate is

$$\frac{1}{N}\left\{ -E\left[\frac{\partial^2 \ln(1+\theta X)}{\partial \theta^2}\right]\right\} = \frac{1}{N}\left\{\frac{1}{\theta^2}\left[-1+\frac{1}{2\theta}\ln\left(\frac{1+\theta}{1-\theta}\right)\right]\right\}^{-1}$$

$$= \frac{\theta^2}{N}\left[-1+\frac{1}{2\theta}\ln\left(\frac{1+\theta}{1-\theta}\right)\right]^{-1}$$

$$\simeq \frac{1}{N}\left(3 - \frac{9}{5}\theta^2 + \cdots\right). \tag{8.8}$$

Comparison of Eqs. (8.7) and (8.8) shows that the variance of the m.l. estimate is smaller than the moments method variance.

8.3. The Maximum Likelihood Method

In Section 7.2.3 we have defined the m.l. estimate $\hat{\theta}$ of a parameter θ as that value, for which the log-likelihood function

$$\ln L(\mathbf{X}|\theta) = \sum_{i=1}^{N} \ln \mathbf{f}(\mathbf{X_i}, \theta)$$

has its maximum given the particular set of independent observations $X_1, \ldots, X_N$ from a distribution $f(\mathbf{X}, \theta)$. Unless the maximum occurs at a limit of the range of θ, a necessary condition for the maximum is the *likelihood equation*

$$\frac{\partial \ln L(\mathbf{X}|\theta)}{\partial \theta} = 0. \tag{8.9}$$

It is assumed that the function L is normalized with respect to $\mathbf{X}$, that is

$$\int L(\mathbf{X}|\theta)d\mathbf{X} = 1.$$

The essential property of $L(\mathbf{X}, \theta)$ is that this integral should be independent of θ.

For r parameters θ_j one has a set of r likelihood equations

$$\partial \ln L(\mathbf{X}|\boldsymbol{\theta})/\partial \theta_j = 0, \quad j = 1, \ldots, r.$$

8.3.1. *Summary of properties of maximum likelihood*

It is important to distinguish between the *asymptotic properties* which hold for sufficiently large N, and the *finite sample* properties.

The maximum likelihood estimator $\hat{\theta}$ is consistent: *Asymptotically* one of the maxima will be arbitrarily close to the true value (Section 7.2.3 and [Fourgeaud, p. 212]). This implies asymptotic unbiasedness.

Under very general conditions, $\hat{\theta}$ is *asymptotically* Normally distributed with minimum variance, the variance being given by the Cramer–Rao lower bound [Section 7.4.1],

$$V(\hat{\theta}) \xrightarrow[N \to \infty]{} \left\{ E\left[\left(\frac{\partial \ln L}{\partial \theta} \right)^2 \right] \right\}^{-1}. \tag{8.10}$$

When the range of $\mathbf{X}$ is independent of θ, one may estimate this by

$$\hat{V}(\hat{\theta}) = \left\{ \left(-\frac{\partial^2 \ln L}{\partial \theta^2} \right) \Big|_{\theta = \hat{\theta}} \right\}^{-1}.$$

The estimate $\sqrt{N}(\hat{\theta} - \theta)$ is distributed as $N[0, I_1^{-1}(\theta)]$.

In many cases, the asymptotic limit where these optimal properties hold, will be approached quite slowly. In particular, the *likelihood function* for *finite* N may have more than one maximum, and there is no way of telling which maximum corresponds to the true value of θ.

For N *finite*, there exists only one case when the m.l. estimate has optimal properties. This is the case when the parent distribution of the observations $\mathbf{X}$ is of the exponential form (5.8),

$$f(X|\theta) = \exp[\alpha(X)a(\theta) + \beta(X) + c(\theta)],$$

which admits sufficient statistics (Section 5.3.4). It has then been shown (Section 7.4.2), that the function

$$r(\theta) = -(dc/d\theta)/(da/d\theta)$$

is estimated efficiently and without bias by the m.l. estimate (7.45),

$$\hat{r} = r(\hat{\theta}) = N^{-1} \sum_{i=1}^{N} \alpha(X_i).$$

See further the example in Section 8.6.1.

An important property is *invariance*: the m.l. estimate $\hat{\tau}$ of a function $\tau(\theta)$ is

$$\hat{\tau} = \tau(\hat{\theta}). \tag{8.11}$$

For example, the m.l. estimate of θ^2 is the square of the estimate of θ.

However, only one function $\tau(\theta)$ will have an unbiased estimate for *finite* N [Section 7.3.2]. This is because the usual definition of bias is itself not invariant. If we had chosen to define bias in terms of the median rather than the mean expectation of the estimates, then it would be invariant and a maximum likelihood estimate which is unbiased in one variable would also be unbiased under all transformations of that variable.

The m.l. estimates can sometimes be shown to converge faster to the asymptotic limit than certain classes of asymptotically efficient estimators. We shall encounter an example later [Section 8.2.6], when the m.l. estimate (obtained by using a histogram) converges faster than the asymptotically efficient methods of least squares and modified least squares. Even then it is not always true that the m.l. estimate will be better or even sensible, given a fixed, small number N of observations.

8.3.2. *Example: determination of the lifetime of a particle in a restricted volume*

To determine the *lifetime of an unstable particle* decaying with time, one uses N events characterized by the flight length, ℓ_i, of the particles from the production point to the decay. Unfortunately this length has an upper bound $\ell_{\text{pot},i}$, and a lower bound $\ell_{\text{min},i}$, due to the finite volume and resolution limiting the experiment.

The actual observation measured is the proper time

$$t_i = \frac{\ell_i}{\gamma_i \beta_i c}$$

where γ_i, β_i are the usual Lorentz quantities, and c is the velocity of light. Thus the limits of t_i are

$$t_{\text{pot},i} = \frac{\ell_{\text{pot},i}}{\gamma_i \beta_i c}$$

and

$$t_{\text{min},i} = \frac{\ell_{\text{min},i}}{\gamma_i \beta_i c} .$$

The likelihood of the N observations t_i is

$$L = \prod_{i=1}^{N} \frac{\dfrac{1}{\tau} \exp\left(\dfrac{-t_i}{\tau}\right)}{\exp\left(\dfrac{-t_{\text{min},i}}{\tau}\right) - \exp\left(\dfrac{-t_{\text{pot},i}}{\tau}\right)} ,$$

where τ is the lifetime to be determined. Then

$$\ln L = \sum_{i=1}^{N} \left\{ -\frac{t_i}{\tau} - \ln \left[\exp\left(-\frac{t_{\min,i}}{\tau} \right) - \exp\left(-\frac{t_{\text{pot},i}}{\tau} \right) \right] \right\} - N \ln \tau .$$

The maximum likelihood estimate $\hat{\tau}$ is obtained by solving

$$\frac{\partial \ln L}{\partial \tau} = \frac{1}{\tau^2} \left\{ \sum_{i=1}^{N} [t_i - g_i(\tau)] - N\tau \right\} = 0 , \tag{8.12}$$

where

$$g_i(\tau) = \frac{t_{\min,i} \exp\left(-\dfrac{t_{\min,i}}{\tau} \right) - t_{\text{pot},i} \exp\left(-\dfrac{t_{\text{pot},i}}{\tau} \right)}{\exp\left(-\dfrac{t_{\min,i}}{\tau} \right) - \exp\left(-\dfrac{t_{\text{pot},i}}{\tau} \right)} .$$

One may solve Eq. (8.12) by Newton's method as follows:
Let us define $F(\tau)$ by

$$F(\tau) = \sum_{i=1}^{N} [t_i - g_i(\tau)] - N\tau .$$

Starting with an initial approximation $\tau = \tau_0$ [e.g. a previously known best value], the next approximation is given by

$$\tau_1 = \tau_0 - \frac{F(\tau_0)}{F'(\tau_0)} ,$$

where

$$F'(\tau) = -\sum_{i=1}^{N} g_i'(\tau) - N$$

and

$$g_i'(\tau) = -\frac{1}{\tau^2} \left[g_i(\tau)^2 - \frac{t_{\min,i}^2 \exp\left(\dfrac{-t_{\min,i}}{\tau} \right) - t_{\text{pot},i}^2 \exp\left(\dfrac{-t_{\text{pot},i}}{\tau} \right)}{\exp\left(\dfrac{-t_{\min,i}}{\tau} \right) - \exp\left(\dfrac{-t_{\text{pot},i}}{\tau} \right)} \right] .$$

Thus one obtains

$$\tau_1 = \frac{N\tau_0 + F(\tau_0) + \tau_0 \sum_{i=1}^{N} g_i'(\tau_0)}{N + \sum_{i=1}^{N} g_i'(\tau_0)} .$$

Successive approximations are given by

$$\tau_{j+1} = \frac{\tau_j \cdot \overline{g'(\tau_j)} + \overline{t} - \overline{g(\tau_j)}}{\overline{g'(\tau_j)} + 1},$$

where

$$\overline{g(\tau_j)} = \frac{1}{N} \sum_{i=1}^{N} g_i(\tau_j), \quad \text{etc}.$$

(The approximation τ_1 will be sufficient if $F'(\tau_1)$ is of the order of $\sqrt{F''(\tau_1)}$, or equivalently, if $\tau_0 - \tau_1$ is of the order of the error on τ_1.)

To determine the variance, we take the second derivative of the log-likelihood

$$\frac{\partial^2 \ln L}{\partial \tau^2} = -\frac{2}{\tau^3} F + \frac{1}{\tau^2} F'$$

$$\simeq \frac{1}{\tau_0^2}\left(F' - \frac{2}{\tau_0} F\right).$$

The inverse is the approximate asymptotic variance on the estimate $\hat{\tau}$ of τ, and

$$\sigma(\hat{\tau}) \simeq \frac{\tau_0}{\sqrt{N[-1 + \bar{g}' + \frac{2}{\tau_0}(\bar{t} - \bar{g})]}}.$$

8.3.3. *Academic example of a poor maximum likelihood estimate*

All of the foregoing discussion of the likelihood method has been based on the assumption that the *range of observations* does not depend on the parameter θ. Let us consider a counter-example. Suppose that one observes N events X_i chosen randomly from a uniform distribution between 0 and θ, where the upper bound θ is the unknown parameter. Since $\theta \geq X_i$ for all i, the likelihood function $L = \theta^{-N}$ will have its maximum at $\hat{\theta} = X_N$, where X_N is the largest observed value of X. Smaller values of θ are not allowed, and larger values would give smaller likelihood. The distributions of $\hat{\theta}$ for $N = 1$, 2, 3 are shown dashed in Fig. 8.1. One can see that this estimator always gives a result which is too small, and in fact one could have done better by simple common sense without using any likelihood. One knows that $\theta_0 \geq X_N$, and a good common-sense estimate would be:

$$\hat{\theta}_{\text{cs}} = X_N + \frac{X_N}{N}.$$

This estimate in fact turns out to be unbiased, as can easily be verified.

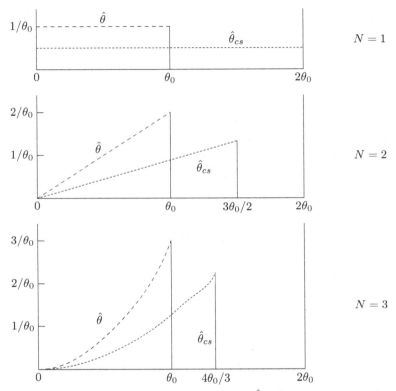

Fig. 8.1. Distribution of maximum-likelihood estimates $\hat{\theta}$ and common-sense estimates $\hat{\theta}_{cs}$ for $N = 1, 2, 3$.

Plotting the distributions of $\hat{\theta}_{cs}$, one gets the lower curves in Fig. 8.1, each of which has the mean value θ_0. But the distribution is still not perfect, being always *peaked* to the right of the true value. In particular, the sharp discontinuity will never disappear for any N, so that neither $\hat{\theta}$ nor $\hat{\theta}_{cs}$ becomes asymptotically Normal. This is due to the fact that the range of observations depends on the value of θ.

This example is largely academic, since in practice one always has to fold the resolution and acceptance functions of the apparatus into the theoretical distribution, which smears the uniform distribution, so that θ is no longer the upper bound. In practice, the actual range of observations may be independent of the parameter. For further comments on this example, see Section 8.6.2.

8.3.4. *Constrained parameters in maximum likelihood*

It often happens in practical problems that the parameters to be fitted are constrained, for instance by some physical law. In this section we are dealing with *constraints* of the form

$$\mathbf{g}(\widehat{\boldsymbol{\theta}}) = 0 . \tag{8.13}$$

The problem of imposing inequalities on functions of the parameters (for instance, positivity of a given parameter) is mostly left aside.

Imposing constraints always implies adding some information, and therefore the errors of the parameters in general get reduced. One should therefore be careful not to add "wrong" information: one should test that the data are not incompatible with the constraints. For example, before setting a parameter to its theoretical value, it should be checked that this value is consistent with the data: even if the theory is true, there may be biases in the experiment, and testing the validity of the constraints is usually a good method to find them. Such tests will be described in Chapter 10.

The most economical method to deal with constraints is to change variables in such a way that Eq. (8.13) becomes trivial. For instance, if

$$g(\boldsymbol{\theta}) = \theta_1 + \theta_2 - 1 = 0 ,$$

replace θ_2 by $1 - \theta_1$ in the likelihood function and maximize with respect to θ_1. This makes use of the invariance property (8.11). Such methods may also be used when θ is bounded by simple inequalities,

$$\theta' < \theta < \theta'' .$$

The variable transform is

$$\theta = \theta' + \frac{1}{2}(\sin \xi + 1)(\theta'' - \theta') , \tag{8.14}$$

and one maximizes the likelihood with respect to ξ.

In the case when the θ_i are proportions, subject to the constraints

$$0 \le \theta_i \le 1 , \quad i = 1, \dots, n$$

$$\sum_{i=1}^{N} \theta_i = 1$$

there is a well-known recipe for changing variables. For $n = 4$ it is

$$\theta_1 = \xi_1$$
$$\theta_2 = (1 - \xi_1)\xi_2$$
$$\theta_3 = (1 - \xi_1)(1 - \xi_2)\xi_3$$
$$\theta_4 = (1 - \xi_1)(1 - \xi_2)(1 - \xi_3),$$

where the ξ_i are bounded by 0 and 1 by the method in Eq. (8.14).

In this way one loses the symmetry of the problem, which may sometimes be a drawback.

In general, such methods may be difficult to apply, and one naturally turns to the method of *Lagrange multipliers*. Given the likelihood function $L(\mathbf{X}|\boldsymbol{\theta})$, one takes the extremum of

$$F = \ln L(\mathbf{X}|\boldsymbol{\theta}) + \boldsymbol{\alpha}^T \mathbf{g}(\boldsymbol{\theta})$$

with respect to $\boldsymbol{\theta}$ and $\boldsymbol{\alpha}$. The *likelihood equation* (8.9) is replaced by the system of equations

$$\frac{\partial}{\partial \boldsymbol{\theta}} \ln L(\mathbf{X}, \hat{\boldsymbol{\theta}}) + \hat{\boldsymbol{\alpha}}^T \frac{\partial \mathbf{g}(\hat{\boldsymbol{\theta}})}{\partial \boldsymbol{\theta}} = 0$$

$$\mathbf{g}(\hat{\boldsymbol{\theta}}) = 0. \tag{8.15}$$

The consistency and the asymptotic Normality of such estimates may be proved in a way similar to the proofs of Section 7.3.

In order to write down the variance of $\hat{\boldsymbol{\theta}}$ and $\hat{\boldsymbol{\alpha}}$, let us define $\mathcal{L}$, the negative of the matrix of second derivatives of F, and its submatrices $\mathcal{A}$ and $\mathcal{B}$, by

$$\mathcal{L} \equiv \begin{pmatrix} \mathcal{A} & \mathcal{B} \\ \mathcal{B}^T & 0 \end{pmatrix} \equiv -E \begin{pmatrix} \dfrac{\partial^2 F}{\partial \boldsymbol{\theta} \partial \boldsymbol{\theta}'} & \dfrac{\partial^2 F}{\partial \boldsymbol{\theta} \partial \boldsymbol{\alpha}} \\ \left(\dfrac{\partial^2 F}{\partial \boldsymbol{\theta} \partial \boldsymbol{\alpha}}\right)^T & \dfrac{\partial^2 F}{\partial \boldsymbol{\alpha}^2} \end{pmatrix}$$

$$= -E \begin{pmatrix} \dfrac{\partial^2 \ln L}{\partial \boldsymbol{\theta} \partial \boldsymbol{\theta}'} & \dfrac{\partial \mathbf{g}}{\partial \boldsymbol{\theta}} \\ \left(\dfrac{\partial \mathbf{g}}{\partial \boldsymbol{\theta}}\right)^T & 0 \end{pmatrix}.$$

The inverse of $\underline{I}$ is

$$
\underline{I}^{-1} = \begin{pmatrix} \underline{A}^{-1} - \underline{A}^{-1}\underline{B}(\underline{B}^T\underline{A}^{-1}\underline{B})^{-1}\underline{B}^T\underline{A}^{-1} & \underline{A}^{-1}\underline{B}(\underline{B}^T\underline{A}^{-1}\underline{B})^{-1} \\ (\underline{B}^T\underline{A}^{-1}\underline{B})^{-1}\underline{B}^T\underline{A}^{-1} & -(\underline{B}^T\underline{A}^{-1}\underline{B})^{-1} \end{pmatrix} .
$$

From the Taylor expansion of the vector

$$
\begin{pmatrix} \partial F/\partial\boldsymbol{\theta} \\ \partial F/\partial\boldsymbol{\alpha} \end{pmatrix}_{\substack{\boldsymbol{\theta}=\widehat{\boldsymbol{\theta}} \\ \boldsymbol{\alpha}=\widehat{\boldsymbol{\alpha}}}} = \begin{pmatrix} \partial F/\partial\boldsymbol{\theta} \\ \partial F/\partial\boldsymbol{\alpha} \end{pmatrix}_{\substack{\boldsymbol{\theta}=\boldsymbol{\theta}_0 \\ \boldsymbol{\alpha}=0}} - \begin{pmatrix} \underline{A} & \underline{B}^T \\ \underline{B} & 0 \end{pmatrix}_{\substack{\boldsymbol{\theta}=\boldsymbol{\theta}_0 \\ \boldsymbol{\alpha}=0}} \begin{pmatrix} \widehat{\boldsymbol{\theta}}-\boldsymbol{\theta}_0 \\ \widehat{\boldsymbol{\alpha}} \end{pmatrix} + \dots ,
$$

and using the fact that this vector is zero, by Eq. (8.15), one can write down the expression for the covariance matrix:

$$
E\left[\begin{pmatrix} \widehat{\boldsymbol{\theta}}-\boldsymbol{\theta}_0 \\ \widehat{\boldsymbol{\alpha}} \end{pmatrix} \begin{pmatrix} \widehat{\boldsymbol{\theta}}-\boldsymbol{\theta}_0 \\ \widehat{\boldsymbol{\alpha}} \end{pmatrix}^T \right] = \underline{I}^{-1} \begin{pmatrix} \underline{A} & 0 \\ 0 & 0 \end{pmatrix} \underline{I}^{-1} .
$$

After tedious but simple algebra one obtains

$$
\underline{V}(\widehat{\boldsymbol{\theta}}-\boldsymbol{\theta}_0) = \underline{A}^{-1} - \underline{A}^{-1}\underline{B}(\underline{B}^T\underline{A}^{-1}\underline{B})^{-1}\underline{B}^T\underline{A}^{-1} \tag{8.16}
$$

and

$$
\underline{V}(\widehat{\boldsymbol{\alpha}}) = (\underline{B}^T\underline{A}^{-1}\underline{B})^{-1} .
$$

The first term on the right-hand side of Eq. (8.16), $\underline{A}^{-1}$, is the ordinary unconstrained variance matrix, whereas the second term represents the reduction in variance obtained by the addition of information by the constraints. The covariance between $\widehat{\boldsymbol{\theta}}$ and $\widehat{\boldsymbol{\alpha}}$ is zero: the parameter estimates are uncorrelated with $\widehat{\boldsymbol{\alpha}}$.

Note that the extremum in F is a *saddle point*. F is maximal with respect to $\widehat{\boldsymbol{\theta}}$ but minimal with respect to $\widehat{\boldsymbol{\alpha}}$. This may completely upset the usual *optimization* methods, based on comparison of successive values of F ("hill climbing" methods). It is, however, outside the scope of this book to discuss the virtues of different optimization methods. The interested reader is referred to the specialized literature [Gill].

One way to overcome this difficulty is to minimize some weighted sum of the squares of the derivatives in Eq. (8.15) instead of maximizing F.

In all that was said above, $\underline{I}$ was assumed to be non-singular. In particular,

$$
\frac{\partial^2 \ln L}{\partial\boldsymbol{\theta}\partial\boldsymbol{\theta}'}
$$

should not be singular for $\boldsymbol{\theta} = \boldsymbol{\theta}_0$.

This assumption is not valid in the practical case when the constraint $g(\boldsymbol{\theta}) = 0$ is necessary to define the parameters unambiguously. For example, if the theoretical distribution is a homogeneous function of the $\boldsymbol{\theta}$, one has to impose some condition, such as

$$\sum_{i=1}^{n} \theta_i = 1.$$

If for symmetry considerations one does not want to impose this restriction by a change of variables, the method of Lagrangian multipliers is valid provided that one replaces F by

$$F' = \ln L(\mathbf{X}|\boldsymbol{\theta}) - \mathbf{g}^2(\boldsymbol{\theta}) + \boldsymbol{\alpha}^{\mathrm{T}} g(\boldsymbol{\theta}).$$

Equations (8.15) become

$$\frac{\partial \ln L(\mathbf{X}|\boldsymbol{\theta})}{\partial \boldsymbol{\theta}} - 2g(\boldsymbol{\theta})\frac{\partial \mathbf{g}}{\partial \boldsymbol{\theta}} + \widehat{\boldsymbol{\alpha}}^{\mathrm{T}}\frac{\partial \mathbf{g}}{\partial \boldsymbol{\theta}}(\widehat{\boldsymbol{\theta}}) = 0$$

$$\mathbf{g}(\widehat{\boldsymbol{\theta}}) = 0$$

and $\underline{\underline{I}}$ is replaced by

$$\underline{\underline{I}}' = -E \begin{pmatrix} \dfrac{\partial^2 \ln L}{\partial\boldsymbol{\theta}\partial\boldsymbol{\theta}'} + \dfrac{\partial\mathbf{g}}{\partial\boldsymbol{\theta}}\dfrac{\partial\mathbf{g}}{\partial\boldsymbol{\theta}'} & \dfrac{\partial\mathbf{g}}{\partial\boldsymbol{\theta}'} \\[2ex] \dfrac{\partial\mathbf{g}^{\mathrm{T}}}{\partial\boldsymbol{\theta}} & 0 \end{pmatrix}_{\boldsymbol{\theta}=\boldsymbol{\theta}_0}.$$

The quantity

$$\frac{\partial^2 \ln L}{\partial\boldsymbol{\theta}\partial\boldsymbol{\theta}'} + \frac{\partial\mathbf{g}}{\partial\boldsymbol{\theta}}\frac{\partial\mathbf{g}}{\partial\boldsymbol{\theta}'}$$

is usually no more singular and all our results (including the variance matrix, replacing F by F') are valid.

8.4. The Least Squares Method (Chi-Square)

Another general method of estimation is that of *least squares*, often referred to as the *chi-square method*. Some confusion in terminology arises between the sum of squares (considered here) and the χ^2 function of Section 4.2.3. Although an important class of least squares functions does behave as the χ^2 function of Section 4.2.3, this is not generally true for any sum of residual squares. (For a discussion of why one uses least squares rather than, for example, least cubes or least absolute values, see Section 8.7.3.)

Consider a set of observations $Y_1, \ldots, Y_N$ from a distribution with expectations $E(Y_i, \boldsymbol{\theta})$ and covariance matrix $\underline{V}$. The $\boldsymbol{\theta}$ are unknown parameters and the $E(Y_i, \boldsymbol{\theta})$ and $V_{ij}(\theta)$ are known functions of $\boldsymbol{\theta}$.

In the method of least squares the estimates of the θ_k are those values $\hat{\theta}_k$ which minimize the covariance form (7.18), or

$$Q^2 = \sum_{i=1}^{N} \sum_{j=1}^{N} [Y_i - E(Y_i, \boldsymbol{\theta})](\underline{V}^{-1})_{ij}[Y_j - E(Y_j, \boldsymbol{\theta})]$$

$$= [\mathbf{Y} - E(\mathbf{Y}, \boldsymbol{\theta})]^{\mathrm{T}} \, \underline{V}^{-1} \, [\mathbf{Y} - E(\mathbf{Y}, \boldsymbol{\theta})] \,. \qquad (8.17)$$

In Section 7.2.4 we have shown that the least squares estimator is consistent, and in Section 7.3.3 we have shown that the linear least squares estimator is unbiased.

In general, the covariance matrix $\underline{V}$ is non-diagonal. When the observations Y_i are independent, it follows that they are uncorrelated, and the covariance matrix is diagonal, with elements

$$V_{ii} = \sigma_i^2(\boldsymbol{\theta}) \,.$$

The covariance form (8.17) then simplifies to the familiar sum of squares

$$Q^2 = \sum_{i=1}^{N} \sigma_i^{-2}(\boldsymbol{\theta})[Y_i - E(Y_i, \boldsymbol{\theta})]^2 \,.$$

The observations may be, for example, numbers of events in histogram bins, or measured coordinates of points along a track.

8.4.1. *The linear model*

The method of *linear* least squares [Section 7.2.4] is applicable when the variances σ_i^2 are independent of the r parameters $\boldsymbol{\theta} = (\theta_1, \ldots, \theta_r)$, and the expectations $E(Y_i, \boldsymbol{\theta})$ are linear in the θ_j's,

$$E(Y_i, \boldsymbol{\theta}) = \sum_{j=1}^{r} a_{ij}\theta_j \,, \quad i = 1, \ldots, N$$

or in matrix notation

$$E(\mathbf{Y}, \boldsymbol{\theta}) = \underline{A} \, \boldsymbol{\theta} \qquad (8.18)$$

The elements a_{ij} of the *design matrix* $\underline{A}$ are given by a model.

As an example, suppose that one wishes to fit the shape of a particle track to a third-degree polynomial in X. For several fixed points X_i, let the measured Y-coordinates be Y_i with variance σ^2.

The assumed polynomial

$$Y_i = Y(X_i) = \theta_0 + \theta_1 X_i + \theta_2 X_i^2 + \theta_3 X_i^3 \tag{8.19}$$

is clearly of the form (8.18). To find the matrix A one only needs to evaluate the $(j-1)^{\text{th}}$ power of X_i.

Solving the *Normal equations*

$$\partial Q^2 / \partial \boldsymbol{\theta} = 0 \, ,$$

we find the linear least squares estimator $\widehat{\boldsymbol{\theta}}$ of Eq. (7.23),

$$\widehat{\boldsymbol{\theta}} = (A^{\mathrm{T}} V^{-1} A)^{-1} A^{\mathrm{T}} V^{-1} \mathbf{Y} \, . \tag{8.20}$$

Note that this "minimization" is done by a simple matrix inversion, involving no approximations, searches, or iterative procedures. It is exact and unique as long as $A^{\mathrm{T}} V^{-1} A$ is non-singular. The unweighted case corresponds to taking $V = I$, the unit matrix, and the solution (8.20) becomes

$$\widehat{\boldsymbol{\theta}} = (A^{\mathrm{T}} A)^{-1} A^{\mathrm{T}} \mathbf{Y} \, . \tag{8.21}$$

It follows from Eqs. (8.20) or (8.21) that the $\widehat{\boldsymbol{\theta}}$ are *linear estimators*.

We have seen in Section 7.3.3 that the estimators (8.20) are unbiased to all orders of N. In fact, the *Gauss–Markov theorem* [Section 7.4.4] states that among all linear unbiased estimators,

$$\mathbf{t} = L \mathbf{Y} \, ,$$

the particular estimators (8.20) have minimum variance. These variances are given by the diagonal elements of

$$\text{cov}(\widehat{\boldsymbol{\theta}}) = E[(\widehat{\boldsymbol{\theta}} - \boldsymbol{\theta})(\widehat{\boldsymbol{\theta}} - \boldsymbol{\theta})^{\mathrm{T}}]$$
$$= \sigma^2 [(A^{\mathrm{T}} V^{-1} A)^{-1}] \, . \tag{8.22}$$

For the determination of $\widehat{\boldsymbol{\theta}}$, the matrix V must be known up to a constant multiplicative factor, σ^2.

For the determination of the *variance* of $\widehat{\boldsymbol{\theta}}$, however, σ^2 must be known also. If it is not known, it can be estimated from the minimum value of Q^2,

$$Q_{\min}^2 = (\mathbf{Y} - A\widehat{\boldsymbol{\theta}})^{\mathrm{T}} V^{-1} (\mathbf{Y} - A\widehat{\boldsymbol{\theta}}) \, , \tag{8.23}$$

called the *residual sum of squares*. A straightforward application of matrix algebra gives the expectation of $Q^2_{\min}$,

$$E(Q^2_{\min}) = \sigma^2(N - r).$$

Therefore, the quantity

$$\hat{\sigma}^2 = \frac{Q^2_{\min}}{N - r} \tag{8.24}$$

is an unbiased estimator of σ^2. In the unweighted case, when $\underline{V} = \underline{I}$, the Eq. (8.24) gives an unbiased estimate of the common variance of the observations.

Note that no assumptions have been made about the parent distribution of the data $\mathbf{Y}$. The optimal properties of least squares estimates are a consequence of the linearity alone.

When the distribution of the data is Gaussian, it follows from the linearity that the least squares estimates are also Gaussian, with covariance matrix given by Eq. (8.22). Furthermore, the least squares method is then equivalent to the maximum likelihood method.

8.4.2. *The polynomial model*

When the power of the polynomial (8.19) is greater than six or seven, the matrix $\underline{A}$ tends to become ill-conditioned and inverting the matrix $\underline{A}^{\mathrm{T}}\underline{A}$ accurately may be difficult. It is then better to rewrite the model in terms of orthogonal polynomials $\xi_j(X)$ of order $j = 1, \ldots, r$, which is always possible. The orthogonality property is expressed by

$$\sum_{i=1}^{N} \xi_j(X_i)\xi_k(X_i) = \delta_{jk}. \tag{8.25}$$

In this case the elements of the matrix $\underline{A}$ are

$$A_{ij} = \xi_j(X_i),$$

and $\underline{A}^{\mathrm{T}}\underline{A} = \underline{I}$, the unit matrix. Thus one is no longer limited by the numerical accuracy in matrix inversion. However, some restriction on the accuracy obtainable is introduced in forming the orthogonal polynomials $\xi_j(X)$.

The model is now $Y(X) = \Sigma_j \cdots \psi_j \xi_j(X)$. The estimates of the parameters become

$$\hat{\psi}_j = \sum_{i=1}^{N} \xi_j(X_i)Y(X_i)$$

with covariance matrix $\sigma^2 \mathbb{1}$. Thus the estimates are independently distributed. The *residual sum of squares* (8.23) then takes the form

$$Q^2_{\min} = \sum_{i=1}^{N} Y^2(X_i) - \sum_{j=1}^{r} \hat{\psi}_j^2,$$

and an unbiased estimate of the variance σ^2 is given by Eq. (8.24). The important property is that the $\hat{\psi}_j$'s are *independently* distributed.

In the more general case, when the observations Y_i, $i = 1, \ldots, N$ have a covariance matrix $\sigma^2 \underline{V}$, the polynomials $\xi_j(X)$ should have the property

$$\sum_{i=1}^{N} \xi_j(X_i)\xi_k(X_i)(\underline{V}^{-1})_{jk} = \delta_{jk},$$

instead of Eq. (8.25). In this case, the residual sum of squares (8.23) takes the form:

$$Q^2_{\min} = \sum_{i=1}^{N}\sum_{j=1}^{N} Y_iY_j(\underline{V}^{-1})_{ij} - \sum_{j=1}^{r} \hat{\psi}_j^2.$$

For polynomial models, the *orthogonal functions* $\xi_j(X_i)$ can be generated most directly by *Forsythe's method*. In general, it is numerically possible to go to appreciably higher order using this technique. If any model can be expressed in terms of orthogonal functions (for example, Legendre or associated Legendre functions, where the functions are orthogonal over the observations), then the useful properties of the above analysis hold, in particular the independence of the estimates of the parameters.

It is important to note that the significance of the kth power in the fit is measured by $\hat{\psi}_k$, and *not* by the coefficient $\hat{\theta}_k$ in the ordinary polynomial form. This will be dealt with in more detail in chapter 10.

8.4.3. *Constrained parameters in the linear model*

(i) *The method of estimation*

Suppose one has a model $Y = \underline{A}\,\theta + \varepsilon$ for N observations Y_i, and r parameters θ_j. Then the expectation is given by Eq. (8.18). The independent variables A_{ij}, forming the matrix $\underline{A}$, may take any form, for example $A_{ij} = X_i^{j-1}$ to give an r-polynomial fit to the observations Y_i, taken at the points X_i, as in the two previous sections.

Suppose that the ε_i have covariance matrix $\sigma^2 \underset{\sim}{V}$, and that the parameters $\boldsymbol{\theta}$ are subject to m linear *constraints*

$$\sum_{j=1}^{r} \ell_{kj}\theta_j = R_k, \quad k = 1,\ldots,m$$

or in matrix notation

$$\underset{\sim}{L}\,\boldsymbol{\theta} = \mathbf{R}\,.$$

To obtain the *least squares estimation of* θ we use a vector of Lagrange multipliers, $\boldsymbol{\lambda}$, to minimize

$$Q^2 = (\mathbf{Y} - \underset{\sim}{A}\,\boldsymbol{\theta})^{\mathrm{T}}\,\underset{\sim}{V}^{-1}\,(\mathbf{Y} - \underset{\sim}{A}\,\boldsymbol{\theta}) + (\underset{\sim}{L}\,\boldsymbol{\theta} - \mathbf{R})^{\mathrm{T}}\boldsymbol{\lambda}\,.$$

Differentiating with respect to $\boldsymbol{\theta}$, $\boldsymbol{\lambda}$, we obtain the *Normal equations*

$$\underset{\sim}{A}^{\mathrm{T}}\underset{\sim}{V}^{-1}\underset{\sim}{A}\,\boldsymbol{\theta} + \underset{\sim}{L}^{\mathrm{T}}\,\boldsymbol{\lambda} = \underset{\sim}{A}^{\mathrm{T}}\underset{\sim}{V}^{-1}\,\mathbf{Y}$$

$$\underset{\sim}{L}\,\boldsymbol{\theta} = \mathbf{R}\,. \tag{8.26}$$

Writing this as

$$\begin{bmatrix} \underset{\sim}{C} & \vdots & \underset{\sim}{L}^{\mathrm{T}} \\ \cdots & \cdots & \cdots \\ \underset{\sim}{L} & \vdots & \underset{\sim}{\phi} \end{bmatrix} \begin{bmatrix} \boldsymbol{\theta} \\ \cdots \\ \boldsymbol{\lambda} \end{bmatrix} = \begin{bmatrix} \mathbf{S} \\ \cdots \\ \mathbf{R} \end{bmatrix},$$

with $\underset{\sim}{C} = \underset{\sim}{A}^{\mathrm{T}}\underset{\sim}{V}^{-1}\underset{\sim}{A}$, $\mathbf{S} = \underset{\sim}{A}^{\mathrm{T}}\underset{\sim}{V}^{-1}\,\mathbf{Y}$, and $\underset{\sim}{\phi} =$ zero matrix, we can solve for $\boldsymbol{\theta}$ and $\boldsymbol{\lambda}$:

$$\begin{bmatrix} \boldsymbol{\theta} \\ \cdots \\ \boldsymbol{\lambda} \end{bmatrix} = \begin{bmatrix} \underset{\sim}{F} & \vdots & \underset{\sim}{G}^{\mathrm{T}} \\ \cdots & \cdots & \cdots \\ \underset{\sim}{G} & \vdots & \underset{\sim}{H} \end{bmatrix} \begin{bmatrix} \mathbf{S} \\ \cdots \\ \mathbf{R} \end{bmatrix}. \tag{8.27}$$

Straightforward matrix algebra leads to the results

$$\underset{\sim}{F} = \underset{\sim}{C}^{-1} - \underset{\sim}{C}^{-1}\underset{\sim}{L}^{\mathrm{T}}\,(\underset{\sim}{L}\underset{\sim}{C}^{-1}\underset{\sim}{L}^{\mathrm{T}})^{-1}\,\underset{\sim}{L}\underset{\sim}{C}^{-1}$$

$$\underset{\sim}{G} = (\underset{\sim}{L}\underset{\sim}{C}^{-1}\underset{\sim}{L}^{\mathrm{T}})^{-1}\,\underset{\sim}{L}\underset{\sim}{C}^{-1}$$

$$\underset{\sim}{H} = -(\underset{\sim}{L}\underset{\sim}{C}^{-1}\underset{\sim}{L}^{\mathrm{T}})^{-1}\,.$$

Using Eqs. (8.27) and the relations between the matrices $\underset{\sim}{C}$, $\underset{\sim}{L}$, $\underset{\sim}{F}$, $\underset{\sim}{G}$, and $\underset{\sim}{H}$, we can study the properties of the estimates of the parameters. These relations are

$$\underset{\sim}{C}\underset{\sim}{F} + \underset{\sim}{L}^{\mathrm{T}}\underset{\sim}{G} = \underset{\sim}{I}, \quad \underset{\sim}{L}\underset{\sim}{F} = \underset{\sim}{\phi},$$

$$\underset{\sim}{C}\underset{\sim}{G}^{\mathrm{T}} + \underset{\sim}{L}^{\mathrm{T}}\underset{\sim}{H} = \underset{\sim}{\phi}, \quad \underset{\sim}{L}\underset{\sim}{G}^{\mathrm{T}} = \underset{\sim}{I}\,.$$

Thus, we have the result

$$\hat{\theta} = \underset{\sim}{F}\, \mathbf{S} + \underset{\sim}{G}^{\mathrm{T}}\, \mathbf{R} = \underset{\sim}{F}\, \underset{\sim}{A}^{\mathrm{T}}\, \underset{\sim}{V}^{-1}\, \mathbf{Y} + \underset{\sim}{G}^{\mathrm{T}}\, \mathbf{R}. \tag{8.28}$$

Since we have assumed that the errors have mean zero, the expectations become

$$E(\hat{\theta}) = E[\underset{\sim}{F}\, \underset{\sim}{A}^{\mathrm{T}}\, \underset{\sim}{V}^{-1}\,(\underset{\sim}{A}\,\theta + \varepsilon) + \underset{\sim}{G}^{\mathrm{T}}\, \underset{\sim}{L}\, \theta]$$

$$= E(\underset{\sim}{F}\, \underset{\sim}{A}^{\mathrm{T}}\, \underset{\sim}{V}^{-1}\,\varepsilon + \theta) = \theta \tag{8.29}$$

showing the estimates to be unbiased.

Also, for the *Lagrange multipliers* $\hat{\lambda}$,

$$\hat{\lambda} = \underset{\sim}{G}\, \mathbf{S} + \underset{\sim}{H}\, \mathbf{R}$$

$$= \underset{\sim}{G}\, \underset{\sim}{A}^{\mathrm{T}}\, \underset{\sim}{V}^{-1}\, \underset{\sim}{A}\, \theta + \underset{\sim}{H}\, \underset{\sim}{L}\, \theta + \underset{\sim}{G}\, \underset{\sim}{A}^{\mathrm{T}}\, \underset{\sim}{V}^{-1}\, \varepsilon$$

$$= \underset{\sim}{G}\, \underset{\sim}{A}^{\mathrm{T}}\, \underset{\sim}{V}^{-1}\, \varepsilon.$$

Therefore,

$$E(\hat{\lambda}) = \phi \equiv 0.$$

Thus the mean value of the Lagrange multipliers estimated in this way is zero. This is reasonable, since it has been assumed that the true values of the parameters satisfy the constraints.

(ii) *Covariance matrix of the estimates*

The covariance matrix of the estimates $\hat{\theta}$ can be found using Eq. (8.29) and the fact that $\mathrm{cov}(\varepsilon) = \sigma^2\, \underset{\sim}{V}$ by definition,

$$\mathrm{cov}(\hat{\theta}) = E[\underset{\sim}{F}\, \underset{\sim}{A}^{\mathrm{T}}\, \underset{\sim}{V}^{-1}\, \varepsilon\varepsilon^{\mathrm{T}}\, \underset{\sim}{V}^{-1}\, \underset{\sim}{A}\, \underset{\sim}{F}]$$

$$= \sigma^2\, \underset{\sim}{F}\, \underset{\sim}{C}\, \underset{\sim}{F}$$

$$= \sigma^2\, \underset{\sim}{F}$$

$$= \sigma^2[\underset{\sim}{C}^{-1} - \underset{\sim}{C}^{-1}\, \underset{\sim}{L}^{\mathrm{T}}\,(\underset{\sim}{L}\,\underset{\sim}{C}\,\underset{\sim}{L}^{\mathrm{T}})^{-1}\, \underset{\sim}{L}\,\underset{\sim}{C}^{-1}]. \tag{8.30}$$

In the unconstrained case, the covariance matrix of the estimates of θ was $\sigma^2\, \underset{\sim}{C}^{-1}$. The diagonal elements of the second term measure the reduction in variances of the estimates, obtained by adding the constraints. The covariances between the parameter estimates may increase or decrease, depending on the particular problem.

The covariance matrix of the Lagrange multipliers is given by

$$\text{cov}(\widehat{\lambda}) = E[\underset{\sim}{G}\underset{\sim}{A}^{\mathrm{T}}\underset{\sim}{V}^{-1}\,\varepsilon\varepsilon^{\mathrm{T}}\,\underset{\sim}{V}^{-1}\underset{\sim}{A}\,\underset{\sim}{G}^{\mathrm{T}}]$$

$$= \sigma^2\,\underset{\sim}{G}\underset{\sim}{C}\underset{\sim}{G}^{\mathrm{T}}$$

$$= \sigma^2\,\underset{\sim}{H}$$

$$= \sigma^2(\underset{\sim}{L}\underset{\sim}{C}^{-1}\underset{\sim}{L}^{\mathrm{T}})^{-1}.$$

Similarly, it can be shown that the covariance between $\widehat{\theta}$ and $\widehat{\lambda}$ is zero.

It is interesting to note that the covariance matrix of $\widehat{\theta}$ is the appropriate submatrix of the inverse of the "Normal equations" matrix of Eq. (8.26), but the covariance of $\widehat{\lambda}$ is the *negative* of the corresponding submatrix.

(iii) Estimation of σ^2

The residual sum of squares is given by Eq. (8.23),

$$Q^2_{\min} = (\mathbf{Y} - \underset{\sim}{A}\,\widehat{\theta})^{\mathrm{T}}\,\underset{\sim}{V}^{-1}\,(\mathbf{Y} - \underset{\sim}{A}\,\widehat{\theta})$$

$$= \mathbf{Y}^{\mathrm{T}}\,\underset{\sim}{V}^{-1}\,\mathbf{Y} - \widehat{\theta}^{\mathrm{T}}\,\underset{\sim}{S} - \underset{\sim}{S}^{\mathrm{T}}\,\widehat{\theta} + \widehat{\theta}^{\mathrm{T}}\,\underset{\sim}{C}\,\widehat{\theta}$$

$$= \mathbf{Y}^{\mathrm{T}}\,\underset{\sim}{V}^{-1}\,\mathbf{Y} - 2\widehat{\theta}^{\mathrm{T}}(\underset{\sim}{C}\,\widehat{\theta} + \underset{\sim}{L}^{\mathrm{T}}\,\widehat{\lambda}) + \widehat{\theta}^{\mathrm{T}}\,\underset{\sim}{C}\,\widehat{\theta}$$

$$= \mathbf{Y}^{\mathrm{T}}\,\underset{\sim}{V}^{-1}\,\mathbf{Y} - (\widehat{\theta}^{\mathrm{T}}\,\underset{\sim}{C}\,\widehat{\theta} + 2\widehat{\lambda}^{\mathrm{T}}\mathbf{R}),$$

or

$$\begin{matrix} \text{Residual} \\ \text{sum of squares} \end{matrix} = \begin{matrix} \text{Total} \\ \text{sum of squares} \end{matrix} - \begin{matrix} \text{Fitted} \\ \text{sum of squares} \end{matrix}.$$

Taking expectations we have:

$$E(\mathbf{Y}^{\mathrm{T}}\,\underset{\sim}{V}^{-1}\,\mathbf{Y}) = E[(\underset{\sim}{A}\,\theta + \varepsilon)^{\mathrm{T}}\,\underset{\sim}{V}^{-1}\,(\underset{\sim}{A}\,\theta + \varepsilon)]$$

$$= \theta^{\mathrm{T}}\,\underset{\sim}{C}\,\theta + E(\varepsilon^{\mathrm{T}}\,\underset{\sim}{V}^{-1}\,\varepsilon),$$

since for any matrix $\underset{\sim}{B}$,

$$E(\varepsilon^{\mathrm{T}}\,\underset{\sim}{B}\,\varepsilon) = E\left[\sum_i\sum_j \varepsilon_i B_{ij}\varepsilon_j\right]$$

$$= \sigma^2 \sum_i\sum_j B_{ij}V_{ij}$$

$$= \text{tr}(\underset{\sim}{B}\underset{\sim}{V}),$$

it follows that

$$E(\mathbf{Y}^{\mathrm{T}} \underset{\sim}{V}^{-1} \mathbf{Y}) = \boldsymbol{\theta}^{\mathrm{T}} \underset{\sim}{\mathcal{G}} \boldsymbol{\theta} + \sigma^2 \operatorname{tr} (\underset{\sim}{V}^{-1} \underset{\sim}{V})$$
$$= \boldsymbol{\theta}^{\mathrm{T}} \underset{\sim}{\mathcal{G}} \boldsymbol{\theta} + N\sigma^2 \,.$$

Similarly, it can be shown that

$$E(\widehat{\boldsymbol{\theta}}^{\mathrm{T}} \underset{\sim}{\mathcal{G}} \widehat{\boldsymbol{\theta}} + 2\widehat{\boldsymbol{\lambda}}^{\mathrm{T}} \mathbf{R}) = \boldsymbol{\theta}^{\mathrm{T}} \underset{\sim}{\mathcal{G}} \boldsymbol{\theta} + (r - m)\sigma^2 \,.$$

Therefore we have the result that

$$E(Q^2_{\min}) = N\sigma^2 - (r - m)\sigma^2$$

and that

$$\hat{\sigma}^2 = Q^2_{\min}/(N - r + m) \tag{8.31}$$

given an unbiased estimate of σ^2.

Thus the least squares estimators (8.28), (8.30), and (8.31) of parameters subject to constraints have similar properties to the corresponding estimators (8.21), (8.22) and (8.24) for unconstrained parameters. In particular, these point estimation results are independent of the distribution of the observations. However, interval estimation requires knowledge about the parent distribution. This will be dealt with in Chapter 9.

8.4.4. *Normally distributed data in nonlinear models*

Suppose that the observations Y_i are Normally distributed about $E(Y_i)$, uncorrelated, and with variances σ_i^2. Let the parameters $\boldsymbol{\theta}$ defined in some nonlinear model, which predicts

$$E(Y_i) = f_i(\boldsymbol{\theta}) \,. \tag{8.32}$$

Then the likelihood function is given by

$$L(\mathbf{Y}, \boldsymbol{\theta}) = \prod_{i=1}^{N} (2\pi\sigma_i^2)^{-\frac{1}{2}} \exp \left\{ - \frac{[Y_i - f_i(\boldsymbol{\theta})]^2}{2\sigma_i^2} \right\},$$

from which

$$\ln L(\mathbf{Y}, \boldsymbol{\theta}) = -\frac{1}{2} \sum_{i=1}^{N} \ln(2\pi\sigma_i) - \sum_{i=1}^{N} \frac{[Y_i - f_i(\boldsymbol{\theta})]^2}{2\sigma_i^2} \,.$$

Since the first term is constant, $\ln L$ is maximal when

$$Q^2 = \sum_{i=1}^{N} \frac{[Y_i - f_i(\boldsymbol{\theta})]^2}{\sigma_i^2} \tag{8.33}$$

is minimal. Thus the least squares method is equivalent to the maximum likelihood method in this case. It is optimal for all N since in this case sufficient statistics exist (7.4.2).

When the parameters $\boldsymbol{\theta}$ are known, Q^2 is the sum of squares of N standard Normal variables, and hence it has a $\chi^2(N)$ distribution. But when the $\boldsymbol{\theta}$ have to be estimated by those values which minimize Q^2, the distribution of Q^2 is no longer $\chi^2(N)$. The minimum value of Q^2 is then the sum of the squares of correlated, non-central (of nonzero mean) random variables, which are not, in general, Normally distributed. Its exact distribution is thus very difficult to evaluate, and the usual procedure is to consider the asymptotic properties, as described in Chapter 7.

When the individual observations Y_i are *not* Normally distributed, the least squares method is still used for estimating parameters, but then it does not have such general optimal properties as to be useful for small N. Even asymptotically, the estimators need not be of minimum variance (in nonlinear models).

8.4.5. *Estimation from histograms; comparison of likelihood and least squares methods*

Consider the case when the observations Y_i are numbers of events in *histogram* bins. In this case, as we have seen, the Y_i will have a multinomial distribution [Section 4.1.2]. As we know, in the limit of large numbers of events, this distribution becomes asymptotically Normal, so that we expect this to be a reasonable application of the least squares method. Now the model (8.32) will tell how many events to expect in the ith bin as a function of the parameters $\boldsymbol{\theta}$.

Assuming many bins (N large) so that $p \ll 1$, we have the case of example (ii) in Section 4.1.2

$$\sigma^2(Y_i) \approx f_i(\boldsymbol{\theta}) . \tag{8.34}$$

Neglecting correlations between bins, the quantity to be minimized is Eq. (8.33),

$$Q^2 = \sum_{i=1}^{N} \frac{[Y_i - f_i(\boldsymbol{\theta})]^2}{f_i(\boldsymbol{\theta})} . \tag{8.35}$$

This is the usual *minimum chi-square method* applied to "fits" of histograms. The reason for the name is that Q^2 *asymptotically* has a $\chi^2(N)$ distribution when $\boldsymbol{\theta}$ is known and when there are many events in each bin (Normality assumption). Note that when there are only very few events in a bin, Eq. (8.34) still holds, since the variance of the binomial distribution is $Np(1-p) \approx Np$ for all N. But in this case, the Y_i are no longer approximately Normally distributed so that one cannot make any general predictions about the way Q^2 is distributed.

Since we have neglected the correlation and made an approximation in forming Eq (8.35), there is no reason to expect any interesting properties. However, it turns out that the estimators obtained by this method have optimal *asymptotic* properties [Kendall II, p. 93]: they are consistent, asymptotically Normal, and efficient (have minimum variance). These are the properties of a *best asymptotically Normal* estimator (*BAN*).

For practical reasons, use is often made of the so-called *modified minimum chi-square* method in the case of histogram fitting. This method consists in minimizing the quantity

$$Q^2 = \sum_{i=1}^{N} \frac{[Y_i - f_i(\boldsymbol{\theta})]^2}{Y_i} . \tag{8.36}$$

Since $E(Y_i) = f_i(\boldsymbol{\theta})$, one might expect the modified method to be a reasonable approximation to the usual Eq. (8.35). In fact it can be shown that *asymptotically* for large N, this estimator coincides with Eq. (8.35), and is also a best asymptotically Normal estimator.

A third method is to apply the maximum likelihood method to the multinomial distribution of the binned observations, rather than to the original observations, as in Section 8.3. The *multinomial maximum likelihood* or *binned likelihood* estimators are found by maximizing

$$\ln L = \sum_{i=1}^{N} Y_i \ln f_i(\boldsymbol{\theta}) , \tag{8.37}$$

when there are N bins.

From the general properties of maximum likelihood (Chapter 7), one can see that the estimates from Eq. (8.37) are asymptotically identical to the estimates from Eqs. (8.35) and (8.36), and they are also BAN.

These three methods are therefore equivalent asymptotically. Nevertheless, it can be shown [Kendall, Section 19.28], [Rao] that in a certain sense the maximum likelihood method converges faster to efficiency under regularity conditions, and that the modified minimum chi-square method is the worst one from that point of view. Therefore, unless very special conditions make one of the other methods preferable (faster computing, for instance, because of linearity), it is recommended to use the maximum likelihood as defined in Eq. (8.37).

In particular, the m.l. method has usually no problems with bins with zero or few events, whereas the other methods either lose efficiency, because of the non-Normality of the distribution of the number of events, or blow up because some terms in Eq. (8.36) are infinite. These drawbacks are usually cured by grouping bins but at the price of loss of information. In contrast, the m.l. method for binned events becomes equivalent to the ordinary m.l. method when the width of the bins goes to zero.

The case of weighted events will be treated in Section 8.5.4.

8.5. Weights and Detection Efficiency

The experimental physicist is usually not able to observe directly the phenomenon he wishes to study. The apparatus generally introduces some distortion or bias, whose effect must be taken account of.

In high-energy physics the imperfect *detection efficiency* of particle detectors is a source of such bias. A good example is the study of events involving the *decay of unstable particles*. An event is observed only when one of the particles decays within the detector into detectable particles. This introduces three different kinds of bias:

(i) Events are not observed when the detector is not sensitive to certain particles. If this sensitivity is independent of the particle momentum, angles, etc., it introduces only a constant bias in the over-all normalization.
(ii) Events are not observed when the decay occurs outside the detector. This loss will be more pronounced for fast particles than for slow ones, and more pronounced also for particles whose production angle is such that they traverse a shorter path within the detector.
(iii) Events are often missed when the particle decays very close to its production point, making it indistinguishable from other reactions without a decaying particle. This not only causes a momentum-dependent loss of events, but may also be a source of contamination when studying other reactions.

The method used to account for these biases may differ according to their importance. If the effect is large (detection efficiency very small for certain classes of events), it will prove necessary to treat the problem exactly, as described in the next section, in order to avoid losing a great deal of information. On the other hand, if the detector is not heavily biased (say $< 20\%$) some approximations may be used which require less detailed knowledge of the form of the bias. Stated briefly, the exact treatment consists in modifying the underlying *physical* p.d.f. to obtain a distorted *observable* p.d.f., which may then be compared directly with the observed data. The approximate methods retain the physical p.d.f., and instead correct the data by assigning *weights* to events according to the probability of observing an event in that class. In the latter case the problem is to estimate correctly the covariance matrix of the estimates.

8.5.1. *Ideal method — maximum likelihood*

Suppose that the measured variables can be divided into two groups: the variables $\mathbf{X}$ having a p.d.f. depending directly on the parameters $\boldsymbol{\theta}$ to be estimated, and the variables $\mathbf{Y}$ depending on $\mathbf{X}$ and thereby indirectly on $\boldsymbol{\theta}$. Thus

$$P(\mathbf{X}, \mathbf{Y}|\boldsymbol{\theta}) = P(\mathbf{X}|\boldsymbol{\theta})\, Q(\mathbf{Y}|\mathbf{X}),$$

where, as always,

$$\int P(\mathbf{X}|\boldsymbol{\theta})d\mathbf{X} = \int Q(\mathbf{Y}|\mathbf{X})d\mathbf{Y} = 1.$$

Let $e(\mathbf{X}, \mathbf{Y})$ be the probability that an event at $(\mathbf{X}, \mathbf{Y})$ be observed. Then $e(\mathbf{X}, \mathbf{Y})$ is the *detection efficiency* of the equipment, $P(\mathbf{X}|\boldsymbol{\theta})$ is the true distribution of $\mathbf{X}$, and the p.d.f. of the actual observations can be written

$$g(\mathbf{X}, \mathbf{Y}|\boldsymbol{\theta}) = \frac{P(\mathbf{X}|\boldsymbol{\theta})\, Q(\mathbf{Y}|\mathbf{X})\, e(\mathbf{X}, \mathbf{Y})}{\int P(\mathbf{X}|\boldsymbol{\theta})\, Q(\mathbf{Y}|\mathbf{X}) e(\mathbf{X}, \mathbf{Y}) d\mathbf{X} d\mathbf{Y}}.$$

The likelihood function of $\boldsymbol{\theta}$, given a set of observations $(\mathbf{X}_i, \mathbf{Y}_i)$, $i = 1, \ldots, N$, is then

$$L(\mathbf{X}_1, \ldots, \mathbf{X}_N, \mathbf{Y}_1, \ldots, \mathbf{Y}_N|\boldsymbol{\theta}) = \prod_{i=1}^{N} g(\mathbf{X}_i, \mathbf{Y}_i|\boldsymbol{\theta})$$

$$= \prod_{i=1}^{N} \frac{P(\mathbf{X}_i|\boldsymbol{\theta})\, Q(\mathbf{Y}_i|\mathbf{X}_i)\, e(\mathbf{X}_i, \mathbf{Y}_i)}{\int P(\mathbf{X}|\boldsymbol{\theta})\, Q(\mathbf{Y}|\mathbf{X})\, e(\mathbf{X}, \mathbf{Y}) d\mathbf{X}\, d\mathbf{Y}}. \tag{8.38}$$

Example

Suppose that one studies the interaction mechanism for the reaction

$$K^+ p \rightarrow K^0 \pi^+ p$$

in a detector sensitive only to charged particles. Let $\mathbf{X}$ be all the variables defining the kinematics of the event (momenta, direction of all tracks), and $\mathbf{Y}$ the coordinates of the interaction point with respect to the sensitive region. Clearly the interaction mechanism is not correlated with $\mathbf{Y}$, but the detection efficiency depends on the lifetime of the K^0 (part of $\mathbf{X}$) and on the possible track length ($\mathbf{Y}$). Thus the detection efficiency correlates $\mathbf{X}$ and $\mathbf{Y}$, so that the $\mathbf{Y}$-dependence does not factor out of the likelihood function.

From Eq. (8.38) one can write

$$\ln L = W + \sum_{i=1}^{N} \ln (e_i Q_i), \qquad (8.39)$$

where

$$W = \sum_{i=1}^{N} \ln P_i - N \ln \int P e Q, \qquad (8.40)$$

using somewhat abbreviated notation. For the sake of simplicity we consider only one parameter θ, with estimate $\hat{\theta}$ obtained by maximizing. W.

From the results of Section 7.2.3 we know that the variance $\sigma_{\hat{\theta}}^2$ of the m.l. estimate $\hat{\theta}$ is given by the inverse of

$$\frac{1}{\sigma_{\hat{\theta}}^2} = N \left[\frac{\int \frac{1}{P} P'^2 e Q}{\int P e Q} - \left(\frac{\int P' e Q}{\int P e Q} \right)^2 \right], \qquad (8.41)$$

where $P' \equiv \partial P(\mathbf{X}|\theta)/\partial\theta$. Then $1/\sigma_{\hat{\theta}}^2$ may be estimated by

$$\hat{\sigma}_{\hat{\theta}}^{-2} = - \frac{\partial^2 W}{\partial \theta^2}\bigg|_{\theta=\hat{\theta}} . \qquad (8.42)$$

It is easily seen that the expected value of $\hat{\sigma}_{\hat{\theta}}^{-2}$ is $\sigma_{\hat{\theta}}^{-2}$.

In practice, difficulties arise when PeQ is not analytically normalized, but the normalization has to be carried out numerically (requiring time-consuming Monte Carlo). Moreover, the results depend on the form of Q, which is often poorly determined. For these reasons, one looks for a method of excluding Q from the expressions for $\hat{\theta}$ and $\sigma_{\hat{\theta}}^2$. Evidently, by excluding some information, one would expect to reduce the precision of the estimates, but the estimation should be more *robust* with respect to the shape of Q.

8.5.2. *Approximate method for handling weights*

In Eq. (8.39) we replace W, Eq. (8.40), by

$$W' = \sum_i \left(\frac{1}{e_i} \ln P_i \right). \tag{8.43}$$

The intuitive justification for expression (8.43) is that each event observed with detection efficiency e_i corresponds, in some sense, to $w_i = 1/e_i$ events actually occurring. Then the "likelihood" of these latter events would instead of P_i have $P_i^{w_i}$ for each observation, corresponding to a distortion of the observations by function $1/e(X, Y)$, and fitting them with the theoretical distribution $P(X|\theta)$. In this case $Q(\mathbf{Y})$ does not appear in the "likelihood".

Whatever the validity of such an argument, it turns out that the estimate $\hat{\theta}'$ obtained by maximizing W' is, just like $\hat{\theta}$, asymptotically Normally distributed about the true value θ_0. However, care must be taken in evaluating the variance of this estimate. Using the second derivative of W' to estimate the information is wrong since it implies

$$\sum_{i=1}^{N} w_i > N$$

events observed. Sometimes the weights are adjusted to $w_i' = Nw_i/\Sigma w_i$, to avoid this difficulty, but this is not a satisfactory method unless all the weights w_i are close to one.

Using the general methods of Section 7.2.3, we now calculate the correct form for the variance of the estimate $\hat{\theta}'$, obtained by maximizing W' in Eq. (8.43). Finally, we shall show that the variance of the estimate may be reduced by dropping events. This paradoxical result follows since the method is not optimal, from the point of view of information use.

In the notation of Section 7.2.3, we have

$$\xi(\theta) = \frac{1}{N} \frac{\partial W'}{\partial \theta} = \frac{1}{N} \sum_{i=1}^{N} \frac{\partial}{\partial \theta} \left(\frac{1}{e_i} \ln P_i \right). \tag{8.44}$$

One may easily verify that

$$E[\xi(\theta_0)] = 0,$$

thus satisfying the requirements of Section 7.2.3.

It follows that $\hat{\theta}'$ is such that $\xi(\hat{\theta}') = 0$ (that is, the maximum of W'), with variance given by

$$\sigma_{\hat{\theta}'}^2 = \frac{D_2'}{ND_1'^2}$$

where

$$D_2' = \frac{\int \frac{P'^2}{P} \frac{Q}{e}}{\int PeQ}$$

and

$$D_1' = \frac{\int \frac{P'^2}{P} Q}{\int PeQ}.$$

Thus the quantity corresponding to Eq. (8.41) becomes

$$\frac{1}{\sigma_{\hat{\theta}'}^2} = \frac{N \left(\int \frac{P'^2}{P} Q \right)^2}{\left(\int \frac{P'^2}{P} \frac{Q}{e} \right) \left(\int PeQ \right)}. \tag{8.45}$$

Although Q does not appear in Eq. (8.44) it is still not eliminated from the variance (8.45). However, we now pass to the estimation of $1/\sigma_{\hat{\theta}}^2$, removing Q. Clearly D_1' and D_2' can be expressed as expectations

$$D_1' = E\left[\frac{1}{e}\left(\frac{P'}{P}\right)^2\right],$$

$$D_2' = E\left[\left(\frac{P'}{Pe}\right)^2\right].$$

Now we have from Eq. (8.44)

$$\frac{\partial^2 W'}{\partial \theta^2} = \sum_{i=1}^{N} \frac{1}{e_i P_i} \frac{\partial^2 P_i}{\partial \theta^2} - \sum_{i=1}^{N} \frac{1}{e_i P_i^2} \left(\frac{\partial P_i}{\partial \theta}\right)^2$$

and

$$E\left(\frac{1}{eP} \frac{\partial^2 P}{\partial \theta^2}\right) = \frac{\int \frac{P'' PeQ}{eP}}{\int PeQ} = \frac{\int P'' Q}{\int PeQ} = 0$$

since $\int PQ = 1$. Thus one can estimate D_1' and D_2' by

$$\hat{D}_1' = -\frac{\partial^2}{\partial \theta^2} \left[\frac{1}{N} \sum_{i=1}^{N} \frac{\ln P_i}{e_i}\right],$$

$$\hat{D}_2' = -\frac{\partial^2}{\partial \theta^2} \left[\frac{1}{N} \sum_{i=1}^{N} \frac{\ln P_i}{e_i^2}\right].$$

In the multidimensional case, the following result holds [Solmitz]. Let us define the matrix elements

$$H_{k\ell} = E\left[\frac{1}{e}\left(\frac{\partial \ln P}{\partial \theta_k}\right)\left(\frac{\partial \ln P}{\partial \theta_\ell}\right)\right]$$

and

$$H'_{h\ell} = E\left[\frac{1}{e^2}\left(\frac{\partial \ln P}{\partial \theta_k}\right)\left(\frac{\partial \ln P}{\partial \theta_\ell}\right)\right].$$

These may be estimated by

$$H_{k\ell} = \frac{1}{N}\sum_{i=1}^{N}\left\{\frac{1}{eP^2}\left(\frac{\partial P}{\partial \theta_k}\right)\left(\frac{\partial P}{\partial \theta_\ell}\right)\right\}_i \tag{8.46}$$

and

$$H'_{k\ell} = \frac{1}{N}\sum_{i=1}^{N}\left\{\frac{1}{e^2 P^2}\left(\frac{\partial P}{\partial \theta_k}\right)\left(\frac{\partial P}{\partial \theta_\ell}\right)\right\}_i.$$

Alternatively, if the first derivatives are not known analytically, one can estimate the elements numerically from

$$H_{k\ell} = -\frac{\partial^2 W'}{\partial \theta_k \partial \theta_\ell} = -\frac{\partial^2}{\partial \theta_k \partial \theta_\ell}\left[\frac{1}{N}\sum_{i=1}^{N}\frac{\ln P_i}{e_i}\right] \tag{8.47}$$

and

$$H'_{k\ell} = -\frac{\partial^2}{\partial \theta_k \partial \theta_\ell}\left[\frac{1}{N}\sum_{i=1}^{N}\frac{\ln P_i}{e_i^2}\right].$$

Then the covariance matrix of $\widehat{\boldsymbol{\theta}}'$ is estimated by

$$\mathrm{cov}(\widehat{\boldsymbol{\theta}}') = \underset{\sim}{H}^{-1}\,\underset{\sim}{H}'\,\underset{\sim}{H}^{-1}. \tag{8.48}$$

Clearly, when $e = 1$, $\underset{\sim}{H}' = \underset{\sim}{H}$, and Eq. (8.48) reduces to the result obtained by the usual maximum likelihood method.

Let us now summarize the approximate method in the case of $\boldsymbol{\theta}$ many-dimensional: find the estimates $\widehat{\boldsymbol{\theta}}'$ by minimizing Eq. (8.43). If possible, compute $\underset{\sim}{H}$ and $\underset{\sim}{H}'$ by Eq. (8.46). If the first derivatives entering Eqs. (8.46) are not known analytically, use Eqs. (8.47). Then the covariance matrix is given by Eq. (8.48).

8.5.3. *Exclusion of events with large weight*

The appearance of one event with a very *large weight* obviously ruins the approximate method, in contrast to the ideal method. Reducing the number of events certainly decreases the precision, but *excluding events* with the largest weights should improve the precision. Thus there must be an optimum number of events.

Let us introduce a step-function $H(e - e_0)$ which excludes events below a minimum efficiency e_0. Then, from Eq. (8.45), the variance becomes

$$\sigma^2(e_0)\left(\frac{1}{N}\int PeQ\right)\frac{\int \frac{P'^2}{P}\frac{Q}{e}H}{\left(\int \frac{P'^2}{P}QH\right)^2}.$$

Taking the derivative with respect to e_0

$$\frac{d\sigma^2(e_0)}{de_0} = \left(\frac{1}{N}\int PeQ\right)\frac{\int \frac{P'^2}{P}Q\delta}{\left(\int \frac{P'^2}{P}QH\right)^2}\left[2\frac{\int \frac{P'^2}{P}\frac{QH}{e}}{\int \frac{P'^2}{P}QH} - \frac{1}{e_0}\right],$$

where $\delta \equiv \delta(e - e_0)$. Solving for $d\sigma^2/de_0 = 0$ one finds

$$\sigma^2_{\min} = \left(\frac{1}{N}\int PeQ\right)\left[2e_{\min}\int \frac{P'^2}{P}QH(e - e_{\min})\right]^{-1}.$$

Example

The choice of functions

$$P = \frac{1}{2}(1 + \alpha X), \quad Q = \frac{1}{2}(1 + \beta Y), \quad e = \left|\frac{Y - X}{2}\right|^\mu,$$

$$-1 \le X, Y \le 1, \quad -1 \le \alpha, \beta \le 1$$

permits an analytic comparison between the ideal and the approximate methods. Instead of the variance σ^2 we study $N\sigma^2$ (the variance having a trivial N^{-1} dependence).

Given α, β, and μ, the ideal method leads to

$$N\sigma^2,$$

whereas the approximate method leads to

$$N\sigma'^2 = f(e_0),\tag{8.49}$$

a function of the threshold efficiency. Let us define the points

$$N\sigma'^2_{\min} = \text{minimum of } f(e_0)\text{, corresponding to threshold } e_{\min}\text{, and}$$
$$r_{\min}\% \text{ events excluded};$$
$$N\sigma'^2_1 = \text{the point at } e_0 = 0.1$$
$$N\sigma'^2_2 = \text{the point at } r = 1\%.$$

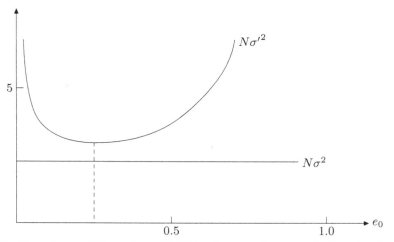

Fig. 8.2. Dependence of the variance (8.49) in the approximate method on the threshold efficiency e_0. Numerical constants: $\alpha = \beta = 0.3$, $\mu = 0.99$.

The ratios

$$R_0 \equiv \frac{\sigma'^2_{\min}}{\sigma^2}\,, R_1 \equiv \frac{\sigma'^2_1}{\sigma'^2_{\min}}\,, R_2 \equiv \frac{\sigma'^2_2}{\sigma'^2_{\min}}$$

turn out to depend almost only on μ but not on α or β. Also $e_{\min}$ and $r_{\min}$ are functions practically only of μ. In Fig. 8.2 we display the function (8.49) for $\alpha = \beta = 0.3$ and $\mu = 0.99$. This demonstrates the existence of the optimum, at

$$e_{\min} = 0.25\,, \quad r_{\min} = 15\% \text{ rejected}$$
$$N\sigma'^2_{\min} = 3.18\,, \quad N\sigma^2 = 2.50\,.$$

(The curve in Fig. 8.2 stays nearly unchanged, up to a multiplicative factor, for different values of α and β. The value of μ is dictated by the fact that $N\sigma'^2$ diverges for $e_0 = 0$ and $\mu \geq 1$).

Thus e_{min}, r_{min} leads to $R_0 = 1.25$, $e_0 = 0.1$ leads to $R_1 = 1.15$, and $r = 1\%$ leads to $R_2 = 1.38$.

8.5.4. *Least squares method*

Consider a histogram with k_i events in the ith bin, $i = 1, \ldots, n$. Suppose a model predicts the normalization as well as the shape of the distribution. Let the expected number of events in the ith bin be

$$a_i(\theta) = N(\theta) \frac{\int_i PeQ}{\int PeQ}, \qquad (8.50)$$

where $N(\theta)$ ($\neq \sum_{i=1}^n k_i$) is the total number of events predicted, and $\int_i$ indicates integration over the bin i. At our disposal we now have the two methods (8.36) and (8.35), for which

$$Q_1^2 = \sum_{i=1}^n \frac{(k_i - a_i)^2}{k_i}, \qquad \frac{\partial Q_1^2}{\partial \theta} = -2 \sum_{i=1}^n \left(1 - \frac{a_i}{k_i}\right) \frac{\partial a_i}{\partial \theta},$$

$$Q_2^2 = \sum_{i=1}^n \frac{(k_1 - a_i)^2}{a_i}, \qquad \frac{\partial Q_2^2}{\partial \theta} = -\sum_{i=1}^n \left[\left(\frac{k_i}{a_i}\right)^2 - 1\right] \frac{\partial a_i}{\partial \theta}.$$

Using the method and notation of Section 7.3.3 (ii), one can verify that asymptotically

$$E[\xi(\theta_0)] = 0$$

for both Q_1^2 and Q_2^2, where

$$\xi = \frac{1}{n} \frac{\partial Q^2}{\partial \theta},$$

since

$$E\left(\frac{1}{k_i}\right) \to \frac{1}{a_i}, \qquad E\left(\frac{1}{k_i^2}\right) \to \frac{1}{a_i^2}.$$

Let us now introduce the approximate method, which avoids using Q in the integrals. Let b_i be the predicted number of events in the ith bin when $e = 1$ (a_i is the predicted number when $e < 1$). Correcting the numbers b_i with the known experimental efficiency, one gets numbers c_i, such that

$$E(c_i) = a_i. \qquad (8.51)$$

Then

$$b_i = N'(\theta)\frac{\displaystyle\int_i PQ}{\displaystyle\int PQ} = N(\theta)\frac{\displaystyle\int_i PQ}{\displaystyle\int PeQ} , \qquad (8.52)$$

where N' is the total number of events predicted when $e = 1$. From Eqs. (8.50) and (8.52)

$$a_i = b_i\frac{\displaystyle\int_i PeQ}{\displaystyle\int_i PQ} .$$

This ratio of integrals, which must be estimated, can be considered as the expectation $E_i(e)$ of the efficiency e of the ith bin, or as $[E_i(1/e)]^{-1}$ of the weight $w = 1/e$. To estimate it we shall take the arithmetic mean of events which were *effectively* obtained in the ith bin. In other words, the normalization comes from $\int_i PeQ$, thus

$$\frac{\displaystyle\int_i PQ}{\displaystyle\int_i PeQ} = E_i\left(\frac{1}{e}\right) \approx \frac{\displaystyle\sum_{j=1}^{k_i} w_{ij}}{k_i} ,$$

where $w_{ij} = 1/e_i$ for the j_{th} event in the i_{th} bin (the summation goes over all the events in that bin).

The problem is clearly not symmetrical between the estimators $E_i(e)$ and $[E_i(1/e)]^{-1}$, because one would need a sample of events observed with detection efficiency $e = 1$ in order to be able to use $E_i(e)$.

Finally, we take

$$c_i = \frac{b_i k_i}{\displaystyle\sum_{j=1}^{k_i} w_{ij}} ,$$

where b_i is completely known, and where Eq. (8.51) holds.

The function to be minimized is then

$$Q^2 = \sum_{i=1}^{n} \frac{1}{\sigma_i^2}\left(k_i - b_i\frac{k_i}{\displaystyle\sum_j w_{ij}}\right)^2 = \sum_{i=1}^{n} \frac{1}{\sigma_i^2}\left(\sum_{j=1}^{k_i} w_{ij} - b_i\right)^2 , \qquad (8.53)$$

with

$$\frac{1}{\sigma_i'^2} = \frac{1}{\sigma_i^2} \left(\frac{k_i}{\sum_j w_{ij}} \right)^2 .$$

We shall use the second expression on the right-hand side of Eq. (8.53). For $\sigma_i'^2$ we take

$$\sigma_i'^2 = E\left[\left(\sum_{j=1}^{k_i} w_{ij} - b_i \right)^2 \right] = E\left[\left(\sum_{j=1}^{k_i} w_{ij} \right)^2 \right] - b_i^2 ,$$

since

$$E\left[\sum_j w_{ij} \right] = b_i .$$

Using the fact that the k_i are Poisson-distributed, one can show that

$$E\left[\left(\sum_{j=1}^{k_i} w_{ij} \right)^2 \right] = E(k_i)E(w_i^2) + b_i^2 .$$

$E(w_i^2)$ can be estimated as the arithmetic mean

$$E(w_i^2) \approx \frac{1}{k_i} \sum_{j=1}^{k_i} w_{ij}^2 .$$

$E(k_i)$ can be estimated by the two methods (8.36) and (8.35):

$$E(k_i) = k_i \text{ giving } Q_1'^2 = \sum_{i=1}^{n} \left[\frac{\left(\sum_{j=1}^{k_i} w_{ij} - b_i \right)^2}{\sum_{j=1}^{k_i} w_{ij}^2} \right]$$

or

$$E(k_i) = a_i \text{ giving } Q_2'^2 = \sum_{i=1}^{n} \left[\frac{\left(b_i \sum_{j=1}^{k_i} \frac{1}{w_{ij}} - n_i^2 \right)^2}{n_i b_i \sum_{j=1}^{k_i} \frac{1}{w_{ij}}} \right] .$$

Clearly Q' approaches Q as $w \to 1$.

As in the usual (unweighted) case $Q_2'^2$ is better justified, but one may prefer $Q_1'^2$ for practical reasons (minimization easier on the computer). If b_i is a linear function of the parameters this is the case, since the solution of the system can be written explicitly. Otherwise, one has to resort to iterative procedures, equally slow for $Q_1'^2$ as $Q_2'^2$.

Application of the general formula of Section 7.2.3 gives a complicated formula for the variance of the parameter estimates.

8.6. Reduction of Bias

In Section 7.4.2 we have discussed the connection between bias and minimum variance (maximum information), and we have given an example of reduction of bias by a change of variable. We concluded that, in general, a reduction of bias could only be achieved at the cost of some information. Below we shall discuss two general methods of reduction of bias.

8.6.1. *Exact distribution of the estimate known*

When the p.d.f. of the estimate is completely known, the bias b is also known, from Eq. (7.1). Instead of a biased estimator t, one can then use the unbiased estimator

$$t_1 = t - b. \tag{8.54}$$

The variance is unchanged

$$V(t_1) = V(t - b) = V(t).$$

More usually, b depends on unknown parameters of the p.d.f. of the experimental observations, and must therefore also be estimated. Using an unbiased estimate $\hat{b}$ in Eq. (8.54) increases the variance of the final estimate. But such a loss of precision may be quite tolerable.

Example

Let us estimate the variance of a Normal distribution of unknown mean μ, by maximum likelihood, from a set of N independent observations $X_1, \ldots, X_N$. Then

$$f(X, \sigma^2) = \frac{1}{\sqrt{2\pi}\sigma} \exp\left[-\frac{(X-\mu)^2}{2\sigma^2}\right].$$

The likelihood function is

$$L(X, \sigma^2) = \frac{1}{(2\pi)^{\frac{N}{2}} \sigma^2} \exp\left[-\frac{1}{2\sigma^2} \sum_{i=1}^{N} (X_i - \mu)^2 \right].$$

The maximum likelihood estimate is easily found to be

$$\hat{\sigma}^2 = \frac{1}{N} \sum_{i=1}^{N} X_i^2 - \left(\frac{\sum X_i}{N} \right)^2 = \frac{1}{N} \sum_{i=1}^{N} (X_i - \bar{X})^2,$$

denoted S^2 in Eq. (4.18). The bias is found from

$$E(\hat{\sigma}^2) = \frac{1}{N} N E (X_i - \bar{X})^2 = E[(X_i - \mu)^2 + (\bar{X} - \mu)^2 - 2(X_i - \mu)(\bar{X} - \mu)]$$

$$= \sigma^2 + \frac{\sigma^2}{N} - \frac{2\sigma^2}{N}$$

$$= \sigma^2 - \frac{\sigma^2}{N}.$$

Therefore the bias in the maximum likelihood estimate is $-\sigma^2/N$. An unbiased estimate of σ^2 (for *any* distribution) is then [Eq. (4.27)]

$$s^2 = \frac{N}{N-1} \hat{\sigma}^2 = \frac{1}{N-1} \sum_{i=1}^{N} (X_i - \bar{X})^2$$

$$= \hat{\sigma}^2 + \frac{\hat{\sigma}^2}{N-1}.$$

For the Normal distribution, however, the variance of $\hat{\sigma}^2$ can be calculated [Kendall II, p. 57] as

$$V(\hat{\sigma}^2) = 2\sigma^4 \frac{(N-1)}{N^2},$$

giving

$$V(s^2) = \left(\frac{N}{N-1} \right)^2 V(\hat{\sigma}^2)$$

$$= \frac{2\sigma^4}{(N-1)}.$$

Thus removal of the bias has increased the variance of the estimate by a quantity of the order of $1/N^2$, namely

$$\frac{2\sigma^2}{N^2} \left(1 + \frac{1}{N-1} \right).$$

Removing the bias of σ^2/N increases the standard error by

$$\sqrt{V(s^2)} - \sqrt{V(\hat{\sigma}^2)} = \sqrt{V(\hat{\sigma}^2)}\left[\frac{N}{N-1} - 1\right]$$

$$= \frac{\sigma^2}{N}\sqrt{\frac{2}{N-1}}.$$

For large N, this loss in precision is very much smaller than the bias.

8.6.2. *Exact distribution of the estimate unknown*

Without detailed knowledge of the p.d.f. there are several methods of reducing the bias to order $1/N^2$ [Quenouille]. We give here the most straightforward of these methods.

Suppose that t_N is a biased estimate of θ and suppose that we express it as a power series in $1/N$ for large N, then the leading term will be θ (independent of N) and the next term will be the variance, falling off as $N^{-\frac{1}{2}}$, and having expectation zero. The leading bias term (non-zero expectation for finite N) will generally be of the order of N^{-1}, and it is this term which we can approximately eliminate by proper treatment of the data. In the series expansion above, consider the leading terms with non-zero expectation for finite N:

$$E(t_N) = \theta + \frac{1}{N}\beta + O\left(\frac{1}{N^2}\right)$$

$$E(t_{2N}) = \theta + \frac{1}{2N}\beta + O\left(\frac{1}{N^2}\right)$$

where the bias is $b = \beta/N$. Subtracting the two lines one has

$$E(2t_{2N} - t_N) = \theta + O\left(\frac{1}{N^2}\right).$$

By a suitable combination of t_{2N} (from all the data) and t_N (from half the data) we have eliminated the leading bias term. A slightly better method is to divide the data into two equal halves, yielding estimates t_N and t'_N, and use

$$2t_{2N} - \frac{1}{2}(t_N + t'_N).$$

As in the previous method, the variance is, in general, increased by a term of order $1/N$. These exists a method which increases the variance only by a term of order $1/N^2$ [Kendall II, p. 66]. Unfortunately this method is not very useful as it requires N estimations.

Example

Applying this method to the example in Section 8.3.3, Fig. 8.1, we divide the sample randomly into two parts, and take

$$t_M = t_{2M} = X_N .$$

On the average

$$t'_M = \frac{1}{2}X_{N-1} + \frac{1}{4}X_{N-2} + \frac{1}{8}X_{N-3} + \cdots ,$$

where we have explicitly chosen $\frac{1}{2}$ for the probability that X_{N-1} is the largest value of X in the primed sample, $\frac{1}{4}$ for the probability that X_{N-2} is the largest value, etc. On the average, the separation between neighbouring values of X is X_N/N. We then get for large N

$$\hat{\theta}_{\text{corr}} \simeq 2X_N - \frac{X_N + \left(X_N - \dfrac{2X_N}{N} \right)}{2} \simeq X_N + \frac{X_N}{N} ,$$

which is unbiased.

8.7. Robust (Distribution-free) Estimation

In the previous sections we have dealt with parametric estimation, assuming the p.d.f. to be of known form. When the p.d.f. is not exactly known, the following questions arise:

(i) What kinds of parameters can be estimated without any assumptions about the form of the p.d.f.?

(ii) How reliable are the parameter estimates if the form assumed for the p.d.f. is not quite correct, for example if the data include misinterpreted observations.

   Traditionally, estimates of type (i) are known as *distribution-free*. However, this term may be misleading [Huber] since properties of such estimates (notably the variance) may depend strongly on the actual underlying distribution. A more proper term is therefore *robust estimation*. In general, *robustness* is taken to imply insensitivity to small deviations from the underlying distribution assumed.

Robust estimation is of considerable practical importance [Tukey], but a full treatment is beyond the scope of this book. We are therefore limited to describing a few methods which lead to more robust estimators than the usual ones. When the underlying distribution and the kinds of estimators considered can be limited to some general classes, it will sometimes be possible to determine "optimally robust" estimators.

There is a close relationship between the methods of robust estimation, presented here, and the distribution-free tests of hypothesis, discussed in Chapter 11. In particular, although both estimates and tests may be formulated in a distribution-free way, nevertheless the variances of such estimates and the power of such tests depend strongly on the underlying distribution.

8.7.1. *Robust estimation of the centre of a distribution*

The only case of robust estimation we will treat here is the estimation of the centre of an unknown, symmetric distribution. The centre of a distribution is defined by a *location parameter*. Some examples of location parameters are:

- The *mean* is the expectation of the variable X.

- The *median* is that value X for which the cumulative distribution has $F(X) = 0.5$.

- The *mode* is that value of X for which the p.d.f. has a maximum.

- The *midrange* is defined when the possible values of X are limited to the range $[X_{\min}, X_{\max}]$. Then the *midrange* is $(X_{\min} + X_{\max})/2$.

For some distributions, not all of the above parameters are defined, e.g. the Cauchy distribution does not have a mean (Section 4.2.11). However, these parameters are all well-defined for any particular sample of (finite) data X_i:

- The *sample mean* is the mean or average of the X_i.

- The *sample median* is the value X such that half the X_i lie above it and half below. If the number of data values is odd, it is the central value. If the number is even, it is usually taken as halfway between the two central values.

- The *sample mode* is the value of X halfway between the two nearest values of X_i.

- The *sample midrange* is halfway between the smallest and largest values of X_i, that is $(X_{i\,\min} + X_{i\,\max})/2$.

The sample mean is the most obvious and most often used estimator of location, because

(i) it is consistent whenever the variance of the underlying distribution is finite (law of large numbers, Section 3.3);

(ii) it is optimal (minimum variance, unbiased) when the underlying distribution is Normal.

However, if the distribution of X is not Normal, the sample mean is not the best estimator of the mean of the distribution, even when the mean of the distribution exists. Below we list the best estimator of location for some important distributions:

Distribution	Minimum-variance location estimator
Normal	sample mean
Uniform	midrange (mean of extreme values)
Cauchy	maximum-likelihood estimate
Double-exponential	median (middle value)

Of course all of these estimators are maximum-likelihood estimators, but in the case of the Cauchy distribution this cannot be expressed in a simpler (distribution-free) way. As in Section 7.4.2, let us define the *efficiency* of an estimator as the ratio of the minimum-variance bound to the estimate of the variance. Then each of the estimators above has (asymptotic) efficiency $= 1$ for the corresponding distribution. One may study the robustness of these estimators when applied to other distributions, by comparing their efficiencies, Table 8.1. [All efficiencies here are asymptotic values. The problem has also been solved for small N [Crow 1967].]

None of the three estimators considered here has non-zero asymptotic efficiency for all four distributions. The reason is that the mean of symmetric finite-range distributions (uniform, triangular, beta, etc.) is determined by specifying the end-points, so that most of the information comes from the tails: the midrange is then an excellent location estimator. On the other hand, for infinite-range distributions, the midrange always has a zero asymptotic efficiency.

Table 8.1. Asymptotic efficiencies of location estimators.

Distribution	Sample median	Sample mean	Sample midrange
Normal	0.64	1.00	0.00
Uniform	0.00	0.00	1.00
Cauchy	0.82	0.00	0.00
Double-exponential	1.00	0.25	0.00

(For asymptotic variances, see Table 8.2)

Since one normally knows at least whether the underlying distribution is of finite or infinite range, it is no serious restriction to consider only infinite-range distributions. Suppose that the underlying distribution is Normal, with an unknown (but non-zero) admixture of Cauchy-distributed entries. Then the sample mean has infinite variance, but the median is (from Table 8.1) at least 64% efficient no matter what the admixture of Cauchy. In general, for infinite-range distributions the sample median is a more robust estimator of location than the sample mean. We discuss below some other estimators which are even more robust for this class of distributions.

8.7.2. *Trimming and Winsorization*

Let us define two new estimators:

(i) *The trimmed mean* of the total of N observations, remove the $n/2$ highest values and the $n/2$ lowest values, and compute the mean of the remaining $N - n$ observations.

(ii) *The Winsorized mean* of the total of N observations, replace the $n/2$ highest values by the highest remaining value, and the $n/2$ lowest by the lowest remaining value, and compute the mean of the new sample of N values.

For both estimators there is one free parameter, usually taken as half the fraction of remaining (unchanged, or not rejected) values,

$$r = \frac{N - n}{2N} .$$

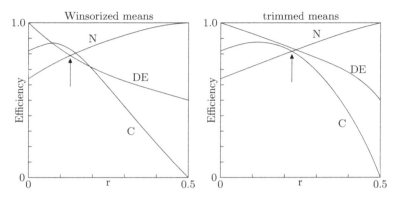

Fig. 8.3. Asymptotic efficiencies of trimmed and Winsorized means for Normal (N), double-exponential (DE) and Cauchy (C) distributions. Arrows indicate minimax optima.

Note that for $r = 0.5$, both estimators are in fact the sample mean, and as $r \to 0$, both become equivalent to the sample median. A comparison of asymptotic efficiencies of trimmed and Winsorized means as functions of r, is shown in Fig. 8.3. Choosing the best value of r is analogous to the problem of choosing decision rules (Section 6.2) since one does not know *a priori* what kind of distribution is present. For optimum robustness, a minimax rule should be used: to maximize the minimum possible efficiency.[c] For Fig. 8.3, this choice gives $r = 0.23$ for the trimmed mean, where an efficiency of at least 82% is assured, no matter which of the three distributions is present. The Winsorized mean is seen to be less robust since the minimax point assures an efficiency of only 77%.

8.7.3. *Generalized p^{th}-power norms*

In Section 8.4 and in Chapter 7 we have discussed in detail the properties of least squares estimators. In particular we know that they are optimal for estimating the mean (or for fitting a line to measured points) whenever the distribution in question is (or the measurement errors are) Normal (Section 8.4.4). In this section we view these estimators in a wider context by considering two related questions:

(a) What are the properties of the least squares estimators if the underlying distributions are not Normal?

---

[c] In using a minimax rule, one is taking the most pessimistic point of view of one's assumptions about the distributions involved, which is precisely the requirement of asking for robustness.

(b) What are the properties of other estimators which take the minimum not of squares of deviations, but of some other powers? Should one sometimes use least cubes, least absolute values, etc.?

Consider again the estimation of a location parameter; the extension to other problems normally attacked by least squares methods will be obvious. We call a_p the L_p estimate of location if a_p minimizes the quantity

$$L_p(a) = \sum_{i=1}^{N} |X_i - a|^p \,.$$

For all $p \geq 1$, the L_p estimate is well-defined, and the properties of these estimates have been studied [Rice]. When $p = 2$, the L_p estimate is the sample mean, and when $p = 1$, it is the sample median. When $p = \infty$, sometimes called the *Tchebycheff norm*, it is equivalent to the midrange since only the extreme values contribute. In Table 8.2 we give the variances of these estimates for several distributions (the variances in boxes are asymptotically the minimum possible).

Table 8.2. Asymptotic variances of L_p location estimates.

Distribution	L_1 (median)	L_2 (mean)	L_∞ (midrange)
Uniform	$1/4N$	$1/12N$	$\boxed{1/(2N^2 + 6N + 4)}$
Triangular		$1/6N$	$(4 - \pi)/4N$
Normal	$\pi/2N$	$\boxed{1/N}$	$\pi^2/12 \ \log\ N$
Double-exponential	$\boxed{1/2N}$	$2/N$	$\pi^2/12$
Cauchy	$\pi^2/4N$	∞	∞

It has often been pointed out that since the efficiency of estimators is defined in terms of the variance of its estimates, we are using an L_2 criterion to compare all L_p estimators. This seemingly arbitrary procedure can, however, be justified in most cases, because of the asymptotic Normality of the estimators.

It is clear from Table 8.2 that part of the popularity of the L_2 estimator is due to the importance of the Normal distribution (see Section 4.3.1). Moreover, among all estimators which are linear functions of the observations, the L_2 estimator is optimal for any distribution with finite variance. For distributions with longer "tails" than the Normal distribution, $p < 2$ is, in general, more efficient, and for distributions with shorter "tails", $p > 2$ is more efficient. Note however, the peculiar N-dependence of the L_∞ estimator. In view of the interesting properties of L_p estimators in the range $1 \leq p \leq 2$, it may be tempting to try to extend the definition of L_p to $p < 1$. Such estimators are found to be computationally inconvenient and they are not generally of interest except for the important class of problems considered in the next section.

8.7.4. *Estimates of location for asymmetric distributions*

Consider the estimation of the centre of a narrow *signal* distribution super-imposed on an unknown but wider *asymmetric background* distribution. The asymmetry of the background makes it difficult to apply trimming or Winsorization.

A common non-robust technique is to parametrize the signal and back-ground in some arbitrary way and carry out a maximum-likelihood or least squares fitting procedure to obtain optimum values for the parameters, includ-ing the location parameter. This is non-robust because the location estimate depends on the background parametrization and on the correlations.

A robust technique for this problem is to estimate the *mode* of the ob-served distribution since this is a well-defined location parameter also for an asymmetric distribution, and it is nearly invariant under a smooth background variation.

A robust technique for estimating the mode is to use $a_{-\infty}$, the L_p estimate for $p = -\infty$. We have seen earlier that as p becomes smaller, a_p tends to ignore information in the tails of the distribution and to give more weight to observations in the region of highest density. The $L_{-\infty}$ estimator is defined by the following procedure: find the two observations which are separated by the smallest distance, and choose the one which has the closer remaining neighbour. The position of this observation is then taken as the estimate of the mode.

The asymptotic efficiency of the $L_{-\infty}$ estimator is about six times smaller than that of the L_2 estimator when applied to the Normal distribution, and it is about three times smaller than L_1 when applied to the Cauchy distribution. On the other hand, $a_{-\infty}$ is quite insensitive to the asymmetry of the distribution. As an example, it has been applied to the $\chi^2(6)$ distribution (see Section 4.2.3) for which the mode $= 4$, the median $= 5.348$, and the mean $= 6$. For $N = 80$ observations, the expectations and variances of the L_p estimates are

$$E(a_{-\infty}) \approx 4.1, \quad V(a_{-\infty}) \approx 1.3$$
$$E(a_1) \approx 5.0, \quad V(a_1) \quad \approx 1.3$$
$$E(a_2) \approx 5.9, \quad V(a_2) \quad \approx 1.9.$$

However, the exact distribution of $a_{-\infty}$ is not known even for simple distributions, and it may be badly behaved for small numbers of observations.

Chapter 9

INTERVAL ESTIMATION

In *point estimation*, Chapters 7 and 8, we have discussed methods to estimate *the values* of unknown parameters. In *interval estimation* we want to find *the range*

$$\theta_a \leq \theta \leq \theta_b \,,$$

which contains the true value θ_0 with probability β. Such an interval is called a *confidence interval* for θ with *probability content* β. It corresponds, in physicists' language, to the "error on the parameter θ".

If one wants to be reasonably confident that the interval indeed contains the true value θ_0, one chooses β to be large, for example 90% or 99%. In experimental physics one often chooses $\beta = 68.3\%$ or 95.5%, and the corresponding "errors" (confidence intervals) are called 1 standard deviation errors or 2 standard deviation errors. This is can be confusing because it is true, in general, only for the Normal distribution.

Given an observation X from a p.d.f. $f(X|\theta)$, the probability content β of the region $[a, b]$ in X-space is

$$\beta = P(a \leq X \leq b) = \int_a^b f(X|\theta)dX \,. \tag{9.1}$$

When the density f is known, as well as the parameter θ, one can always calculate β, given a and b.

When the parameter θ is unknown, one has to find another variable

$$Z = Z(X, \theta) \,,$$

215

a function of the observation X and the parameter θ, such that its p.d.f. is independent of the unknown θ. If this can be found, it may then be possible to re-express Eq. (9.1) as a problem of interval estimation: given β, find the optimal range $[\theta_a, \theta_b]$ in θ-space such that

$$P(\theta_a \leq \theta_0 \leq \theta_b) = \beta. \tag{9.2}$$

The interval (θ_a, θ_b) is then called a *confidence interval*. A method which yields an interval (θ_a, θ_b) satisfying Eq. 9.2 is said to possess the property of *coverage*. Formally, if an interval does not possess the property of *coverage*, it is not a confidence interval, although we will consider later in this chapter approximate confidence intervals, which have only approximate coverage.

Our approach to interval estimation is that of classical statistics: θ_0 is an unknown *constant*, and θ_a and θ_b are functions of the random variable X. However, we shall also briefly illustrate the Bayesian interpretation of confidence intervals in the last Section.

9.1. Normally distributed data

9.1.1. *Confidence intervals for the mean*

Let $f(X|\theta)$ in Eq. (9.1) be the Normal distribution $N(\mu, \sigma^2)$, Eq. (4.16). When μ and σ^2 are known, Eq. (9.1) can be evaluated:

$$\beta = P(a \leq X \leq b) = \int_a^b N(\mu, \sigma^2) dX$$

$$= \Phi\left(\frac{b-\mu}{\sigma}\right) - \Phi\left(\frac{a-\mu}{\sigma}\right),$$

where Φ is the *Normal probability integral*, Eq. (4.17).

When μ is unknown, one can no longer calculate the probability content of the interval $[a, b]$. Instead, one calculates the probability β that X lies in some interval relative to its unknown mean, say $[\mu+c, \mu+d]$. Letting $Y = (X'-\mu)/\sigma$, we have:

$$\beta = P(\mu + c \leq X \leq \mu + d) = \int_{\mu+c}^{\mu+d} N(\mu, \sigma^2) dX'$$

$$= \int_{c/\sigma}^{d/\sigma} \frac{1}{\sqrt{2\pi}} \exp\left[-\frac{1}{2}Y^2\right] dY$$

$$= \Phi\left(\frac{d}{\sigma}\right) - \Phi\left(\frac{c}{\sigma}\right). \tag{9.3}$$

One can now invert the probability statement in Eq. (9.3), to take the form of the statement (9.2):

$$\beta = P(X - d \leq \mu \leq X - c).$$ (9.4)

Note that this is a probability statement about the random variable X, although it looks as if one had put bounds on the variation of μ. But, in fact, μ is an unknown constant.

When both μ and σ are unknown, one chooses to use the *standardized variable*

$$Z = \frac{X - \mu}{\sigma},$$

which, when X is Normal, is known to have the *standard Normal distribution* $N(0,1)$ (Section 4.2.1). Thus the probability statement about Z becomes

$$\beta = P(c \leq Z \leq d) = \int_c^d \frac{1}{\sqrt{2\pi}} \exp\left[-\frac{1}{2}Z^2\right] dZ$$

$$= \Phi(d) - \Phi(c).$$ (9.5)

The confidence interval (9.4) then follows by converting the inequality in Z to an inequality in μ. For the Normal distribution this conversion happens to be algebraically simple, due to the symmetry of the function between X and μ. But it need not, in general, be so simple, and it is not even always algebraically possible. Techniques for handling the general case will be discussed in Section 9.2.

Let us now define a useful concept: given a random variable X with p.d.f. $f(X)$ and cumulative distribution $F(X)$, the α-*point* X_α is defined by

$$\int_{-\infty}^{X_\alpha} f(X)dX = F(X_\alpha) = \alpha.$$

A solution to the problem of estimating the interval $[c, d]$ in Eq. (9.5), given β, is obviously the interval

$$[Z_\alpha, Z_{\alpha+\beta}].$$ (9.6)

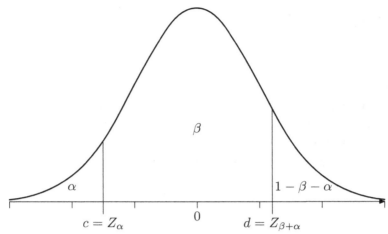

Fig. 9.1. $N(0, 1)$ with regions of probability content α, β, and $1 - \beta - \alpha$. c is the α-point and d the $(\alpha + \beta)$-point.

This is illustrated in Fig. 9.1 for the standard Normal distribution. However, since the interval (9.6) has probability content β for any α, there are obviously infinitely many confidence intervals which satisfy Eq. (9.5). The most usual choice is $\alpha = (1 - \beta)/2$, which gives the *central* interval, symmetric about zero.

Example: Central Intervals for N(0,1).

$\beta = (1 - \alpha)/2$	Z_α	$Z_{\alpha+\beta}$
0.6827	-1.00	1.00
0.9000	-1.65	1.65
0.9500	-1.96	1.96
0.9545	-2.00	2.00
0.9900	-2.58	2.58
0.9973	-3.00	3.00

9.1.2. *Confidence intervals for several parameters*

We will now extend the above results to the case of more than one parameter, to establish *confidence regions* in many parameters, with a given probability level β, in the sense of Eq. (9.5).

Suppose that one has estimators $\mathbf{t}$ of N parameters $\boldsymbol{\theta}$, and that the distribution of $\mathbf{t}$ is Normal many-dimensional with mean $\boldsymbol{\theta}$ and covariance $\underset{\sim}{V}$. Then the p.d.f., from Eq. (4.19), is

$$f(\mathbf{t}|\boldsymbol{\theta}) = \frac{1}{(2\pi)^{N/2}| \underset{\sim}{V} |^{1/2}} \exp\left[-\frac{1}{2}(\mathbf{t} - \boldsymbol{\theta})^T \underset{\sim}{V}^{-1} (\mathbf{t} - \boldsymbol{\theta}) \right]. \tag{9.7}$$

It follows from the Normality of the **t** that the *covariance form*

$$Q(\mathbf{t}, \boldsymbol{\theta}) = (\mathbf{t} - \boldsymbol{\theta})^T \underset{\sim}{V}^{-1} (\mathbf{t} - \boldsymbol{\theta})$$

has a $\chi^2(N)$ distribution (Section 4.2.2). This means that the distribution of Q is independent of $\boldsymbol{\theta}$. We can therefore write down a probability statement similar to Eq. (9.5), namely

$$P[Q(\mathbf{t}, \boldsymbol{\theta}) \le K_\beta^2] = \beta \tag{9.8}$$

where K_β^2 is the β-point of the $\chi^2(N)$ distribution.

The region in **t**-space defined by

$$Q(\mathbf{t}, \boldsymbol{\theta}) \le K_\beta^2 \tag{9.9}$$

is a hyperellipsoid of constant probability density for the function (9.7). From Eq. (9.8), the probability content of this region is β.

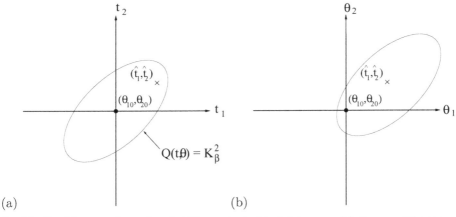

(a)   (b)

Fig. 9.2. Confidence regions of probability content β in (a) **t**-space, (b) $\boldsymbol{\theta}$-space. The true values are $(\theta_{10}, \theta_{20})$ and the estimates $(\hat{t}_1, \hat{t}_2)$.

The inversion of the probability statement (9.8) about **t** to a confidence region for $\boldsymbol{\theta}$ is now evident, because of the symmetry of $Q(\mathbf{t}, \boldsymbol{\theta})$ in the variables

t and $\boldsymbol{\theta}$. The confidence region in $\boldsymbol{\theta}$ space of content β is a hyperellipsoid, centred at the point **t**, and defined by Eq. (9.9).

The two-dimensional case is sufficiently general to allow one to visualize the properties of many-dimensional confidence regions. The Eq. (9.8) is a probability statement about the *simultaneous* values of all the parameters $\boldsymbol{\theta}$, and in two dimensions it defines an elliptical region, shown in Fig. 9.2a. Here the estimates are $\hat{t}_1$, $\hat{t}_2$ and the true values of the unknown parameters θ_{10}, θ_{20}. Figure 9.2b shows the corresponding region in $\boldsymbol{\theta}$-space, resulting from converting the probability statement (9.8).

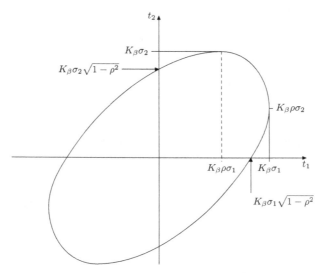

Fig. 9.3. Confidence region of probability content β for the estimators t_1, t_2. Shown here is the case $\rho = 0.5$. If $\rho = 0$, the axes of the ellipse are horizontal and vertical. If $\rho = 1$, the ellipse degenerates to a diagonal line.

In Fig. 9.3 we show the elliptical region (Eq. 9.9) in **t**-space, for two Normally-distributed variables with covariance matrix

$$\underset{\sim}{V} = \begin{pmatrix} \sigma_1^2 & \rho\sigma_1\sigma_2 \\ \rho\sigma_1\sigma_2 & \sigma_2^2 \end{pmatrix}.$$

The probability content of this ellipse is the probability that the ellipse will contain both parameters simultaneously. Alternatively, one may wish to make a different probability statement about two parameters, corresponding to different regions in **t**-space.

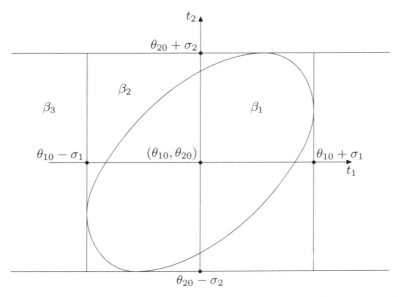

Fig. 9.4. Confidence regions for the Normal estimators t_1, t_2, with $K_\beta = 1$, $\rho = 0.5$. The probability content is β_1 for the elliptic regions Eq. (9.9), β_2 for the circumscribed rectangle Eq. (9.10), and β_3 for the horizontal band Eq. (9.11).

In Fig. 9.4 we show three different confidence regions for the case $K_\beta = 1$, $\rho = 0.5$. Region 1, the ellipse, has probability content given by Eq. (9.8) with $\beta = \beta_1$. Region 2, the circumscribing square, has probability content given by

$$P[(\theta_1 - \sigma_1 \le t_1 \le \theta_1 + \sigma_1) \text{ and } (\theta_2 - \sigma_2 \le t_2 \le \theta_2 + \sigma_2)] = \beta_2 . \qquad (9.10)$$

And region 3, the long horizontal band, defined by the lines

$$t_2 = \theta_{20} \pm \sigma_2$$

gives a one-parameter confidence interval $[t_2 - \sigma_2, \, t_2 + \sigma_2]$ corresponding to the statement that, *whatever the value of t_1*

$$P[\theta_2 - \sigma_2 \le t_2 \le \theta_2 + \sigma_2] = \beta_3 . \qquad (9.11)$$

In table 9.1, we give the probability contents of the three regions for different values of K_β and the correlation ρ, for two Gaussian-distributed variables. For the inner ellipse (region 1) and the infinite band (region 3), the probability content β does not depend on ρ.

Table 9.1. Probability content of different regions in two variables.

	$K_\beta = 1$	$K_\beta = 2$	$K_\beta = 3$
inner ellipse β_1	0.393	0.865	0.989
square β_2 for $\rho = 0.00$	0.466	0.911	0.995
for $\rho = 0.50$	0.498	0.917	0.995
for $\rho = 0.80$	0.561	0.929	0.996
for $\rho = 0.90$	0.596	0.936	0.996
for $\rho = 0.95$	0.622	0.941	0.996
for $\rho = 1.00$	0.683	0.954	0.997
infinite band β_3	0.683	0.954	0.997

Example

Consider the estimation of the mass M and width Γ of a resonance from an observed mass spectrum. Suppose that one has obtained Normally distributed estimates with a known covariance matrix. Wishing to quote confidence intervals for M and Γ, one may proceed in several different ways.

For instance, if one wants to give a probability that the point (M, Γ) lies in some region, one must use Eq. (9.9) for an elliptical region, or Eq. (9.10) for rectangular regions. Alternatively, if one wants to use the data to determine a confidence region only for M, one can use Eq. (9.11) which, given β, allows one to establish tighter bounds on M because one says nothing about Γ. (This does not prevent a collaborator from doing the same thing for Γ alone, using the same data, and thereby creating an incompatible results!).

In this way one gives a 68% confidence region for M, *and* a 68% confidence region for Γ, but the probability is only 46% that both regions include the true values simultaneously (assuming the estimates to be uncorrelated).

A third possibility is to give a confidence interval for M, conditional on knowing a value for Γ. Suppose that the estimates of M and Γ have a covariance matrix

$$\underset{\sim}{V} = \begin{pmatrix} \sigma_M^2 & \rho\sigma_M\sigma_\Gamma \\ \rho\sigma_M\sigma_\Gamma & \sigma_\Gamma^2 \end{pmatrix} .$$

Then, conditional on a value Γ, the distribution of M is Normal, with mean $M_0 + (\sigma_M/\sigma_\Gamma)\rho(\Gamma - \Gamma_0)$ and variance $\sigma_M^2(1 - \rho^2)$, where M_0 is the true value of M. Therefore the standardized variable is

$$Z = \frac{M - \left[M_0 + \dfrac{\rho\sigma_M}{\sigma_\Gamma}(\Gamma - \Gamma_0) \right]}{\sigma_M\sqrt{1 - \rho^2}}$$

and one can write the probability statement

$$P(Z_\alpha \leq Z \leq Z_{\beta+\alpha}|\Gamma) = \beta .$$

Then a symmetric β confidence interval for M_0 (of $\frac{1}{2}z_{(1+\beta)/2}$ standard deviations), conditional on $\Gamma = \hat{\Gamma}$, is

$$\widehat{M} - \frac{\rho\sigma_M}{\sigma_\Gamma}(\hat{\Gamma} - \Gamma_0) - Z_{(1-\beta)/2}\,\sigma_M\sqrt{1-\rho^2} \leq M_0 \leq$$

$$\widehat{M} - \frac{\rho\sigma_M}{\sigma_\Gamma}(\hat{\Gamma} - \Gamma_0) - Z_{(1+\beta)/2}\,\sigma_M\sqrt{1-\rho^2} .$$

This interval is a function of the true value Γ_0. The purpose of constructing such an interval should now be clear. If, from some source other than $\mathbf{X}$, the data at hand, one knows a value Γ^* for Γ_0, then the information in $\hat{\Gamma}$ is not required for constructing a confidence interval for Γ_0. This information can instead be applied to improving one's knowledge about M_0, by constructing the conditional confidence interval

$$\widehat{M} - \frac{\rho\sigma_M}{\sigma_\Gamma}(\hat{\Gamma} - \Gamma^*) - Z_{(1-\beta)/2}\,\sigma_M\sqrt{1-\rho^2} \leq M_0 \leq$$

$$\widehat{M} - \frac{\rho\sigma_M}{\sigma_\Gamma}(\hat{\Gamma} - \Gamma^*) - Z_{(1+\beta)/2}\,\sigma_M\sqrt{1-\rho^2} .$$

Note that it would be wrong to construct a conditional interval for M, setting $\Gamma_0 = \hat{\Gamma}$. One can only use the conditional distribution when some additional information is added, namely the known value Γ^*.

9.1.3. *Interpretation of the covariance matrix*

The general many-dimensional problem is to find a confidence region for a subset of parameters, $\boldsymbol{\theta}_{(r)} = (\theta_1, \ldots, \theta_r)$, either

(i) *independently* of the estimates $\mathbf{t}_{(s)}$ of the other $s = N - r$ parameters, or

(ii) *conditional* on the estimates $\mathbf{t}_{(s)}$, given that the parameters

$\boldsymbol{\theta}_{(s)} = (\theta_{r+1}, \ldots, \theta_N)$ have known values, $\boldsymbol{\theta}^*_{(s)}$.

The estimates $\mathbf{t}_{(N)}$ have the many-dimensional Normal distribution with mean $\boldsymbol{\theta}_{(N)}$ and covariance matrix

$$\underset{\sim}{V} = \begin{pmatrix} \sigma_1^2 & \rho_{12}\sigma_1\sigma_2 & \cdots & & & \rho_{1N}\sigma_1\sigma_N \\ \rho_{12}\sigma_1\sigma_2 & \sigma_2^2 & & & & \vdots \\ \vdots & & \ddots & & & \\ \vdots & & & \ddots & & \vdots \\ \rho_{1N}\sigma_1\sigma_N & \cdots & & \cdots & \cdots & \sigma_N^2 \end{pmatrix}.$$

This may be partitioned as

$$\underset{\sim}{V} = \begin{pmatrix} \underset{\sim}{V}_{(rr)} & \vdots & \underset{\sim}{V}_{(rs)} \\ & \vdots & \\ \cdots & \cdots & \cdots & \cdots & \cdots & \cdots \\ \underset{\sim}{V}^T_{(rs)} & \vdots & \underset{\sim}{V}_{(ss)} \\ & \vdots & \end{pmatrix}.$$

In case (i) one uses the fact that the distribution of the estimates $\mathbf{t}_{(r)} = (\mathbf{t_1}, \ldots, \mathbf{t_r})$ is Normal r-dimensional with mean $\boldsymbol{\theta}_{(r)}$ and covariance matrix $\underset{\sim}{V}_{(rr)}$, obtained by retaining the rows $(1, \ldots, r)$ and columns $(1, \ldots, r)$ of $\underset{\sim}{V}$.

In case (ii), one uses the fact that the distribution of the $\mathbf{t}_{(r)}$, conditional on observing the values $\mathbf{t}_{(s)}$, is Normal r-dimensional, with mean

$$E\left(\mathbf{t}_{(r)} | \mathbf{t}_{(s)}\right) = \boldsymbol{\theta}_{(r)} \underset{\sim}{V}_{(rs)} \underset{\sim}{V}^{-1}_{(ss)} \left(\mathbf{t}_{(s)} - \boldsymbol{\theta}_{(s)}\right).$$

and covariance matrix $\underset{\sim}{V}_{(rr|s)}$, obtained from the inverse matrix $\underset{\sim}{V}^{-1}$, deleting the rows $(r+1, \ldots, N)$ and columns $(r+1, \ldots, N)$, and inverting the remaining submatrix. The conditional confidence regions are then defined by

$$Q\left(\mathbf{t}_{(r)}, \boldsymbol{\theta}_{(r)} | \mathbf{t}_{(s)}, \boldsymbol{\theta}^*_{(s)}\right) \leq \mathbf{K}_\beta^2$$

where K_β^2 is the β-point of the $\chi^2(r)$ distribution, and

$$Q\left(\mathbf{t}_{(r)}, \boldsymbol{\theta}_{(r)} | \mathbf{t}_{(s)}, \boldsymbol{\theta}^*_{(s)}\right)$$
$$= \left[\mathbf{t}_{(r)} - \left\{\boldsymbol{\theta}_{(r)} + \underset{\sim}{V}_{(rs)} \underset{\sim}{V}^{-1}_{(ss)} \left(\mathbf{t}_{(s)} - \boldsymbol{\theta}^*_{(s)}\right)\right\}\right]^{\mathbf{T}}$$
$$\times \underset{\sim}{V}^{-1}_{(rr|s)} \left[\mathbf{t}_{(r)} - \left\{\boldsymbol{\theta}_{(r)} + \underset{\sim}{V}_{(rs)} \underset{\sim}{V}^{-1}_{(ss)} \left(\mathbf{t}_{(s)} - \boldsymbol{\theta}^*_{(s)}\right)\right\}\right].$$

Again, it should be noted that this region depends both on the observed estimates $\mathbf{t}_{(s)}$, and on the known value of the parameters $\boldsymbol{\theta}^*_{(s)}$.

As an example, consider the case $N = 3$. The conditional distribution of t_1, given the values t_2, t_3 is Normal with mean

$$E[t_1|t_2, t_3, \theta_2, \theta_3] = \theta_1 - \frac{\sigma_1}{\sigma_2}\left(\frac{\rho_{13}\rho_{23} - \rho_{12}}{1 - \rho_{23}^2}\right)(t_1 - \theta_2)$$

$$- \frac{\sigma_1}{\sigma_3}\left(\frac{\rho_{12}\rho_{23} - \rho_{13}}{1 - \rho_{23}^2}\right)(t_1 - \theta_3)$$

and variance

$$V(t_1) \equiv V(t_1|t_2, t_3) = \sigma_1^2\left(1 - \frac{\rho_{12}^2 + \rho_{13}^2 - 2\rho_{12}\rho_{13}\rho_{23}}{1 - \rho_{23}^2}\right).$$

The confidence region of probability content β, conditional on observing t_2, t_3, and given the values θ_2^* and θ_3^* are of the form

$$E(t_1|t_2, t_3, \theta_2^*, \theta_3^*) - Z_{\beta+\alpha}\sqrt{V(t_1)} \leq \theta_1 \leq E(t_1|t_2, t_3, \theta_2^*, \theta_3^*) - Z_\alpha\sqrt{V(t_1)}.$$

In case (ii), the parameters $\boldsymbol{\theta}_s$ are "frozen" at known values $\boldsymbol{\theta}_s^*$. Then the size of the confidence region (the "error") is always smaller than in case (i), since by fixing $\boldsymbol{\theta}_{(s)}$, one is adding information about $\boldsymbol{\theta}_{(r)}$. This information is transmitted by means of the correlations between $\mathbf{t}_{(r)}$ and $\mathbf{t}_{(s)}$.

9.2. The General Case in One Dimension

9.2.1. *Confidence intervals and belts*

Here we give the classical method for "inverting" the probability statement (9.1). This method is known as the *Neyman construction* since the general solution is due to Jerzy Neyman [Neyman]. Suppose that $t(\mathbf{X})$ is some function of the data with the p.d.f. $f(t|\theta)$, then Eq. (9.1) can be written

$$\beta = P(t_1 \leq t \leq t_2|\theta) = P[t_1(\theta) \leq t \leq t_2(\theta)]$$

$$= \int_{t_1}^{t_2} f(t|\theta)dt. \tag{9.12}$$

The solution to Eq. (9.12), an interval in t-space, $[t_1(\theta), t_2(\theta)]$, is not unique. One way to obtain a unique solution is to specify the *central interval*, defined by

$$\int_{-\infty}^{t_1} f(t|\theta)dt = \frac{1 - \beta}{2} = \int_{t_2}^{\infty} f(t|\theta)dt.$$

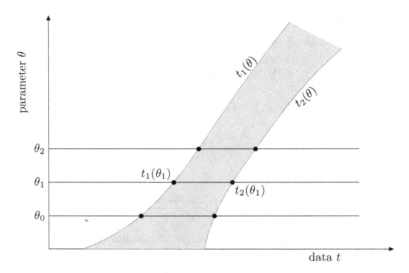

Fig. 9.5. The confidence belt is constructed horizontally. For each hypothetical value of θ, the points $t_1(\theta)$ and $t_2(\theta)$ are determined such that $\int_{t_1}^{t_2} P(t|\theta)dt = \beta$.

Some other possibilities are the *upper limit*, the *lower limit* and the *unified approach*, all discussed in the following sections. For the time being, assume simply that we have a way of determining t_1 and t_2, which satisfy Eq. (9.12), for all values of θ. These values can be plotted as in Fig. 9.5 to form the curves $t_1(\theta)$, $t_2(\theta)$. The region between $t_1(\theta)$ and $t_2(\theta)$ is known as a *confidence belt*. Along any horizontal line of fixed θ, the belt has a probability content β, by construction.

After the confidence belt is constructed horizontally, the experimental data are used to read the *confidence interval* vertically. Given the value t_0 observed in the experiment, the required confidence interval for θ is given by the intersections of the vertical $t = t_0$ with $t_1(\theta)$ and $t_2(\theta)$, as shown in Fig. 9.6.

That the interval $[\theta_L,\ \theta^U]$ is a confidence interval can be seen as follows. Suppose θ_0 is the true, unknown value of θ. Then, in a large series of experiments, the proportion of values of $t(X)$ falling in the range $t_1(\theta_0) \leq t \leq t_2(\theta_0)$ would be β, by definition. But it is clear from Fig. 9.6 that the intervals $[\theta_L, \theta^U]$, corresponding to values of t in $[t_1(\theta_0), t_2(\theta_0)]$ will cover the true value θ_0, that is, the proportion of intervals covering the true value is β, so we can make the statement

$$P(\theta_L \leq \theta_0 \leq \theta^U) = \beta\,.$$

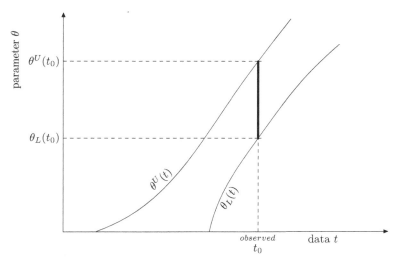

Fig. 9.6. The confidence interval is read vertically. From the observed data t_0, the intersection of the vertical line with the confidence belt determines the confidence interval (θ_L, θ^U).

In highly non-linear problems, the confidence belt may wiggle in such a way that the resulting confidence interval consists of several disconnected pieces. This is still a confidence interval, even if it is not simply-connected.

9.2.2. *Upper limits, lower limits and flip-flopping*

The Neyman construction, as long as Eq. (9.12) is satisfied, assures that the method has exact coverage. However, as we have already remarked, coverage alone does not make the interval unique, so we may apply different methods to produce different confidence intervals for the same problem. Apart from the *central interval* discussed above, two other common choices are the *upper limit* and the *lower limit*. These are the extremely non-central intervals, where the upper or lower *confidence bound* is at $\theta = +\infty$ or $-\infty$, for example

$$P(\theta \geq t_1) = \beta\,.$$

Upper limits are often used for a parameter which cannot be negative but is very close to zero. Then it is customary to quote only an upper limit, rather than a two-sided confidence interval.

We illustrate the choice of *confidence belt* with the simple example of a Gaussian-distributed measurement of a parameter which cannot be negative (for example, a particle mass or decay rate). We assume that the variance of the Gaussian measurement is known and is equal to one. For this problem, two different confidence belts are shown in Fig. 9.7. The heavy solid lines give 90% central limits, and the single dashed line gives the 90% upper limit (The other side of this belt is at infinity.). Both confidence belts have exact coverage by construction, so the physicist may quote either an upper limit or a two-sided confidence interval and be sure that whichever method he chooses covers exactly.

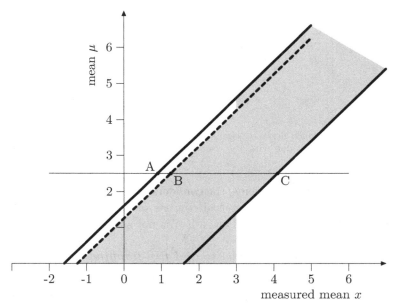

Fig. 9.7. Flip-flopping for a Gaussian measurement. The solid lines delimit the central 90% confidence belt, the dashed line the 90% upper limit, and the shaded area the effective confidence belt resulting from choosing between the two after seeing the data. This effective belt undercovers for $1.2 < \mu < 4.3$, for example at $\mu = 2.5$ where the intervals AC and $B\infty$ each contain 90% probability but BC contains only 85%.

However, the global procedure (certainly used by many physicists) which consists of choosing between the upper limit and the central interval after seeing the data, may not have coverage. This is illustrated in Fig. 9.7 where the shaded area corresponds to quoting an upper limit whenever the measured value is less than three standard deviations above zero, but quoting a two-sided interval when the measurement is more than 3σ above. It can be seen that the effective confidence belt for this choice does not give full coverage for some values of μ. In particular, for $\mu = 2.5$ as shown in the figure, the probability content of the interval BC is only 85%. This phenomenon is known as *flip-flopping* [Feldman]. It can be avoided by using a unified approach, as discussed below.

9.2.3. *Unphysical values and empty intervals*

There is another danger that may be encountered in the use of confidence belts near a boundary of the physical region. This can again be illustrated in Fig. 9.7, where the boundary of the physically allowed region is at $\mu = 0$, since μ cannot be negative. Suppose that we have decided to quote a 90% central interval and we measure a value $x = -2.0$. This is an allowed value of x, even if it is not allowed for μ.

Now it can be seen in Fig. 9.7, that a vertical line drawn at $x = -2.0$ does not intersect the chosen confidence belt anywhere in the physical region for μ. We might then be tempted to extrapolate all the straight lines down into the non-physical region and say that this implies a non-physical confidence interval for μ, but as that region is not really defined, the proper mathematical formulation of this situation is simply to say that the confidence interval is the *empty interval*. It contains no physically allowed values of μ.

When Neyman proposed his construction of confidence belts [Neyman], he was aware of this problem, and stated specifically that a valid confidence belt should not yield an empty interval. In the next section we describe a method which solves the problem of empty intervals.

Although empty intervals are to be avoided, it is instructive to notice that a method that produces such intervals can still have exact coverage. That is, even though we know that the empty confidence interval does not cover the true value of μ, the method that produces it can still have exact coverage, because, if the experiment is repeated many times, 90% of the confidence intervals produced by that method may still cover the true value. It is then clear that coverage is a property of the *method*, not of any particular *interval*. In the frequentist approach, it is the *method* that has well-defined properties like coverage, whereas in the Bayesian approach it is the *interval* that is deemed to have a certain probability of containing the true value (see Section 9.6).

9.2.4. *The unified approach*

Recall that the Neyman construction requires, for each hypothetical value of the parameter θ (or μ), accumulating the elements of probability that will be inside the confidence belt until the total integral between the two end points of the interval is β. This assures coverage, but it still allows the freedom to choose *which elements* will be chosen to be inside the interval, or more precisely, the freedom to choose the order in which elements of probability will be accepted into the interval. This choice, which then makes the intervals unique, is called the *ordering principle*.

We look then for an ordering principle which in some precise sense yields an optimal confidence belt, and hopefully solves at the same time the problems mentioned above: flip-flopping and empty intervals. Such an ordering principle was indeed found by Feldman and Cousins [Feldman] and is known as the *likelihood ratio ordering principle*. It is at the heart of their *unified approach* to the construction of confidence intervals. They point out that it cannot be considered new, because it is based entirely on well-known statistical theory and practice, but the full algorithm does not seem to have been described previously.

The Feldman–Cousins ordering principle is inspired by an important result from *hypothesis testing* which is discussed in the following chapter. There it is shown that the optimal test for distinguishing two simple hypotheses is the *Neyman–Pearson test*, in which the acceptance region is composed of those elements of the observable space which have the highest values of a likelihood ratio. For confidence intervals, no principle can be proved optimal, but the optimality for tests of simple hypotheses strongly suggests that the likelihood ratio ordering principle should also have very good properties for confidence intervals, which indeed it does.

The *likelihood ratio ordering principle* in the Neyman construction works as follows: When determining the interval for $\mu = \mu_0$, include the elements of probability $P(x|\mu_0)$ which have the largest values of the likelihood ratio

$$R(x) = \frac{P(x|\mu_0)}{P(x|\hat{\mu})} \, ,$$

where $\hat{\mu}$ is the value of μ for which the likelihood $P(x|\mu)$ is maximized within the physical region. For details and examples of this construction, please refer to the original paper [Feldman] which gives a clear and complete presentation of the method. Fig. 9.8 shows the confidence belt constructed using this ordering principle (labelled F-C) for the example of the Gaussian measurement considered previously. For values of $\mu > 1.65$, it coincides with the confidence belt for central intervals, but for smaller values of μ, the ordering principle begins to differ between the two, as $\hat{\mu}$ remains equal to zero for smaller values of μ. This results in the long left-hand tail which prevents the empty intervals

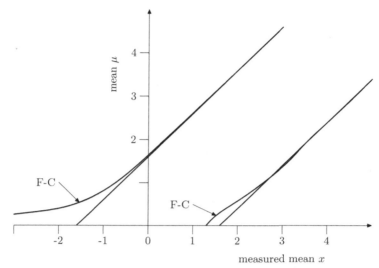

Fig. 9.8. Belts of 90% confidence for a Gaussian measurement showing the effect of using different ordering principles. The Feldman–Cousins belt is labelled "F-C", and the straight lines give central intervals.

that occur in central intervals for measurements $x < -1.65$. In order to compensate the extra probability under the left-hand tail, the right-hand edge of the belt must also move a little to the left.

Thus the Feldman–Cousins unified approach optimizes the confidence intervals in the sense of minimizing the error of the second kind, and the ordering principle that does this also eliminates the problem of empty intervals. Consistent use of the unified approach of course eliminates also the danger of flip-flopping.

9.2.5. *Confidence intervals for discrete data*

In the Neyman construction, we have so far assumed that Eq. (9.12) could be solved exactly because the data t were continuous. When the data are discrete, which will be the case for Poisson- and binomial-distributed data for example, we must replace in Eq. (9.12) the continuous t with discrete t_i, the integral becomes a summation, and unfortunately the equals sign must also go:

$$\int_{t_1}^{t_2} f(t|\theta)dt = \beta \quad \rightarrow \quad \sum_{i=L}^{U} P(t_i|\theta) \geq \beta. \tag{9.13}$$

The Neyman construction is performed in the same way as for continuous data, except that instead of accepting infinitesimal elements of probability into the confidence belt, one has to accept finite chunks of probability, with the obvious consequence that it will in general not be possible to attain exactly β probability. One must then accept some *overcoverage*, as indicated by the $\geq$ sign in Eq. (9.13). The degree of overcoverage depends in general on the (unknown) parameter θ.

A typical problem which produces unavoidable overcoverage is the determination of confidence intervals for the parameter μ of a Poisson distribution in a counting experiment without background. For this case, the actual coverage of the 90% Feldman–Cousins confidence intervals is shown in Fig. 9.9. It can be seen that the overcoverage is considerable for small values of μ, but it is still less than that of other methods, for example simple 90% upper limits.

There is however one aspect of discrete data which is advantageous compared with continuous data. For continuous data, the concept of *shortest interval* is not well defined (although it is often used by statisticians) since it is not invariant under transformations $\theta \rightarrow f(\theta)$. However, for discrete data one obtains the shortest interval in a well-defined way using the maximum-probability ordering, that is accepting t_i into the interval in order of decreasing $P(t_i|\theta)$. For the Poisson distribution, the shortest interval is the *Crow–Gardner interval*.

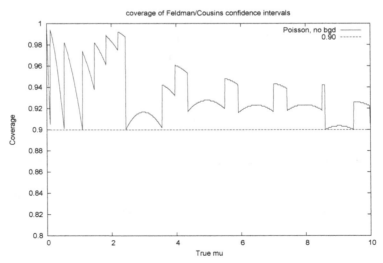

Fig. 9.9. Actual coverage of nominal 90% confidence intervals for the parameter of a Poisson distribution as given by the unified approach of Feldman and Cousins, as a function of the unknown true value μ.

There is in fact a way to avoid the overcoverage which seems to be inherent when data are discrete, but this involves using a trick which goes beyond conventional statistical practice. The idea is to introduce an element of randomness into the determination of the confidence interval, choosing a given interval only with a certain probability calculated to compensate its overcoverage so that the global algorithm has exact coverage. This method, due to W. L. Stevens [Stevens], is described in sections 19.10 and 19.11 of the sixth edition of [Stuart]. Since it requires the use of a random number to choose the confidence interval, it has the unfortunate effect that two experiments with exactly the same data, using this method, would in general obtain different confidence intervals. This can however be avoided by using an additional trick proposed in [Cousins 1994]. In this variant, one uses instead of the random number some otherwise unused but potentially relevant aspect of the data, such as the (normalized) distance to the last event removed by the cuts.

9.3. Use of the Likelihood Function

9.3.1. *Parabolic log-likelihood function*

Consider the case of a Normal distribution with known variance σ^2, and unknown mean μ. Then the likelihood for a single observation X is given by Eq. (5.7) with $N = 1$.

As a function of μ, $L(X|\mu)$ has the same bell-shaped form as the Normal probability density function (of X). The log-likelihood is a parabola in μ,

$$\ln L(X|\mu) = \ln c - \frac{(\mu - X)^2}{2\sigma^2}$$

with maximum at $\mu = X$. In Fig. 9.10, we plot this log-likelihood function, after a change of origin, to have $\ln L = 0$ at the maximum. A line drawn at $\ln L = -0.5$ gives the interval $(X - \sigma \le \mu \le X + \sigma)$, from $(\mu - X)^2/2\sigma^2 = 1/2$. From the properties of the Normal distribution one has

$$P[(X - \mu)^2 \le \sigma] = 68.3\%$$

$$P[-\sigma \le X - \mu \le \sigma] = 68.3\%$$

$$P[X - \sigma \le \mu \le X + \sigma] = 68.3\%.$$

Similarly, a line drawn at $\ln L = -2$ yields the 95.5% confidence interval for μ. As before, the probability statements are based on the properties of the random variable X, not on any properties of the parameter μ.

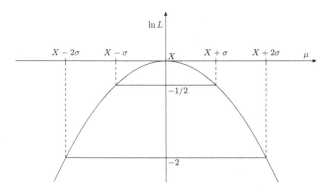

Fig. 9.10. Parabolic log-likelihood function.

9.3.2. *Non-parabolic log-likelihood functions*

We have seen in Eq. (8.11) that the *maximum likelihood estimate* has the *invariance* property, that is, instead of estimating parameter θ one can estimate any (monotonic) function of θ and make the same inference. This invariance is not only a property of the maximum of the likelihood, but applies to all the relative values of the likelihood function. Thus the *likelihood ratio* is an invariant, which helps explain its importance in *hypothesis testing* as well as in interval estimation.

Suppose, for example, that we have a non-linear problem for which the log-likelihood function takes a non-parabolic shape as in Fig. 9.11b. It is clearly possible to find a transformation of variables $\theta \to g(\theta)$ such that the log-likelihood curve will take on a parabolic shape in the new variable g. Now one can make the same inferences about g as were made about the Normal parameter μ in the previous section. And by the *invariance* property, these inferences also apply to the corresponding values of θ.

In Fig. 9.11a, the intersection of $\ln L_g$ with $\ln L = -2$ determines the interval $[g_L, g^U]$ which corresponds to a 95.5% confidence interval for g. By the invariance of likelihood ratios, this translates in Fig. 9.11b, to the interval θ_L, θ^U which is then a 95.5% confidence interval for θ.

Thus, by using the invariance properties of likelihood ratios it is possible to make inferences about a parameter of a non-Normal likelihood function *without actually finding the transformation* to a Normally distributed likelihood. This is extremely useful and elegant procedure is has been known to physicists for many years; they often call it the method of *minos*, since it is invoked by the command MINOS in the program Minuit [James 1975]. Statisticians use the

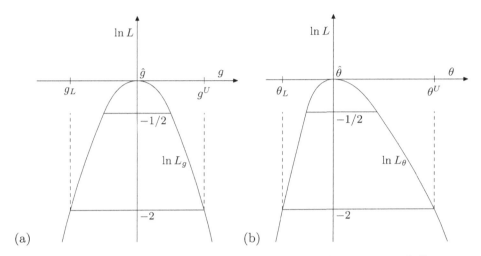

Fig. 9.11. Transformation of non-parabolic log-likelihood function to parabolic.

term *likelihood based confidence intervals*, and the multidimensional extension (where it is even more useful) is called *profile likelihood*.

Likelihood-based confidence intervals are in general only approximate, since the method is exact *only to order* $1/N$. It removes the primary bias term in the likelihood expansion but leaves other terms of higher order in $1/N$. The point is that we have made the *experimental* likelihood distribution Normal in the parameter instead of finding a transformation that would have made the *theoretical* distribution Normal (that is, using the true value of θ). To order $1/N$ this is the same, but for small samples, it is not exact.

The interval obtained is symmetric about $\hat{g}$ for the dummy variable g, but is in general asymmetric in the parameter of physical interest, reflecting the fact that in this parameter there may be more uncertainty on one side than on the other.

When the likelihood function has a pathological shape with for example multiple maxima, the procedure may give rise to a multiply-connected confidence interval as shown in Fig. 9.12. This would correspond to a confidence statement of the form

$$P(\theta_1 \le \theta \le \theta_2 \quad \text{or} \quad \theta_3 \le \theta \le \theta_4) = 1 - \beta.$$

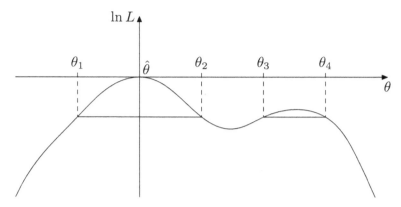

Fig. 9.12. "Pathological" log-likelihood function.

Such a multiply-connected interval is still an approximate confidence interval, although the approximation may be poorer when the likelihood function has such a shape, and it is not so convenient to handle.

The likelihood-based method is particularly useful when there are many parameters being estimated, including *nuisance parameters*, as discussed in the next section.

9.3.3. *Profile likelihood regions in many parameters*

Likelihood-based intervals as defined in the previous section are easily generalized to k-dimensional confidence regions given by the hypersurface

$$\ln L_{\boldsymbol{\theta}}(\mathbf{X}|\boldsymbol{\theta}) = \ln L_{\max} - \frac{1}{2}\chi_{\beta}^2(k)\,.$$

One then treats this region as a confidence region of probability content β. This practice can be justified asymptotically, by making use of the properties of the likelihood ratio

$$\ell = \frac{L(\mathbf{X}|\boldsymbol{\theta})}{L(\mathbf{X}|\widehat{\boldsymbol{\theta}})}\,.$$

In Section 10.5 it will be shown that $-2\ln\ell(\boldsymbol{\theta})$ is distributed asymptotically as $\chi^2(k)$. This leads directly to the probability statement

$$\beta = P[-2\ln\ell(\boldsymbol{\theta}) \le \chi_{\beta}^2(k)]\,,$$

which defines an (approximate) confidence region for θ. In many dimensions, the shape of such a region will often be too complicated to make it very useful *per se*, but we will show how it is used to produce single-parameter confidence intervals with remarkably good coverage in highly non-linear many-dimensional problems.

In two dimensions the situation can be visualized as in Fig. 9.13, where $\hat{\theta}$ is the maximum-likelihood joint estimate of the two parameters $\theta = (\theta_1, \theta_2)$, and two likelihood contours are plotted, at the same values as the horizontal lines in Figs. 9.10 and 9.11.

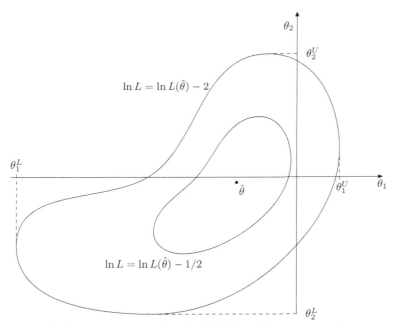

Fig. 9.13. Log-likelihood contours and 95.5% MINOS confidence intervals for a typical non-linear problem in two parameters θ_1, θ_2. See text.

The contours shown here are typical of those observed in practice (in fact these are taken from a real experiment performed by the author). Notice that the inner contour is more nearly elliptical than the outer one. If we plotted contours of $\ln L$ very close $\ln L(\hat{\theta})$, they would be good ellipses, reflecting the fact that all models are linear over a sufficiently small region. In order to take into account the non-linearities, we need to follow the log-likelihood all the way out to the edge of the confidence interval. The program Minuit performs these calculations and finds automatically the values θ^U and θ_L as indicated in the figure, for the contour specified by the user, in any number of dimensions.

In many dimensions, the MINOS (profile likelihood) confidence interval for parameter θ_i is defined by the two values of θ_i for which

$$\max_{\theta_j, j \neq i} \ln L(\theta) = \ln L(\hat{\theta}) - \lambda,$$

where $\lambda = Z_{1-\beta}^2/2$ is equal to $1/2$ for $\beta = .683$ and $\lambda = 2$ for $\beta = .955$.

9.4. Use of Asymptotic Approximations

9.4.1. *Asymptotic Normality of the maximum likelihood estimate*

In Sections 7.3.1 and 7.3.3 we have seen that the maximum likelihood estimate $\hat{\theta}$ is asymptotically distributed as $N(\theta, 1/NI)$, when N observations have been made, and I is the information about θ in one observation.

This immediately suggests a confidence interval, from the calculation used in the Normal case (Section 9.3.1), namely

$$\hat{\theta} \pm \frac{\lambda_{\beta/2}}{\sqrt{NI(\hat{\theta})}} \tag{9.14}$$

where $\lambda_{\beta/2}$ is the $\beta/2$-point of the standard Normal distribution.

However, because of the bias of the estimate $\hat{\theta}$ and the fact that the asymptotic behaviour may be reached only slowly this method can give only an approximate confidence interval.

9.4.2. *Asymptotic Normality of $\partial \ln L / \partial \theta$*

A more accurate confidence interval (in fact, near the true value θ_0, the most accurate interval) is obtained by using the fact that $\partial \ln L/\partial \theta$ is distributed with mean 0 and variance $NI(\theta)$ for all N. The approximation now used is that the distribution of

$$\frac{1}{\sqrt{NI(\hat{\theta})}} \frac{\partial \ln L}{\partial \theta}$$

is standard Normal. The interval is then

$$\left| \frac{1}{\sqrt{NI(\hat{\theta})}} \frac{\partial \ln L}{\partial \theta} \right| < \lambda_{\beta/2}. \tag{9.15}$$

Example

Consider the case of observations $\mathbf{X}$ from the negative exponential distribution (Section 4.2.9) with mean μ

$$f(X, \mu) = \frac{1}{\mu} e^{-X/\mu} \,.$$

The log-likelihood is

$$\ln L = -N \ln \mu - \sum_{i=1}^{N} \frac{X_i}{\mu} = -N \left(\ln \mu + \frac{\bar{X}}{\mu} \right),$$

and its derivatives

$$\frac{\partial \ln L}{\partial \mu} = -\frac{N}{\mu} + \frac{N\bar{X}}{\mu^2}, \quad \frac{\partial^2 \ln L}{\partial \mu^2} = \frac{N}{\mu^2} - \frac{2 \sum_{i=1}^{N} X_i}{\mu^3} \,.$$

The information $I(\hat{\mu})$ of $\hat{\mu} = \bar{X}$ is found by taking the expectation

$$E \left(\frac{-\partial^2 \ln L}{\partial \mu^2} \right) = \frac{N}{\mu^2} = N I(\mu) \,,$$

thus

$$N I(\hat{\mu}) = \frac{N}{\bar{X}^2} \,.$$

The rough confidence interval, Eq. (9.14), is now given by

$$\bar{X} \left(1 - \frac{\lambda_{\beta/2}}{\sqrt{N}} \right) \leq \mu \leq \bar{X} \left(1 - \frac{\lambda_{\beta/2}}{\sqrt{N}} \right) \,.$$

The asymptotically most accurate interval, Eq. (9.15), is given by

$$\frac{\left| \dfrac{N\bar{X}}{\mu^2} - \dfrac{N}{\mu} \right|}{\dfrac{\sqrt{N}}{\mu}} \leq \lambda_{\beta/2} \,,$$

or, expressed as a confidence interval for μ,

$$\frac{\bar{X}}{1 + \frac{\lambda_{\beta/2}}{\sqrt{N}}} \leq \mu \leq \frac{\bar{X}}{1 - \frac{\lambda_{\beta/2}}{\sqrt{N}}} \,.$$

If, instead, one looks for a confidence interval for $1/\mu$, it turns out that Eqs. (9.14) and (9.15) give the same result.

9.4.3. $\partial L/\partial \boldsymbol{\theta}$ confidence regions in many parameters

The generalization of Eq. (9.15) to a confidence region in k parameters is

$$\frac{1}{N}\left(\frac{\partial L(\mathbf{X}|\boldsymbol{\theta})}{\partial \boldsymbol{\theta}}\right)^{T} I^{-1}(\boldsymbol{\theta})\left(\frac{\partial L(\mathbf{X}|\boldsymbol{\theta})}{\partial \boldsymbol{\theta}}\right) \leq \chi_{\beta}^{2}(k), \tag{9.16}$$

where $I(\boldsymbol{\theta})$ is the information matrix, and $\chi_{\beta}^{2}(k)$ is the β-point of the χ^2 distribution with k degrees of freedom. When $I(\boldsymbol{\theta})$ is approximated by

$$\left(-\frac{\partial^{2}\ln L}{\partial \boldsymbol{\theta}\partial \boldsymbol{\theta}'}\right)\bigg|_{\boldsymbol{\theta}=\widehat{\boldsymbol{\theta}}},$$

Eq. (9.16) yields a hyperellipsoidal region in $\partial L/\partial \boldsymbol{\theta}$. The corresponding region in $\boldsymbol{\theta}$-space is not, in general, ellipsoidal.

9.4.4. Finite sample behaviour of three general methods of interval estimation

Three methods of interval estimation in the general case (data not Normally distributed, $\boldsymbol{\theta}$ is p-dimensional) are based on the following three asymptotic properties of the likelihood function:

(i) $\sqrt{N}(\widehat{\boldsymbol{\theta}} - \boldsymbol{\theta})$ is asymptotically Normal with mean $\mathbf{0}$ and covariance matrix $\mathcal{I}^{-1}$ [Sections 7.2.3, 7.3.1 and 7.3.3 (iii)]. $\mathcal{I}$ is the information matrix with elements

$$(I)_{ij} = E\left[\left(\frac{\partial \ln f(X,\boldsymbol{\theta})}{\partial \theta_{i}}\right)\left(\frac{\partial \ln f(X,\boldsymbol{\theta})}{\partial \theta_{j}}\right)\right] = -E\left[\frac{\partial^{2}\ln f(X,\boldsymbol{\theta})}{\partial \theta_{i}\partial \theta_{j}}\right]$$

(ii) $\partial \ln f(X,\boldsymbol{\theta})/\partial \boldsymbol{\theta}$ is distributed with mean $\mathbf{0}$ and covariance matrix $\mathcal{I}(\boldsymbol{\theta})$, and is asymptotically Normal (Section 7.2.3).

(iii) $2\Sigma_{k}[\ln f(X_{k},\widehat{\boldsymbol{\theta}}) - \ln f(X_{k},\boldsymbol{\theta})]$ behaves asymptotically as a χ^2 with p degrees of freedom.

Each of these properties can be used to define a function $Q^2(\mathbf{X}, \boldsymbol{\theta})$ with distribution independent of $\boldsymbol{\theta}$. Thus, each of the functions

$$Q_1^2 = N(\widehat{\boldsymbol{\theta}} - \boldsymbol{\theta})^T \, \mathcal{I}\,(\boldsymbol{\theta})(\widehat{\boldsymbol{\theta}} - \boldsymbol{\theta}) \tag{9.17}$$

$$Q_2^2 = \left[\frac{1}{N} \sum_{k=1}^{N} \frac{\partial \ln f(X_k, \boldsymbol{\theta})}{\partial \boldsymbol{\theta}}\right]^T N \, \mathcal{I}^{-1}\,(\boldsymbol{\theta}) \left[\frac{1}{N} \sum_{k=1}^{N} \frac{\partial \ln f(X_k, \boldsymbol{\theta})}{\partial \boldsymbol{\theta}}\right] \tag{9.18}$$

$$Q_3^2 = 2 \sum_{k=1}^{N} [\ln f(X_k, \widehat{\boldsymbol{\theta}}) - \ln f(X_k, \boldsymbol{\theta})] \tag{9.19}$$

has asymptotically a $\chi^2(p)$ distribution.

In practice, it may not always be possible to calculate the information matrix $\mathcal{I}\,(\boldsymbol{\theta})$ analytically. It must then be replaced in Eqs. (9.17) and (9.18) by estimates from the data. This can be done in two ways. Either one estimates the matrix elements $(I)_{ij}$ by the sample average of the second derivatives,

$$[\mathcal{I}_N\,(\boldsymbol{\theta})]_{ij} = -\frac{1}{N} \sum_{k=1}^{N} \frac{\partial^2 \ln f(X_k, \boldsymbol{\theta})}{\partial \theta_i \partial \theta_j} . \tag{9.20}$$

Or, one assumes that when $\boldsymbol{\theta}$ is near $\widehat{\boldsymbol{\theta}}$, the second derivatives are constant,

$$[\mathcal{I}_N\,(\widehat{\boldsymbol{\theta}})]_{ij} = -\left\{\frac{1}{N} \sum_{k-1}^{N} \left(\frac{\partial^2 \ln f(X_k, \boldsymbol{\theta})}{\partial \theta_i \partial \theta_j}\right)\right\}_{\boldsymbol{\theta}=\widehat{\boldsymbol{\theta}}} . \tag{9.21}$$

Thus, we have seven possible functions $Q^2(\mathbf{X}, \boldsymbol{\theta})$:

1. $Q_{1a}^2 = $ Eq. (9.17) with Information computed analytically
2. $Q_{1b}^2 = $ Eq. (9.17) with Information from Eq. (9.20)
3. $Q_{1c}^2 = $ Eq. (9.17) with Information from Eq. (9.21)
4. $Q_{2a}^2 = $ Eq. (9.18) with Information computed analytically
5. $Q_{2b}^2 = $ Eq. (9.18) with Information from Eq. (9.20)
6. $Q_{2c}^2 = $ Eq. (9.18) with Information from Eq. (9.21)
7. $Q_3^2 = $ Eq. (9.19), the likelihood ratio

Probability statements can then take the form

$$P\left[\chi_{1-\beta/2}^2(p) \le Q^2(\boldsymbol{\theta}) \le \chi_{1+\beta/2}^2(p)\right] = \beta . \tag{9.22}$$

Asymptotically, confidence regions for $\boldsymbol{\theta}$ can be found from Eq. (9.22).

Let us now turn to the finite sample distributions of the $Q^2(\mathbf{X}, \boldsymbol{\theta})$, when they are not $\chi^2(p)$, so that Eq. (9.22) is inaccurate. In particular we shall consider the bias of the distributions of $Q^2(\mathbf{X}, \boldsymbol{\theta})$ relative to $\chi^2(p)$. Clearly, if $P[Q^2(\boldsymbol{\theta})]$ is biased relative to $P[\chi^2(p)]$, the probability content on either side of the interval in Eq. (9.22)

$$\left[\chi^2_{(1-\beta)/2}(p), \quad \chi^2_{(1+\beta)/2}(p)\right]$$

will be unequal for the distribution of $Q^2(\boldsymbol{\theta})$.

Let us expand each of the functions $Q^2(\mathbf{X}, \boldsymbol{\theta})$ in successive orders of $1/N$. Limiting ourselves to the simplest case with only one parameter $(p = 1)$, we can write

$$\frac{1}{N}\sum_k \frac{\partial \ln f(X_k, \theta)}{\partial \theta} = E\left(\frac{\partial \ln f(X, \theta)}{\partial \theta}\right) + \frac{X}{\sqrt{N}}\frac{X}{\sqrt{N}}.$$

$$\frac{1}{N}\sum_k \frac{\partial^2 \ln f(X_k, \theta)}{\partial \theta^2} = E\left(\frac{\partial^2 \ln f(X, \theta)}{\partial \theta^2}\right) + \frac{Y}{\sqrt{N}} = -I + \frac{Y}{\sqrt{N}}.$$

$$\frac{1}{N}\sum_k \frac{\partial^3 \ln f(X_k, \theta)}{\partial \theta^3} = E\left(\frac{\partial^3 \ln f(X, \theta)}{\partial \theta^3}\right) + \frac{Z}{\sqrt{N}} = K + \frac{Z}{\sqrt{N}}.$$

By the Central Limit Theorem X, Y, and Z are asymptotically Normal, with zero mean and finite variance. In Section 7.3.3 (iii) we have seen that

$$(\hat{\theta} - \theta) = \frac{X}{I\sqrt{N}}\left[1 + \frac{Y}{1\sqrt{N}} + \frac{1}{2}\frac{KX}{I^2\sqrt{N}}\right] + \cdots.$$

From this we can deduce that

$$-I_N(\hat{\theta}) = -I_N(\theta) + (\hat{\theta} - \theta)\frac{1}{N}\sum_{k=1}^N \frac{\partial^3 \ln f(X_k, \theta)}{\partial \theta^3} + \cdots$$

$$= -I(\theta) + \frac{Y}{\sqrt{N}} + \frac{KX}{I\sqrt{N}} + \cdots$$

$$= -I(\theta)\left[1 - \frac{Y}{I\sqrt{N}} - \frac{KX}{I^2\sqrt{N}}\right] + \cdots,$$

Similarly we can show that

$$2\sum_k [\ln f(X_k, \hat{\theta}) - \ln f(X_k, \theta)]$$

$$= 2N\left[\frac{1}{N}\sum_k \frac{\partial \ln f(X_k, \theta)}{\partial \theta}(\hat{\theta} - \theta) + \frac{1}{2N}\sum_k \frac{\partial^2 \ln f(X_k, \theta)}{\partial \theta^2}(\hat{\theta} - \theta)^2\right.$$

$$\left. + \frac{1}{6N}\sum_k \frac{\partial^3 \ln f(X_k, \theta)}{\partial \theta^3}(\hat{\theta} - \theta)^3\right] + \cdots$$

$$= \frac{2X^2}{I}\left(1 + \frac{Y}{I\sqrt{N}} + \frac{1}{2}\frac{KX}{I^2\sqrt{N}}\right)$$

$$+ \frac{X^2}{2I}\left(-1 + \frac{Y}{I\sqrt{N}}\right)\left(1 + \frac{2Y}{I\sqrt{N}}\frac{KX}{I^2\sqrt{N}}\right) + \frac{KX^3}{6I^3\sqrt{N}} + O\left(\frac{1}{N}\right)$$

$$= \frac{X^2}{2I} + \frac{X^2Y}{I^2\sqrt{N}} + \frac{KX^3}{3I^3\sqrt{N}} + O\left(\frac{1}{N}\right).$$

We can now construct expansions for each function; these are given in Table 9.2. The mean values are expressed in terms of two constants, a and b, where

$$a = \frac{E(\sqrt{N}X^2Y)}{I^2}$$

and

$$b = \frac{E(\sqrt{N}KX^3)}{I^3}.$$

To show that a and b are finite, we write

$$X^2Y = \frac{1}{N}\left(\sum_i g_k\right)^2\left[\sqrt{N}I_0 + \frac{1}{N}\sum_i h_i\right]$$

where

$$g_i = \frac{\partial \ln f(X_i, \theta)}{\partial \theta}$$

and

$$h_i = \frac{\partial^2 \ln f(X_i, \theta)}{\partial \theta^2}.$$

Table 9.2. Functions for interval estimation in the general case.

Method	$Q^2(\theta)$	Asymptotic Expansion	Mean
(1a) Information computed analytically	Q^2_{1a}	$\dfrac{X^2}{1} + \dfrac{2X^2Y}{I^2\sqrt{N}} + \dfrac{KX^3}{I^3\sqrt{N}}$	$1 + \dfrac{1}{N}(2a + b)$
(1b) Information estimated from data	Q^2_{1b}	$\dfrac{X^2}{I} + \dfrac{X^2Y}{I^2\sqrt{N}} + \dfrac{KX^3}{I^3\sqrt{N}}$	$1 + \dfrac{1}{N}(a + b)$
(1c) Information estimated from data, at $\theta = \hat{\theta}$	Q^2_{1c}	$\dfrac{X^2}{I} + \dfrac{X^2Y}{I^2\sqrt{N}}$	$1 + \dfrac{1}{N}a$
(2a) Information computed analytically	Q^2_{2a}	$\dfrac{X^2}{I}$	1
(2b) Information estimated from data	Q^2_{2b}	$\dfrac{X^2}{I} + \dfrac{X^2Y}{I^2\sqrt{N}}$	$1 + \dfrac{1}{N}a$
(2c) Information estimated from data, at $\hat{\theta}$	Q^2_{2c}	$\dfrac{X^2}{I} + \dfrac{X^2Y}{I^2\sqrt{N}} + \dfrac{KX^3}{I^3\sqrt{N}}$	$1 + \dfrac{1}{N}(a + b)$
(3) Likelihood ratio	Q^2_3	$\dfrac{X^2}{I} + \dfrac{X^2Y}{I^2\sqrt{N}} + \dfrac{KX^3}{3I^3\sqrt{N}}$	$1 + \dfrac{1}{N}\left(a + \dfrac{b}{3}\right)$

Then we have

$$\frac{E(\sqrt{N}X^2Y)}{I^2} = \frac{I}{I^2}E\left[I\left(\sum_i g_i\right)^2\right] + \frac{1}{I^2}E\left[\frac{1}{N}\left(\sum_i g_i\right)^2\sum_i h_i\right]$$

$$= \frac{1}{I}E\left(\sum_i g_i^2\right) + \frac{1}{NI^2}e\left[\sum_i g_i^2 h_i + \sum_{i*j} g_i^2 h_j\right]$$

$$= N + \frac{E(g^2h)}{I^2} + \frac{N(N-1)}{NI^2}I(-I) = \frac{E(g^2h)}{I^2} + 1.$$

Thus

$$a = 1 + \frac{1}{I^2} E\left[\left(\frac{\partial \ln f(X,\theta)}{\partial \theta}\right)^2 \frac{\partial \ln f(X,\theta)}{\partial \theta^2}\right]$$

and similarly

$$b = \frac{1}{I^3} E\left(\frac{\partial^3 \ln f(X,\theta)}{\partial \theta^3}\right) E\left[\left(\frac{\partial \ln f(X,\theta)}{\partial \theta}\right)^3\right].$$

From Table 9.2 it is clear that, with exception for $Q^2_{2a}(\boldsymbol{\theta})$, the functions $Q^2(\boldsymbol{\theta})$ are biased with respect to χ^2. This bias is proportional to $1/N$, and it exists even if the log-likelihood function $\ln f(\mathbf{X}, \boldsymbol{\theta})$ is symmetric in $\boldsymbol{\theta}$, in contradiction to point estimation, where a sufficient condition for the absence of bias was the symmetry of the log-likelihood function.

The bias can be removed by a change of variable on the parameter, except in the likelihood ration method, where the bias is invariant to such transformations.

The function $Q^2_{2a}(\theta)$ is unbiased with respect to the χ^2. However, X^2/I is not a χ^2 variable. The variance of X^2/I is

$$V\left(\frac{X^2}{I}\right) = E\left(\frac{X^4}{I^2}\right) - \left[E\left(\frac{X^2}{I}\right)\right]^2$$

$$= 2 + \frac{1}{N}\left\{\frac{1}{I^2} E\left[\left(\frac{\partial \ln f(X,\theta)}{\partial \theta}\right)^4\right] + 1\right\},$$

the $1/N$ term giving the bias relative to the variance of $\chi^2(1)$. Functions approximating more closely χ^2 can be found [Kendall II, p. 112].

In the other methods, the extra terms will similarly contribute to the variance and higher moments of the distribution. However, for the likelihood ratio method, it may be shown that if one divides by the mean, the function

$$X_4^2 = \frac{2\left[\sum_m \ln f(X_k, \hat{\theta}) - \sum_k \ln f(X_k, \theta)\right]}{1 + \frac{1}{N}\left[a(\theta) + \frac{1}{3}b(\theta)\right]}$$

is distributed as χ^2 to order $1/N$ *included* [Kendall II, p. 233], [Lawley 1956].

In practice, one does not usually want to compute higher order terms. If one can compute the function $\mathcal{L}(\boldsymbol{\theta})$ analytically, then the best method to use is Q_{2a}^2. One has then eliminated the bias relative to the χ^2 distribution. Otherwise, the best method would seem to be the likelihood ratio method (Section 9.4.3 and Chapter 10), since this has the useful property of invariance to change of variable in the parameters.

9.5. Summary: Confidence Intervals and the Ensemble

Faced with the plethora of results and methods given in this chapter, the reader may well feel the need for some practical advice on how to proceed. Fortunately, in the majority of cases, he will probably be in what we may call the "asymptotic region", where the uncertainties in the parameter estimates are small compared with the non-linearities of the model and small compared with the distance to the nearest boundary of the physical region. In this case the Normal theory is sufficient, and all methods (including even Bayesian methods) will give the same numerical values for the confidence interval.

So the problem reduces to how to know when more sophisticated methods are needed, and what to do in that case. A common procedure is to calculate the confidence interval using (1) a simple Normal theory method and (2) the profile likelihood method, since both of these are easily calculated by readily available programs. In the framework of the program Minuit, (1) is called the "parabolic error", and (2) is the "Minos error". If these two are identical, there is no need to proceed further unless there is some other reason to suspect a problem.

If the intervals (1) and (2) are different, it is up to the physicist to judge whether this difference is big enough to warrant further effort. If so, the next step depends mostly on the number of unknown parameters being estimated in the experiment. This number includes not only the parameters of interest, but also the *nuisance parameters*, whose values are not of interest except insofar as they are needed in order to estimate the parameters of interest. A notorious nuisance parameter is the expected rate of background in a counting experiment; when the background expectation is only approximately known, it must be considered as an unknown parameter in the statistical analysis, even if its value is of no direct interest. Real experiments may have many nuisance parameters.

If the total number of parameters is small (one or two), it is worthwhile considering the Feldman–Cousins analysis. This involves some work, but the original publication [Feldman] describes exactly what you have to do, and there are now many examples where it has been used successfully in rather complicated situations. It is the method of choice.

When the number of free parameters is more than two, Feldman–Cousins may prove to be too complicated, and it may also be too conservative. Fortunately, this is just the situation in which profile likelihood happens to give very good coverage according to the Monte Carlo studies which have been performed.

Whatever method is used, if the result is important it should be verified with a Monte Carlo calculation of the coverage. For this calculation, one assumes values for each of the unknown parameters and then calculates the coverage for that point in parameter space by generating a large set of simulated experiments (usually about 10,000) and counting how many times the confidence interval for the parameter(s) of interest covers the assumed true value. This procedure is then repeated for enough different points in the parameter space to get a reasonable idea of the coverage over the region of interest. Fig. 9.9 is an example of the plot you would get if you did this calculation for the simple case of a counting experiment with no background using Feldman–Cousins intervals. The effect of the discrete data is evident in this plot. When data are continuous, the coverage tends to be a smooth function of the position in parameter space. If an approximate method has been used, the coverage may occasionally drop below the target value, in which case the confidence interval should be widened to attain correct coverage.

When doing the Monte Carlo coverage calculation, the physicist may encounter one of the most debated problems in frequentist statistics, that of the *ensemble*, usually referred to in the statistics literature as *conditioning*. This problem occurs when assigning values to certain quantities during the calculation; sometimes it may not be obvious whether the value should be a constant (the value observed in the actual experiment), or whether it should be considered a *random variable*, free to take on a range of values around that observed in the experiment. We consider two examples, an easy one and a hard one.

Suppose that when the experiment (the real one) was designed, it was decided that the main magnet should have a diameter of 2 metres. The design was then sent to the magnet builder, specifying the desired diameter and a tolerance of 1%. When the magnet was delivered, the diameter was measured to be 2.01m, within the specified tolerance, so it was accepted and installed in the experiment. Now the question is: In the Monte Carlo calculation, should the magnet have a constant diameter of 2.01m, or should it be a random variable with a value of 2.00 ± 0.02m, or perhaps even something else? In this case, it is fairly clear that nothing would be gained by making the diameter variable since one really wants to know the coverage for the experiment as it was performed, not as it was designed.

The hard example concerns the famous case where one is trying to measure a small signal in the presence of a significant background. Suppose that the total number of events observed (signal plus background) is less than that expected from background alone. Then it is known that the experimental data contain fewer background events than were expected. The double question is then: (1) Should the statistical analysis take account of this knowledge by *conditioning* on the unexpectedly low background, and (2) if so, how would it be done? To our knowledge, none of the attempts to answer these questions have been entirely successful.

Other problems of a purely practical nature may arise during the coverage calculation. During the analysis of the Monte Carlo data, some catastrophic event may occur (a program crashes or an iterative procedure fails to converge) which makes it impossible to calculate the confidence interval. If such an event happens during the analysis of the real data, the physicist will take the time to fix the programs before continuing. But in the millions of Monte Carlo experiments, such an event can be expected more frequently, and there may not be time to fix all those bugs that have no effect on the real data.

9.6. The Bayesian Approach

The Bayesian approach to interval estimation often produces results numerically identical to frequentist confidence intervals. However, even in these cases, the meaning of the intervals is totally different. This is not surprising since they are based on different definitions of probability.

In the classical approach the unknown parameter has a constant value θ_0, whereas in the Bayesian approach there is a distribution of possible true values, $\pi(\theta|\mathbf{X})$. Recall from Section 2.3.5 that this *posterior distribution* is a distribution of degrees of belief, given the data $\mathbf{X}$ and given one's *prior belief* $\pi(\theta)$,

$$\pi(\theta|\mathbf{X}) = \frac{\displaystyle\prod_{i=1}^{N} f(X_i, \theta)\pi(\theta)}{\displaystyle\int \prod_{i=1}^{N} f(X_i, \theta)\pi(\theta)d\theta}. \tag{9.23}$$

The posterior density (9.23) summarizes one's complete knowledge of the parameter θ. Using this, an interval $[\theta_L, \theta^U]$ with probability content β, can be defined by

$$\int_{\theta_L}^{\theta^U} \pi(\theta|\mathbf{X})d\theta = \beta \tag{9.24}$$

By this definition, the interval $[\theta_L, \theta^U]$ contains a fraction β of one's total belief about θ; that is, one is prepared to bet that the true value lies in this interval, at odds of $\beta/(1-\beta)$. Using P to denote the Bayesian (degree-of-belief) probability, we have:

$$P(\theta_L < \theta < \theta^U) = \beta$$
$$P\{\theta \notin [\theta_L, \theta^U]\} = 1 - \beta.$$

As in the classical approach, the interval $[\theta_L, \theta^U]$ is not unique. And we again have several possibilities: the *central interval*, the *upper limit* and the *shortest interval* are the most common.

Recall that the frequentist central interval is central *in the data* and when transformed to parameter space it may be very non-central or even empty. This cannot happen with Bayesian intervals since they are constructed directly in parameter space, with equal probabilities in the two tails of $\pi(\theta|\mathbf{X})$ outside of the interval. This means that central intervals cannot be empty (which is good), but it also means that central intervals cannot automatically become upper limits when the data indicate that should happen (which is not so good).

The above problem can be avoided by using the Bayesian *shortest interval*, which is anyway the most often used. The algorithm is straightforward: One accepts into the interval the largest values of the p.d.f. $\pi(\theta|\mathbf{X})$. This method suffers from the same problem as frequentist shortest intervals when the parameter θ is continuous. The concept of "largest values" of a *p.d.f.* is metric-dependent; for example, the shortest interval for θ would not correspond to the shortest interval obtained for θ^2 or $1/\theta$. However, there is in principle a remedy for this non-invariance in the Bayesian framework, because there exists now a preferred metric: the one in which the prior belief p.d.f. $\pi(\theta)$ is uniform. Since the *uniform prior* is anyway the most commonly used, this simplifies things considerably, but it makes it even more important to consider the problems of the uniform prior discussed in Section 7.5.1.

9.6.1. *Confidence intervals and credible intervals*

In order to distinguish them from frequentist confidence intervals which by definition have coverage, Bayesian intervals are called *credible intervals* or sometimes (improperly) Bayesian confidence intervals. The point is that credible intervals have a specified *degree of belief*, but do not necessarily cover the true value with any given frequentist probability. This is not surprising because coverage is simply not a meaningful concept in the Bayesian framework. In spite of this, it is widely considered to be of interest to consider the coverage of Bayesian intervals [Cousins 1995]. Our experience listening to physicists discuss their statistical methods is that coverage is considered of prime importance, even by those who use Bayesian methods. One attitude often encountered among physicists is that both frequentist and Bayesian methods are valid, but if your intervals do not have coverage, you are cheating and your intervals should be bigger. We do not try to defend this attitude, only note that it exists.

Many people have studied the frequentist coverage of Bayesian intervals, even though it is not a meaningful property in the Bayesian framework. In one famous case (upper limits for the parameter of a Poisson distribution) Bayesian intervals are conservative (overcover) and this has led some people to believe mistakenly that Bayesian limits are always conservative. In fact, it can be seen that Bayesian credible intervals have a property which we may call *average coverage*, where the average is taken over all values of the parameter being estimated, weighted by the prior density. Since this average coverage is exact, one might expect that Bayesian methods would undercover about half the time, but the situation is more complicated than that because of the prevalence of the uniform prior. Indeed, when the uniform prior is used for parameters of semi-infinite range, all the weight is put at infinity, so the coverage is exact only for $\theta = \infty$ where we are generally not interested. Then the coverage in the region of interest may be anything.

9.6.2. *Summary: Bayesian or frequentist intervals?*

Both Bayesian and frequentist methodologies have many good points and a few bad ones, but it is futile to attempt to demonstrate that one or the other is right or wrong. They are based on different definitions of probability and they answer different questions. The decision which one to use must be based on which question you want to answer, that is, it depends mostly on the consumer of your intervals. What does he or she want to know? Some possibilities are:

(i) The consumer wants to know your personal opinion about the parameter in question, in order to make a decision which depends on the value of the parameter. In this case the Bayesian interval is the most appropriate.

(ii) The consumer is the editor of a scientific journal in which you want to publish your result. Journals are usually interested in an objective study reporting intervals with exact coverage. Then it is the frequentist interval that is appropriate.

(iii) The consumer wants to combine your results with his own, or with several others in order to produce a more accurate combined result. For this purpose, none of the methods described here will be of much use since it is not possible in general to combine either confidence or credible intervals without losing a lot of information. For this purpose there is general agreement [James 2000] that experimenters should make available their *likelihood function*. These are easily combined (just add the log-likelihoods of all the experiments), and since the likelihood function is a *sufficient statistic* for the parameter in question, the *Fisher information* is preserved.

Chapter 10

TESTS OF HYPOTHESES

In the previous three chapters we have seen how experimental information can be used for estimating parameters. In this chapter and the following one we show how experimental data may be used for *testing*: to verify or disprove a theory or hypothesis, or to help to choose between alternative hypotheses.

When the *hypothesis under test* concerns the value of a parameter, the problems of *parameter estimation* and *hypothesis testing* are related; for instance, good techniques for estimation often lead to analogous testing procedures. The two situations, lead, however, to principally different conclusions, and should not be confused (in practice they often are!).

From example, if nothing is known *a priori* about the parameter involved, it is natural to use the data to estimate it, using the techniques of Chapters 7, 8 and 9. On the other hand, if a theoretical prediction has been made that the parameter should have a certain value, it may be more appropriate to formulate the problem as a test of whether the data are consistent with this value. In either case, the nature of the problem (estimation or test) must be clear from the beginning and consistent to the end.

An hypothesis which is completely specified is called a *simple hypothesis*. An ensemble of more than one simple hypothesis is known as a *composite hypothesis*, the most common example being an hypothesis involving a free parameter. To clarify most of the concepts involved in testing it is sufficient to consider simple hypotheses, and we shall do so in most of this Chapter. A most important problem is that of testing a given hypothesis (denoted H_0)

against the ensemble of all other possible hypotheses (not H_0). Chapter 11 is devoted to such tests, known as *goodness-of-fit tests*.

10.1. Formulation of a Test

10.1.1. *Basic concepts in testing*

Suppose that it is desired to test an hypothesis H_0 (known as the *null hypothesis*) against an alternative hypothesis H_1, on the basis of some experimental observations. Let X be some function of the observations, called the *test statistic*, and W be the space of all possible values of X. We divide the space W into a *critical region* w and a *region of acceptance* $W - w$, such that observations X falling into w are regarded as suggesting that the null hypothesis H_0 is not true. Choosing a test of H_0 therefore amounts to choosing a test statistic X and a critical region w.

Usually one adjusts the size of the critical region so as to obtain a desired *level of significance* α, (*size of test*) defined as the probability of X falling in w when H_0 is true:

$$P(X \in w \,|\, H_0) = \alpha. \tag{10.1}$$

That is, α is the probability that H_0 would be rejected even if H_0 was indeed true.

The usefulness of a test depends on its ability to discriminate against the alternative hypothesis H_1. The measure of this usefulness is the *power of the test*, defined as the probability $1 - \beta$ of X falling into the critical region if H_1 is true:

$$P(X \in w \,|\, H_1) = 1 - \beta. \tag{10.2}$$

In other words, β is the probability that X will fall in the acceptance region if the alternative hypothesis is true:

$$P(X \in W - w \,|\, H_1) = \beta.$$

When testing hypotheses, we distinguish between two kinds of erroneous conclusions:

(a) *error of the first kind* or loss: rejecting the null hypothesis when it is true. The probability for making such an error is α.

(b) *error of the second kind* or contamination: accepting the null hypothesis when it is false. The probability for making such an error is β.[*]

---

[*]The symbols α and β are conventionally used by most authors for risks of 1st and 2nd kind, but some authors use $1 - \beta$ where we use β.

10.1.2. *Example: Separation of two classes of events*

Suppose one wants to distinguish elastic proton scattering events

$$pp \to pp \quad \text{(the hypothesis under test, } H_0\text{)}$$

from inelastic scattering events

$$pp \to pp\pi° \quad \text{(the alternative hypothesis, } H_1\text{)},$$

in an experimental set-up which measures the proton trajectories, but where the π^0 cannot be detected directly. The separation of events must then be based on the existing kinematic information (the measured angles and momenta of the protons). This information in turn determines the *missing mass* (the total rest energy of all unseen particles) for the event. This missing mass, M, can be chosen as the test statistic.

Under hypothesis H_0 the true value of M would be $M = 0$ and under H_1, $M = M_{\pi^0} = 135$ MeV/c^2. Thus the acceptance region would be chosen around $M = 0$ and the critical region around $M = 135$ MeV/c^2. True elastic events having M close to 135 MeV/c^2 instead of zero (because of measurement errors) would be classified as inelastic, and would constitute the loss. True inelastic events with M close to zero constitute the contamination to the elastic sample.

The exact choice of the critical region depends on how much loss and contamination one is willing to accept. Clearly it is desirable to minimize both the loss α and the contamination β.

Suppose that the p.d.f. of the random variable M is known under each of the two hypotheses: $p(M|H_0)$ and $P(M|H_1)$, illustrated in Fig. 10.1. Such knowledge comes from understanding the functioning of the measuring apparatus, the measurement errors, etc. These distributions are usually known as the *resolution functions*.

Consider the case when only one event has been observed. Judging from Fig. 10.1(a), it would be reasonable to choose a critical region

$$M > M_c$$

such that

$$P(M > M_c) = \int_{M_c}^{\infty} p(M|H_0)dM = \alpha,$$

where α is the desired significance level. One would then classify the event as H_0 if it had a value $M \le M_c$.

From the distribution in Fig. 10.1(b), the probability of the observation falling in

$$P(M < M_c \,|\, H_1) = \int_{-\infty}^{M_c} p(M|\theta_1) = \beta \,.$$

The *power of the test*, $1 - \beta$, is thus calculable.

The usual procedure is to choose α small, say $\alpha = 0.05$ or 0.01. β is then determined by the particular test chosen. In the example of Fig. 10.1, obviously $\beta \gg \alpha$. Thus the probability of accepting the event as elastic when in fact it is inelastic is large. Such an event is M_1 in Fig. 10.1, which would be accepted, although clearly an inelastic event has a large probability to occur with missing mass close to the value M_1.

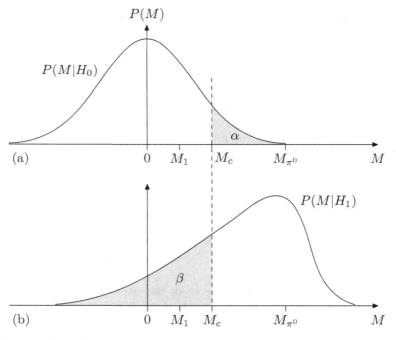

Fig. 10.1. Resolution functions for the missing mass M under the hypotheses H_0 and H_1, with critical region $M > M_c$.

Note that the actual contamination (in number of π^0 events accepted as elastic) is βR, where R is the *a priori* abundance of π^0 events produced. One can therefore accept a large value of β if R is small.

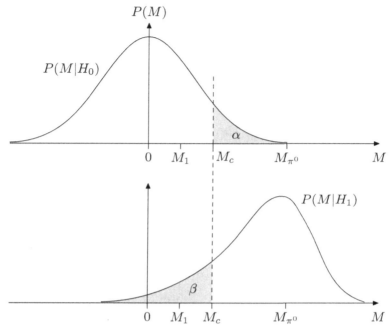

Fig. 10.2. Resolution functions for the missing mass M after modification of the experiment to improve the resolution.

On the basis of the above analysis, one may well decide to improve the apparatus so that the momentum measurements are more precise, to yield a narrower resolution function. Then the situation may improve, as illustrated in Fig. 10.2, where the distributions have become more distinct and the probability β of contamination has become much smaller.

In general, the problem is to construct not only a valid test (where the probabilities α and β are known, as in Figs. 10.1 and 10.2), but the best test. For example, it could be that by using missing momentum or missing energy as a test statistic in place of missing mass, one could obtain better separation between the two hypotheses. In the following sections we shall address ourselves to this problem.

10.2. Comparison of Tests

10.2.1. *Power*

In Section 10.1.1 we defined $p(\theta_1)$, the *power of the test* to discriminate the hypothesis

$$H_0 : \theta = \theta_0$$

against the alternative

$$H_1 : \theta = \theta_1$$

as

$$p(\theta_1) = 1 - \beta.$$

In general (composite hypotheses), we may define the *power function*

$$p(\theta) = 1 - \beta(\theta)$$

for alternative hypotheses specified by the value of θ. By construction, then

$$p(\theta_0) = 1 - \beta(\theta_0) = \alpha.$$

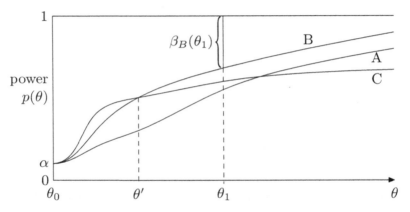

Fig. 10.3. Power functions of tests A, B, and C at significance level α. Of these three tests, B is the best for $\theta > \theta'$. For smaller values of θ, C is better.

Tests may be compared on the basis of their power functions: If the alternative H_1 is simple the best test of H_0 against H_1 at the significance level a is the test with maximum power at $\theta = \theta_1$. In the example shown in Fig. 10.3, test B has maximum power (or minimum contamination β) at all $\theta > \theta'$, and in particular at $\theta = \theta_1$. Test C, again, is best when testing H_0 against an alternative which assumes the true θ to lie in the interval (θ_0, θ').

If, for a given value of θ, a test is at least as powerful as any other possible test, it is called a *most powerful test* at that value of θ. A test which is most powerful for all values of θ under consideration is called *Uniformly Most*

Powerful (UMP). We shall discuss later the conditions under which a *UMP test* exists.

In the example of Fig. 10.4 the test U is UMP. If we construct a most powerful test at a particular value of θ, say θ_1, and this test is independent of θ_1, then such a test is uniformly most powerful. A test other than a UMP

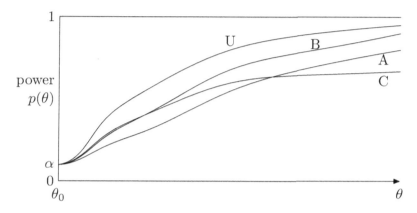

Fig. 10.4. Example of power functions of four tests, one of which (U) is UMP.

test may still be preferred on the grounds of simplicity or *robustness*, i.e. low sensitivity to unimportant changes in the null hypothesis.

10.2.2. *Consistency*

A highly desirable property of a test is that, as the number of observations increases, it should distinguish better between the hypotheses being tested. A test is said to be consistent if the power tends to unity as the number of observations tends to infinity. Mathematically, a test is consistent if

$$\lim_{N \to \infty} P(\mathbf{X} \in w_\alpha | H_1) = 1 \,,$$

where $\mathbf{X}$ is the set of N observations, and w_α is the critical region, of size α, under H_0.

A *consistent test* has a power function which tends to a step function as $N \to \infty$, as in Fig. 10.5.

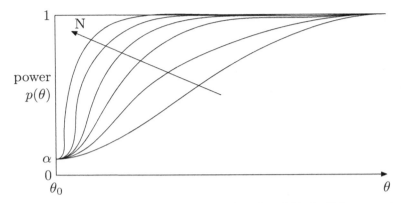

Fig. 10.5. Power function for a consistent test as a function of N. As N increases, it tends to a step function.

10.2.3. *Bias*

Consider the power curve which does not take on its minimum at $\theta = \theta_0$ (curve B in Fig. 10.6). In this case, the probability of accepting $H_0 : \theta = \theta_0$ is greater when $\theta = \theta_1$ than when $\theta = \theta_0$

$$1 - \beta < \alpha.$$

That is, we are more likely to accept the null hypothesis when it is false than when it is true. Such a test is called a *biased* test, and such a property is clearly undesirable in general.

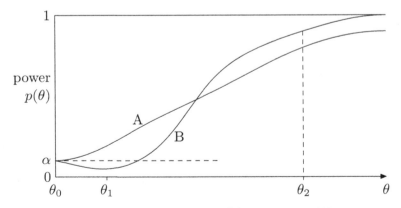

Fig. 10.6. Power functions for biased (B) and unbiased (A) tests.

Particular situations may nonetheless arise when a biased test is preferred because of its power for certain values of θ. In Fig. 10.6 for instance, test B (biased) may be chosen over test A (unbiased) if it is especially important to discriminate against the hypothesis that $\theta = \theta_2$, where test B is more powerful. The user of test B must, however, realize that it does not allow him to distinguish at all between θ_0 and θ_1.

Generalizing to the case where the null hypothesis H_0 includes a range θ_0 out of all possible values θ, the definition of an unbiased test is one for which

$$P(X \in w_\alpha \mid \theta) \leq \alpha \quad \text{for all } \theta \in \theta_0$$

and

$$P(X \in w_\alpha \mid \theta) \geq \alpha \quad \text{for all } \theta \in \theta - \theta_0 .$$

The following corollary follows immediately from the definitions of most powerful and unbiased tests:

If a UMP test exists, and if at least one unbiased test exists, then the UMP test is unbiased. Note that in real-life situations, a uniformly most powerful test usually does not exist, although an unbiased test usually does.

10.2.4. *Choice of tests*

Traditionally, the choice between competing tests is made by comparing the different power curves, as discussed in the previous sections. In following this procedure, it is assumed that the risk of the first kind (or loss) is a given constant α and that the choice is based on minimizing the risk of the second kind (or contamination) β. We feel that this method tends to obscure the basic symmetry between α and β. Moreover it is often unrealistic in experimental situations where, for a given α, the minimum β may still be unacceptable.

Let us therefore take both α and β as variables in comparing different tests between the two hypotheses $H_0 : \theta = \theta_0$ and $H_1 : \theta = \theta_1$. For a given test and a given value of $\alpha = p(\theta_0)$, one can determine $\beta = p(\theta_1)$ as in Section 10.2.1. Then for each test, a curve, giving β as a function of α, can be established as in the example of Fig. 10.7.

The dotted line in Fig. 10.7 corresponds to $1 - \beta = \alpha$, so that all unbiased test curves will lie entirely below this line, passing through the two points $(1,0)$ and $(0,1)$. Since the goal is to have both α and β small, test C is clearly inferior to all others in the figure, for all values of α and β. We will show in Section 10.3.1 that if the hypotheses are simple (completely specified), there exists a test (Neyman–Pearson) which is at least as good as any other test for all α

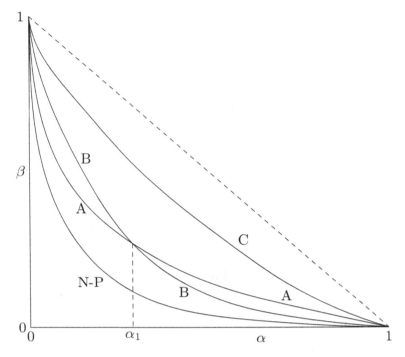

Fig. 10.7. $\alpha - \beta$ symmetric comparison of tests.

and β. If this test is too complicated or does not exist (composite hypotheses) one may be in the position of choosing, for example, between tests A and B. Clearly test A should be chosen for $\alpha < \alpha_1$ and test B for $\alpha > \alpha_1$.

If the hypotheses involved are not completely specified, or if more than two values of θ are possible, Fig. 10.7 becomes a multidimensional diagram with new axes corresponding to θ or to other unspecified parameters in the problem, or each test may be represented by a family of curves in the $\alpha - \beta$ plane.

A minor difficulty arises when θ is discrete, since only a discrete set of α's are then available, and the $\alpha - \beta$ curves are discontinuous.

The above techniques allow one to choose the best test and calculate β if α is given (and if other possible parameters are specified) or to calculate α if β is given. To take the next step and decide on a value of α or β requires knowing the cost of making a wrong decision (see Section 10.6 and Chapter 6).

10.3. Test of Simple Hypotheses

10.3.1. *The Neyman–Pearson test*

The problem of finding the *most powerful test* of hypothesis H_0 against hypothesis H_1 can be stated as the problem of finding the *best critical region* in X-space. Suppose that the random variable $\mathbf{X} = X_1, \ldots, X_N)$ has p.d.f. $f_N(\mathbf{X}|\theta)$, and that the parameter space consists of only two points, θ_0 and θ_1. Suppose also that the functions $f_N(\mathbf{X}|\theta_0)$, $f_N(\mathbf{X}|\theta_1)$ are absolutely continuous with respect to each other.

From the definition (10.2) of the power $(1 - \beta)$ of a test of significance α, we have

$$\int_{w_\alpha} f_N(\mathbf{X}|\theta_0)d\mathbf{X} = \alpha \tag{10.3}$$

$$1 - \beta = \int_{w_\alpha} f_N(\mathbf{X}|\theta_1)d\mathbf{X} . \tag{10.4}$$

We want to find the region w_α, given α, which maximizes $1 - \beta$. Equation (10.4) can be rewritten

$$1 - \beta = \int_{w_\alpha} \frac{f_N(\mathbf{X}|\theta_1)}{f_N(\mathbf{X}|\theta_0)} f_N(\mathbf{X}|\theta_0)d\mathbf{X}$$

$$= E_{w_\alpha} \left(\frac{f_N(\mathbf{X}|\theta_1)}{f_N(\mathbf{X}|\theta_0)} \bigg|_{\theta=\theta_0} \right) .$$

Clearly this will be maximal if and only if w_α is that fraction α of X-space containing the largest values of $f_N(\mathbf{X}|\theta_1)/f_N(\mathbf{X}|\theta_0)$. Thus the best critical region w_α consists of points satisfying

$$\ell_N(\mathbf{X}, \theta_0, \theta_1) \equiv \frac{f_N(\mathbf{X}|\theta_1)}{f_N(\mathbf{X}|\theta_0)} \geq c_\alpha , \tag{10.5}$$

c_α being so chosen that the significance level condition (10.3) is satisfied.

The procedure leads to the criteria:

if $\quad \ell_N(\mathbf{X}, \theta_0, \theta_1) > c_\alpha \quad$ choose $\quad H_1 : \quad f_N(\mathbf{X}|\theta_1)$

if $\quad \ell_N(\mathbf{X}, \theta_0, \theta_1) \leq c_\alpha \quad$ choose $\quad H_0 : \quad f_N(\mathbf{X}|\theta_0) .$

This is known as the *Neyman–Pearson test*. The test statistic ℓ_N is essentially the ratio of the likelihoods for the two hypotheses, and this ratio must be calculable at all points $\mathbf{X}$ of the observable space. The two hypotheses H_0

and H_1 must therefore be completely specified simple hypotheses, and under these conditions, this gives the *best test*.

The corresponding test for non-simple hypotheses is not necessarily optimal, as discussed in Sections 10.4 and 10.5.

10.3.2. *Example: Normal theory test versus sign test*

Suppose that one is measuring a quantity, as for instance beam intensity X. After a modification of the apparatus, one wants to know whether the intensity has increased or not. To find out, one measures the intensity N times and takes the mean $\bar{X}$. Assuming that the measuring errors are Normal and of variance σ^2, this *Normal theory test* can be formulated as follows:

Null hypothesis, $H_0 : \mu = \mu_0$, distribution $N(\mu_0, \sigma^2)$.

Alternative hypothesis, $H_1 : \mu = \mu_1$, distribution $N(\mu_1, \sigma^2)$.

Then the most powerful test is based on the test statistic

$$t = \frac{\exp\left[-\dfrac{1}{2\sigma^2} \sum_{i=1}^{N} (X_i - \mu_1)^2 \right]}{\exp\left[-\dfrac{1}{2\sigma^2} \sum_{i=1}^{N} (X_i - \mu_0)^2 \right]}$$

or

$$\sigma^2 \ln t = \frac{1}{2} N(\mu_0^2 - \mu_1^2) + (\mu_1 - \mu_0) \sum_{i=1}^{N} X_i \, .$$

One sees that t is a monotonic function of

$$z = \frac{\bar{X} - \mu_0}{\dfrac{\sigma}{\sqrt{N}}}$$

which is distributed according to a Normal law $N(\mu_1 - \mu_0, 1)$, if H_1 is true, and $N(0, 1)$ under H_0.

The Neyman–Pearson test is therefore equivalent to a cut on z: if α is the significance level, the critical region is that region of X-space where $z > \lambda_\alpha$,

$$\{X; z > \lambda_\alpha\} \, .$$

λ_α is the α-point of the standard Normal distribution. The *power* of the test is

$$1 - \beta = P[z > \lambda_\alpha | X_i \text{ are } N(\mu_1, \sigma^2)]$$
$$= P[z > \lambda_\alpha | z \text{ is } N(d, 1)]$$
$$= \int_{\lambda_\alpha}^{\infty} \frac{1}{\sqrt{2\pi}} \exp\left[-\frac{(z-d)^2}{2}\right] dz$$
$$= \phi(d - \lambda_\alpha),$$

where

$$d = \frac{\sqrt{N}}{\sigma}(\mu_1 - \mu_0).$$

The power $1 - \beta$ is an increasing function of $(\mu_1 - \mu_0)$ and N. Also $P(\mu_1 = \mu_0) = \alpha$, as one would expect.

An alternative to the Normal theory test is the *sign test*, which can be constructed as follows.

H_0: The N observations have a mean μ_0, and are distributed symmetrically.
H_1: The N observations have some other mean μ_1.

Let N_+ be the number of times that $(X_i - \mu_0)$ is positive, and let $N_- = N - N_+$. The test statistic N_+ is then distributed symmetrically under H_0, as

$$P(N_+ = r) = \binom{N}{r} \left(\frac{1}{2}\right)^N.$$

Suppose for instance that $N = 10$, and that there are two observations of $(X_i - \mu_0)$ being negative, $N_- \leq 2$. Then the significance level α is

$$\alpha = \frac{1}{2^{10}}\left[\binom{10}{0} + \binom{10}{1} + \binom{10}{2}\right] = \frac{7}{128} = 0.0547.$$

Let q_+ and q_- be the probabilities of a positive and a negative sign, respectively, so that $q_+ + q_- = 1$, and let

$$\mu = \mu_1 - \mu_0.$$

Then $q_- = \phi(\mu/\sigma)$, and the power is

$$p(\mu) = q_+^{10} + 10q_+^9 q_- + 45q_+^8 q_-.$$

For $\sigma = 1$, the power takes on the values in Table 10.1.

Table 10.1. Power of the sign test $p(\mu)$ for $N = 10$, $\sigma = 1$.

μ	0	0.5	1
q_-	0.5	0.6915	0.8413
$p(\mu)$	0.055	0.360	0.797

We can now compare the two tests in the above example. It can be shown that for the same α and the same N, the power of the sign test is always less than that of the Normal test. Since the Normal theory test is a Neyman–Pearson test, it is UMP and its power must be at least as great as that of the sign test for all values of μ.

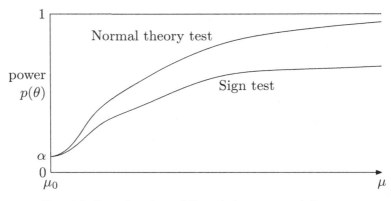

Fig. 10.8. Power functions of Normal theory test and sign test.

10.4. Tests of Composite Hypotheses

In the previous two sections we have seen how to construct the best test between two simple hypotheses. Unfortunately, no such generally optimal technique exists when H_0 or H_1 (or both) is not a completely specified hypothesis. The case of composite hypotheses is, in fact, the more common one, since it arises whenever one is testing a theory with one or more free parameters, or when making a test based on experimental results which involve previously unmeasured phenomena.

In the following we shall distinguish between the case when the hypotheses belong to *one continuous family*, and the case when they belong to *separate families* of hypotheses. In the first case the p.d.f.'s have the same analytic form; the difference between the hypotheses consist in the specification of different regions of the parameter space. Thus the whole *parametric family* can be obtained by continuous variation of the parameters. For example, we may have

$$H_0 : \quad f(\mathbf{X}|\theta) \quad \text{with} \quad \theta < \theta_0 \quad \text{is valid}$$
$$H_1 : \quad f(\mathbf{X}|\theta) \quad \text{with} \quad \theta > \theta_1 \quad \text{is valid}.$$

Hypotheses belonging to separate families cannot be deduced from each other by continuous variation of the parameters. We shall meet an example of such a case in Section 10.5.6. Obviously, one can construct a *comprehensive* parametric family out of two separate families $f(\mathbf{X}|\phi)$ and a new parameter λ,

$$\lambda f(\mathbf{X}|\phi) + (1 - \lambda) g(\mathbf{X}|\psi).$$

Thus, by varying λ from 0 to 1 one goes continuously from g to f. We shall discuss this case in Section 10.5.7.

In the following sections we indicate under what conditions a best (UMP) test does exist for composite hypotheses, and how to construct reasonable tests when no UMP test exists. In view of the importance of composite hypotheses, considerable discussion is then devoted to the properties of the maximum likelihood ratio test, Section 10.5, which is the usual one for such situations.

10.4.1. *Existence of a uniformly most powerful test for the exponential family*

Just as in Darmois' theorem for sufficient statistics (Section 5.3.4), we are again led to the *exponential family* of functions as those which admit a *UMP test*. A general result showing when a UMP test exists, is given by the following theorem:

If $X_1 \cdots X_N$ are independent, identically distributed random variables with a p.d.f. of the form

$$F(X)G(\theta)\exp[A(X)B(\theta)],\tag{10.6}$$

where $B(\theta)$ is strictly monotonic, then there exists a UMP test of $H_0 : \theta = \theta_0$ against $H_1 : \theta > \theta_0$. Note that this test is only *one-sided*: see the following section and [Lehmann, Section 3.6].

Applying the Neyman–Pearson test, the most powerful test of $H_0 : \theta_0$ against $H_1 : \theta_1$, takes the form

$$\frac{G^N(\theta_1)}{G^N(\theta_0)}\exp\left\{\left[\sum_{i=1}^{N}A(X_i)\right][B(\theta_1) - B(\theta_0)]\right\} \gtrless c_\alpha$$

or

$$\sum_{i=1}^{N} A(X_i) \gtrless k_\alpha \,,$$

where k is chosen to give the required significance level when $\theta = \theta_0$. This test is independent of θ_1, and hence the result is UMP.

10.4.2. *One- and two-sided tests*

When the hypotheses under test involve the value of one parameter θ, one usually distinguishes between *one-sided tests* of the form

$$H_0 : \theta = \theta_0 \quad H_1 : \theta > \theta_0$$

and *two-sided tests* of the form

$$H_0 : \theta = \theta_0 \quad H_1 : \theta \neq \theta_0 \,.$$

We have seen for the *exponential family* that a UMP one-sided test exists (Section 10.4.1). For two-sided tests, however, no UMP test generally exists as can be seen from Fig. 10.9. In this figure, curve 1^+ is the power curve for a test which is UMP for $\theta > \theta_0$. Curve 1^- is UMP for $\theta < \theta_0$, and curve 2 is the two-sided test corresponding to the sum of tests 1^+ and 1^-. If the one-sided tests have significance level $\alpha/2$, the two-sided test has level α, and if the size of the two-sided test is reduced to be also $= \alpha/2$, it is seen to be less powerful (on one side) than each of the one-sided tests.

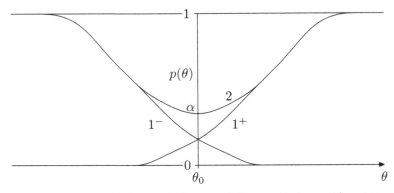

Fig. 10.9. The two-sided test 2 is the sum of the one-sided tests 1^+ and 1^-.

In the case of the exponential family there exists a UMP test for

$$H_0 : \theta \leq \theta_1 \quad \text{or} \quad \theta \geq \theta_2$$

against

$$H_1 : \theta_1 < \theta < \theta_2$$

[Lehmann, Section 3.7]. However, there exists no UMP test of H_1 against H_0.

For a two-sided problem where no UMP test exists, the usual procedure is in fact to combine two one-sided UMP tests.

10.4.3. *Maximizing local power*

If no UMP test exists, an important alternative is to look for a test which is most powerful in the neighbourhood of the null hypothesis. Then we have

$$H_0 : \theta = \theta_0 \quad H_1 : \theta = \theta_0 + \Delta$$

where Δ is small.

Expanding the log-likelihood we have

$$\ln L(\mathbf{X}, \theta_1) = \ln L(\mathbf{X}, \theta_0) + \Delta \frac{\partial \ln L}{\partial \theta}\bigg|_{\theta=\theta_0} + \cdots .$$

If we now apply the Neyman–Pearson lemma [Eq. (10.5)] to H_0 and H_1, the test is of the form:

$$\ln L(\mathbf{X}, \theta_1) - \ln L(\mathbf{X}, \theta_0) \gtrless c_\alpha .$$

or

$$\frac{\partial \ln L}{\partial \theta}\bigg|_{\theta=\theta_0} \gtrless k_\alpha .$$

If the observations are independent and identically distributed, then

$$E\left(\frac{\partial \ln L}{\partial \theta}\bigg|_{\theta=\theta_0}\right) = 0$$

$$E\left[\left(\frac{\partial \ln L}{\partial \theta}\right)^2\right] = +NI ,$$

where N is the number of observations and I is the information matrix (Section 5.2.1). Under suitable conditions (Section 7.3.1), $\partial \ln L/\partial \theta$ is approximately Normal. Hence a locally most powerful test is approximately given by

$$\frac{\partial \ln L}{\partial \theta}\bigg|_{\theta=\theta_0} \gtrless \lambda_\alpha \sqrt{NI} . \tag{10.7}$$

10.5. Likelihood Ratio Test

Let the observations $\mathbf{X}$ have a distribution $f(\mathbf{X}|\boldsymbol{\theta})$, depending on two parameters, $\boldsymbol{\theta} = (\theta_1, \theta_2)$. Then the likelihood function is

$$L(\mathbf{X}|\boldsymbol{\theta}) = \prod_{i=1}^{N} f(X_i|\boldsymbol{\theta}). \tag{10.8}$$

In general, let the total $\boldsymbol{\theta}$-space be denoted θ, and let ν be some subspace of θ, then any test of parametric hypotheses (of the same family) can be stated as

$$H_0 : \boldsymbol{\theta} \in \nu$$

$$\tag{10.9}$$

$$H_1 : \boldsymbol{\theta} \in \theta - \nu.$$

Hypotheses belonging to the same parametric family may, for example, be formulated:

(i) $H_0 : \theta_1 = a, \quad \theta_2 = b$
      $H_1 : \theta_1 \neq a, \quad \theta_2 \neq b$

(ii) $H_0' : \theta_1 = c, \quad \theta_2$ unspecified
      $H_1' : \theta_1 \neq c, \quad \theta_2$ unspecified

(iii) $H_0'' : \theta_1 + \theta_2 = d$
      $H_1'' : \theta_1 + \theta_2 \neq d.$

10.5.1. *Test statistic*

Using the likelihood function (10.8) and the notation in (10.9), we can define the *maximum likelihood ratio*

$$\lambda = \frac{\max\limits_{\boldsymbol{\theta} \in \nu} L(\mathbf{X}|\boldsymbol{\theta})}{\max\limits_{\boldsymbol{\theta} \in \theta} L(\mathbf{X}|\boldsymbol{\theta})}. \tag{10.10}$$

Clearly $0 \leq \lambda \leq 1$. λ is certainly a reasonable test statistic for H_0. For the case of a simple H_0 against a simple H_1, it usually produces the most powerful test (Section 10.3.1). For other than simple case, the success of λ as a test statistic is due to the fact that it is always a function of the sufficient statistic for the problem. The main justification of the method is its past success in producing workable tests with good properties, at least for large sets of observations.

The hypotheses (10.9) usually take the form

$$H_0 : \theta_i = \theta_{i0} \quad i = 1, 2, \ldots, r \quad [\text{say } \boldsymbol{\theta}_r = \boldsymbol{\theta}_{r0}]$$
$$\theta_j \text{ unspecified} \quad j = 1, \ldots, s \quad [\text{say } \boldsymbol{\theta}_s]$$
$$H_1 : \theta_i \neq \theta_{i0}, \quad i = 1, 2, \ldots, r$$
$$\theta_j \text{ unspecified} \quad j = 1, \ldots, s$$

where the null hypothesis specifies fixed values for a subset of the parameters, while the alternative hypothesis H_1 leaves the parameters free to take any values other than those specified in H_0.

The *maximum likelihood ratio* (10.10) is then the ratio between the value of $L(\mathbf{X}|\boldsymbol{\theta})$, maximized with respect to $\theta_j, j = 1, \ldots, s$, while holding fixed $\theta_i = \theta_{io}, i = 1, \ldots, r$, and the value of $L(\mathbf{X}|\boldsymbol{\theta})$, maximized with respect to *all* the parameters.

With this notation the statistic (10.10) becomes

$$\lambda = \frac{\max\limits_{\boldsymbol{\theta}_s} L(\mathbf{X}|\boldsymbol{\theta}_{r0}, \boldsymbol{\theta}_s)}{\max\limits_{\boldsymbol{\theta}_r, \boldsymbol{\theta}_s}{}' L(\mathbf{X}|\boldsymbol{\theta}_r, \boldsymbol{\theta}_s)}$$

or   (10.11)

$$\lambda = \frac{L(\mathbf{X}|\boldsymbol{\theta}_{r0}, \boldsymbol{\theta}_s)}{L(\mathbf{X}|\boldsymbol{\theta}_r', \boldsymbol{\theta}_s)}$$

where $\boldsymbol{\theta}_s''$ is the value of $\boldsymbol{\theta}_s$ at the maximum in the restricted $\boldsymbol{\theta}$ region and $\boldsymbol{\theta}_r', \boldsymbol{\theta}_s'$ are the values of $\boldsymbol{\theta}_r, \boldsymbol{\theta}_s$ at the maximum in the full $\boldsymbol{\theta}$ region.

10.5.2. *Asymptotic distribution for continuous families of hypotheses*

Having seen that the likelihood ratio (10.11) is a reasonable test statistic, it remains to determine the critical region from the distribution of this statistic for the null hypothesis. Often, however, this procedure is difficult, since the distribution is unknown, or is awkward to handle. One can sometimes fall back on Monte Carlo simulation, but this is not always satisfactory. The usual procedure is to consider the *asymptotic distribution* of the likelihood ratio, and use this as an approximation to the true distribution.

Asymptotically, the maximum likelihood estimator $\hat{\theta}$ attains the minimum variance bound (Section 7.4.1). It follows (see Section 7.3.2) that the likelihood function $L(\mathbf{X}|\theta)$ must be of the form

$$\frac{\partial \ln L}{\partial \theta} = -E\left[\frac{\partial^2 \ln L}{\partial \theta^2}\right](\hat{\theta} - \theta)$$

or

$$L(\mathbf{X}|\theta) \propto \exp\left[\frac{1}{2}E\left(\frac{\partial^2 \ln L}{\partial \theta^2}\right)(\hat{\theta} - \theta)^2\right].\tag{10.12}$$

For $\boldsymbol{\theta}$ many-dimensional, Eq. (10.12) takes the form

$$L(\mathbf{X}|\boldsymbol{\theta}) = L(\mathbf{X}|\boldsymbol{\theta}_r, \boldsymbol{\theta}_s)$$

$$\propto \exp\left[-\frac{1}{2}(\hat{\boldsymbol{\theta}} - \boldsymbol{\theta})^T \mathcal{I}(\hat{\boldsymbol{\theta}} - \boldsymbol{\theta})\right]\tag{10.13}$$

where $\mathcal{I}$ is the information matrix for $\boldsymbol{\theta}$,

$$\mathcal{I} = \begin{bmatrix} \mathcal{I}_r & \vdots & \mathcal{I}_{rs} \\ \cdots & \vdots & \cdots \\ \mathcal{I}_{rs}^T & \vdots & \mathcal{I}_s \\ & \vdots & \end{bmatrix}.$$

Thus, Eq. (10.13) can be written

$$L(\mathbf{X}|\boldsymbol{\theta}_r, \boldsymbol{\theta}_s) \propto \exp\Big\{ -\frac{1}{2}[\hat{\boldsymbol{\theta}}_r - \boldsymbol{\theta}_r)^T \mathcal{I}_r (\hat{\boldsymbol{\theta}}_r - \boldsymbol{\theta}_r)$$

$$+ 2(\hat{\boldsymbol{\theta}}_r - \boldsymbol{\theta}_r)^T \mathcal{I}_{rs} (\hat{\boldsymbol{\theta}}_s - \boldsymbol{\theta}_s)$$

$$+ (\hat{\boldsymbol{\theta}}_s - \boldsymbol{\theta}_s)^T \mathcal{I}_s (\hat{\boldsymbol{\theta}}_s - \boldsymbol{\theta}_s)]\Big\}.\tag{10.14}$$

We also know from Section 7.3.1 that the asymptotic distribution of $\hat{\boldsymbol{\theta}} = (\hat{\boldsymbol{\theta}}_r, \hat{\boldsymbol{\theta}}_s)$ is $(r + s)$-dimensional Normal with covariance matrix $\mathcal{I}^{-1}$. Thus we see that the likelihood function tends asymptotically to the p.d.f. of $\hat{\boldsymbol{\theta}}$.

Consider now the value of Eq. (10.13), when $L(\mathbf{X}|\boldsymbol{\theta})$ is maximized with respect to all $\boldsymbol{\theta}$,

$$L(\mathbf{X}|\boldsymbol{\theta}_r', \boldsymbol{\theta}_s').$$

It follows from the fact that L can be written in the form (10.13) that

$$\boldsymbol{\theta}_r' = \hat{\boldsymbol{\theta}}_r \quad \text{and} \quad \boldsymbol{\theta}_s' = \hat{\boldsymbol{\theta}}_s,$$

and that

$$L(\mathbf{X}|\boldsymbol{\theta}'_r, \boldsymbol{\theta}'_s) \propto 1. \tag{10.15}$$

Let the restricted region again be given by $\boldsymbol{\theta}_r = \boldsymbol{\theta}_{r0}$. When $L(\mathbf{X}|\boldsymbol{\theta})$ is maximized over this region, it takes on the value $L(\mathbf{X}|\boldsymbol{\theta}_{r0}, \boldsymbol{\theta}''_s)$. From the fact that L can be written in the form (10.13) it again follows that $\boldsymbol{\theta}''_s = \hat{\boldsymbol{\theta}}_s$. Thus Eq. (10.14) takes the form

$$L(\mathbf{X}|\boldsymbol{\theta}_{r0}, \boldsymbol{\theta}''_s) \propto \exp\left[-\frac{1}{2}(\hat{\boldsymbol{\theta}}_r - \boldsymbol{\theta}_{r0})^T \mathcal{L}_r (\hat{\boldsymbol{\theta}}_r - \boldsymbol{\theta}_{r0})\right]. \tag{10.16}$$

Using Eqs. (10.15) and (10.16) in Eq. (10.11), we write the likelihood ratio

$$\lambda = \frac{L(\mathbf{X}|\boldsymbol{\theta}_r, \boldsymbol{\theta}''_s)}{L(\mathbf{X}|\boldsymbol{\theta}'_r, \boldsymbol{\theta}'_s)}$$

$$= \exp\left[-\frac{1}{2}(\hat{\boldsymbol{\theta}}_r - \boldsymbol{\theta}_{r0})^T \mathcal{L}_r (\hat{\boldsymbol{\theta}}_r - \boldsymbol{\theta}_{r0})\right].$$

It follows from the asymptotic properties of $\hat{\boldsymbol{\theta}}_r$ that the distribution of

$$-2\ln\lambda = [(\hat{\boldsymbol{\theta}}_r - \boldsymbol{\theta}_{r0})^T \mathcal{L}_r (\hat{\boldsymbol{\theta}}_r - \boldsymbol{\theta}_{r0})]$$

is asymptotically a central $\chi^2(r)$ under H_0, and a non-central $\chi^2(r)$ with non-centrality parameter

$$K_1 = (\boldsymbol{\theta}_r - \boldsymbol{\theta}_{r0})^T \mathcal{L}_r (\boldsymbol{\theta}_r - \boldsymbol{\theta}_{r0})$$

under the hypothesis H_1. In other words, if H_0 imposes r constraints on the $s + r$ parameters in H_0 and H_1, then $-2\ln\lambda$ is distributed as a $\chi^2(r)$ under H_0.

10.5.3. *Asymptotic power for continuous families of hypotheses*

The result of the previous section enables one to calculate the power of the test. The critical region is specified by values of $-2\ln\lambda$ greater than the α-point of $\chi^2(r)$, say $\chi^2_\alpha(r)$. After calculating the matrix $\mathcal{L}_r$, the non-centrality parameter K_1 measures the "distance" of the alternative hypothesis from H_0. The power of the test is then given by

$$p(K) = \int_{\chi^2_\alpha(r)}^{\infty} dF_1\left[\chi^2(r, K_1)\right] \tag{10.17}$$

where $F_1[\chi^2(r, K_1)]$ is the cumulative distribution function of the non-central χ^2. One can use the approximation of $\chi^2(N, \Delta)$, given at the end of Section 4.2.3, to yield

$$p(K) \simeq \int_{\left(\frac{r+K_1}{r+2K_1}\right)\chi^2_\alpha(r)}^{\infty} dF_2\left[\chi^2\left(r + \frac{K_1^2}{r + 2K_1}\right)\right]$$

where F_2 is the cumulative distribution of the central χ^2 variable.

It can be shown that the likelihood ratio test is *consistent*. As the number of observations N increases, the power of the test tends to 1, for all alternatives H_1, that is

$$\lim_{N \to \infty} P(\mathbf{X} \in w_\alpha | H_1) = 1,$$

where w_α is the critical region defined by the test.

As a result of the consistency, it follows that the test is *asymptotically unbiased*. However, unbiasedness is a sufficiently desirable property that one would like to have it in small samples as well.

In testing hypotheses about location and scale parameters using the likelihood ratio test, it turns out that if the maximum likelihood estimators used are first adjusted to make them unbiased, the resulting test is also unbiased. This gives another good reason for adjusting estimates for bias.

Note that the "usual" conditions of regularity and independence of the range on the parameters have been assumed. If these do not hold, then under certain conditions [Kendall II, p. 236], [Hogg] the quantity $-2 \ln \lambda$ is distributed as a $\chi^2(2r)$ exactly.

10.5.4. *Examples*

(i) Test of some theories of particle interactions

In this experiment one attempts to determine the ratio X of two decay amplitudes which are complex numbers,

$$X = \frac{A(\text{reaction 1})}{A(\text{reaction 2})}.$$

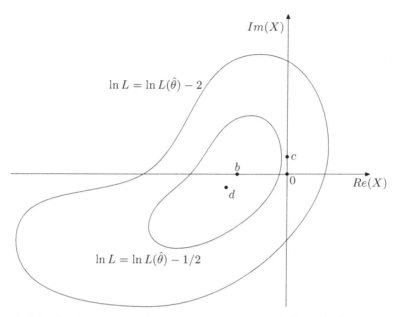

Fig. 10.10. Likelihood contours and constrained maxima for the leptonic decay amplitudes example .

In the general case, X may be any complex number, but there exist three different theories which predict the following for X:

Hypothesis A: If Theory A is valid, $X = 0$.
Hypothesis B: If Theory B is valid, X is real and $Im(X) = 0$.
Hypothesis C: If Theory C is valid, X is purely imaginary and non-zero.

At our current stage of knowledge, the value of X is interesting only in so far as it could distinguish between the hypotheses A, B, C or the general case. Therefore we are concerned here not so much with estimation as with tests of hypotheses.

In this experiment, hypothesis A is simple, and hypothesis B is composite, including hypothesis A as a special case. Hypothesis C is also composite, and separate from A and B. The alternative hypothesis to all these is that Re (X) and Im (X) are both non-zero.

From a set of experimental data the log-likelihood function $\ln L(X)$ is computed and its contours are drawn in the space of $Re(X), Im(X)$ near its maximum. In the example of Fig. 10.10 $X = d$ is the point where $\ln L$ is maximal.

The points $X = b$ and $X = c$ correspond to maxima of L when, respectively, the imaginary or real part of X is constrained to be zero.

The maximum likelihood ratio for hypothesis A versus the general case is

$$\lambda_a = \frac{L(0)}{L(d)}.$$

If hypothesis A is true, $-2\ln\lambda - a$ is distributed asymptotically as a $\chi^2(2)$, and this give the usual test for Theory A.

To test Theory B, the m.l. ratio for hypothesis B versus the general case is

$$\lambda_b = \frac{L(b)}{L(d)}.$$

If B is true, $-2\ln\lambda_b$ is distributed asymptotically as a $\chi^2(1)$. Finally, Theory C can be tested in the same way, using $L(c)$ in place of $L(b)$.

(ii) Test of whether Normally distributed data of unknown variance
     have zero mean

Suppose $X_1,\ldots,X_N$ are $N(\mu,\sigma^2)$; $H_0 : \mu = 0$, σ^2 unknown. $H_1 : \mu \neq 0$, σ^2 unknown. The likelihood function is

$$L(\mu,\sigma^2) = \frac{1}{(2\pi)^{N/2}\sigma^N} \exp\left[-\frac{1}{2}\sum_{i=1}^{N}\frac{(X_i-\mu)^2}{\sigma^2}\right].$$

Under H_0, the maximum likelihood estimate of σ^2 is

$$\hat{\sigma}^2 = \frac{1}{N}\sum_{i=1}^{N}X_i^2.$$

Thus

$$\max_{\sigma^2} L(0,\sigma^2) = \frac{\exp\left(-\frac{N}{2}\right)}{(2\pi)^{N/2}\left(\frac{\Sigma X_i^2}{N}\right)^{N/2}}.$$

Under the general hypothesis, the maximum likelihood estimates are

$$\hat{\mu} = \overline{X}; \quad S^2 = \frac{1}{N}\sum_{i=1}^{N}(X_i-\overline{X})^2$$

and

$$\max_{\mu,\sigma^2} L(\mu,\sigma^2) = \frac{\exp\left(-\dfrac{N}{2}\right)}{(2\pi)^{N/2}\left[\dfrac{1}{N}\sum_{i=1}^{N}(X_i-\overline{X})^2\right]^{N/2}}.$$

Taking the ratio, we have

$$\lambda = \left[\frac{\displaystyle\sum_{i=1}^{N}(X_i-\overline{X})^2}{\displaystyle\sum_{i=1}^{N}X_i^2}\right]^{N/2}.$$

Some algebra gives us

$$\lambda^{2/N} = \frac{1}{\left(1+\dfrac{t^2}{N-1}\right)}$$

where

$$t = \frac{\sqrt{N}\,\overline{X}}{\sqrt{\dfrac{1}{N-1}\sum(X_i-\overline{X})^2}} = \sqrt{N}\frac{\overline{X}}{s}$$

is the Student's t (Section 4.2.4). A test based on λ is therefore equivalent to a test based on the two-sided Student's t-test. The critical region corresponds to small values of λ.

(iii) Test of whether Poisson-distributed data have the same mean

Suppose $X_1,\ldots,X_N$ are independent observations from Poisson distributions with means μ_i. The null hypothesis to be tested is

$$H_0: \quad \mu_1 = \mu_2\cdots = \mu$$

against the alternative

$$H_1: \quad \mu_1 \quad \text{arbitrary, but not all equal}.$$

The likelihood function is

$$L(\mu_1,\ldots,\mu_N) = \prod_{i=1}^{N} e^{-\mu_i}\frac{\mu_i X_i}{X_i!}.$$

Under H_0 the m.l. estimator is $\hat{\mu} = \overline{X}$, and under H_1 the estimates are $\hat{\mu}_i = X_i$. Then

$$\max_{\mu} \, L(\mu) = \frac{e^{-N\overline{X}} \overline{X}^{N\overline{X}}}{\prod\limits_{i=1}^{N} X_i!}$$

and

$$\max_{\mu_i} \, L(\mu_i, \ldots, \mu_N) = e^{-N\overline{X}} \prod_{i=1}^{N} X_i^{X_i} / X_i! \; .$$

Hence the likelihood ratio is

$$\lambda = \frac{\overline{X}^{N\overline{X}}}{\prod\limits_{i=1}^{N} X_i^{X_i}}$$

or

$$\ln \lambda = N\overline{X} \, \ln \, \overline{X} - \sum_{i=1}^{N} X_i \, \ln \, X_i \, . \tag{10.18}$$

This is not the usual test in this situation, which is to consider

$$\sum_{i=1}^{N} (X_i - \overline{X})^2 / \overline{X}$$

which is asymptotically a χ^2 with $(N-1)$ degrees of freedom.

Let us put $X_i = \overline{X} + \sigma_i$ and suppose the σ_i are small. Then from Eq. (10.18) one has

$$\ln \lambda = N\overline{X} \, \ln \, \overline{X} - \sum_{i=1}^{N} (\overline{X} + \delta_i) \, \ln \, (\overline{X} + \delta_i)$$

$$= N\overline{X} \, \ln \, \overline{X} - \sum_{i=1}^{N} (\overline{X} + \delta_i) \, \ln \, \overline{X} - \sum_{i=1}^{N} (\overline{X} + \delta_i) \left(\frac{\delta_i}{\overline{X}} - \frac{\delta_i^2}{2\overline{X}^2} + \cdots \right)$$

$$= -\frac{1}{2} \sum_{i=1}^{N} \frac{\delta_i^2}{\overline{X}} + 0(\delta_i^3) \, .$$

The sum over terms of order δ_i vanishes, by construction. Note now that

$$\frac{\sum\limits_{i=1}^{N} \delta_i^2}{\overline{X}}$$

is the χ^2 statistic. We arrive therefore at the usual result that in the vicinity of the null hypothesis, $-2 \ln \lambda$ is asymptotically equivalent to $\chi^2(N-1)$.

Whether the χ^2 test or the likelihood ratio test is better in small samples is an open question, but probably depends on the alternative hypotheses.

10.5.5. *Small sample behaviour*

Although the asymptotic properties of the likelihood ratio test for continuous families of hypotheses are simple, the small sample behaviour is not so clear.

Under the hypothesis H_0, it is possible to find a closer approximation to the distribution of $-2 \ln \lambda$ than is given by the asymptotic result. It has been shown in Section 9.4.4 that

$$E[-2 \ln \lambda] = r\left(1 + \frac{a}{N} + \cdots\right)$$

where, for the one-parameter case,

$$a = \frac{N}{I^2}\left\{E\left[\left(\frac{\partial \ln L}{\partial \theta}\right)^2 \frac{\partial^2 \ln L}{\partial \theta^2}\right] - \frac{N}{3} \frac{E\left[\left(\frac{\partial \ln L}{\partial \theta}\right)^3\right] E\left[\frac{\partial^3 \ln L}{\partial \theta^3}\right]}{I}\right\} + 1.$$

Then it turns out that

$$-\frac{2 \ln \lambda}{1 + \dfrac{a}{N}}$$

behaves as a $\chi^2(r)$ to order $1/N^2$ [Kendall II, p. 233], [Lawley]. This scaling factor is thus an unequivocal improvement, and should be used when $N < 100$.

A particular case worth attention is *the linear model* (e.g. Sections 8.4.1 to 8.4.3), because then the distribution is known also for finite N. Suppose that some observations Y_i are related to other observations X_i within a random error ε_i,

$$Y_i = \sum_k \theta_k \rho_k(X_i) + \varepsilon_i,$$

where the ε are Normally distributed with mean 0 and constant variance.

Suppose that one wants to test whether the θ_k have specified constant values θ_{0k}, or more generally, whether the components of $\boldsymbol{\theta}$ satisfy some set of r linear constraints

$$\mathcal{A}\,\boldsymbol{\theta} = \mathbf{b},$$

where $\mathcal{A}$ and $\mathbf{b}$ are given. A particular case of this is the test of the significance of the last term of an expansion.

The likelihood for both hypotheses, [H_0: the θ obey the linear constraints, H_1: they are free] is given by

$$L(\mathbf{X}|\boldsymbol{\theta}) = \frac{1}{(2\pi\sigma^2)^{N/2}} \exp\left\{-\frac{1}{2\sigma^2} \sum_{i=1}^{r} \left[Y_i - \sum_{k=1}^{N} \theta_k \rho_k(\mathbf{X}_i)\right]^2\right\}. \qquad (10.19)$$

Let us first take the *variance σ^2 to be known*. The estimates θ_k are given by the least squares solutions, with constraints for H_0, $\hat{\theta}_{0k}$, and without constraints for H_1, $\hat{\theta}_{1k}$. The m.ℓ. ratio is given by

$$-2\ln\lambda = \frac{1}{\sigma^2} \sum_{i=1}^{N} \left[Y_i - \sum_{k=1}^{r} \hat{\theta}_{0k}\rho_k(X_i)\right]^2 - \frac{1}{\sigma^2} \sum_{i=1}^{N} \left[Y_i - \sum_{k=1}^{r} \hat{\theta}_{1k}\rho_k(X_i)\right]^2.$$

It has been shown that the second term can be expressed as a sum of the first term and a term which can also be expressed as a quadratic form in the errors ε. The same result which was true asymptotically in Section 10.5.2, that $-2\ln\lambda$ behaves as a $\chi^2(r)$, now holds exactly [Cochran 1934]. This result is also obtained if the errors ε are not independent, but have a known covariance matrix. Weighted least squares estimates (Section 8.4) are then used in the calculations.

When the variance σ^2 is unknown and has to be estimated from the data, one has to proceed differently. The likelihood function $L(\mathbf{X}|\boldsymbol{\theta},\sigma^2)$ is again given by the right-hand side of Eq. (10.19). The residual sum of squares in the constrained fit is

$$\sum_{i=1}^{N} \left[Y_i - \sum_{k=1}^{r} \hat{\theta}_{0k}\,\rho_k(X_i)\right]^2.$$

This leads to the estimate

$$S_{\omega}^2 = \frac{1}{N} \sum_{i=1}^{N} \left[Y_i - \sum_{k=1}^{r} \hat{\theta}_{0k}\,\rho_k(X_i)\right]^2, \qquad (10.20)$$

where ω denotes the constrained $\boldsymbol{\theta}$ space. Then the maximum likelihood under H_0 becomes

$$\max_{\omega} L(\mathbf{X}|\boldsymbol{\theta},\sigma^2) = \frac{1}{(2\pi)^{N/2}(S_{\omega}^2)^{N/2}} \exp\left(-\frac{N}{2}\right).$$

Similarly, if NS_Ω^2 is the residual sum of squares under the free fit, Ω denoting the full $\boldsymbol{\theta}$ space, the maximum likelihood becomes

$$\max_\Omega L(\mathbf{X}|\boldsymbol{\theta}, \sigma^2) = \frac{1}{(2\pi)^{N/2}(S_\Omega^2)^{N/2}} \exp\left(-\frac{N}{2}\right).$$

The likelihood ratio is then

$$\lambda = \left(\frac{S_\Omega^2}{S_\omega^2}\right)^{N/2}$$

or

$$\lambda^{-\frac{N}{2}} = 1 + \frac{S_\Omega^2 - S_\Omega^2}{S_\Omega^2}.$$

It can then be shown [Cochran 1934] that $(S_\Omega^2 - S_\Omega^2)/\sigma^2$ and S_Ω^2/σ^2 are distributed independently as χ^2 with r and $(N-s)$ degrees of freedom respectively. Therefore

$$F = (\lambda^{-\frac{N}{2}} - 1)\frac{N-s}{r}$$

is distributed as a Fisher–Snedecor $F(r, N-s)$ variable, if the constraints are satisfied. For only one constraint, this simplifies to a Student's test, where [Section 10.5.4 (ii)]

$$t^2 = F(1, N-1).$$

When the constraints do not hold, then the function (10.20) follows a non-central Fisher–Snedecor law which enables us to compute the power of this test.

Example: Representing a curve by a polynomial

As explained in Section 8.4.2, it is best from an estimation point of view, to use polynomials

$$\xi_j(X)$$

orthogonal on the observations X. The polynomial model can then be written

$$Y_i = \sum_j \psi_j \xi_j(X_i) + \varepsilon_i$$

where the ψ are estimated by

$$\hat{\psi}_j = \sum_{i=1}^{N} \xi_j(X_i)Y(X_i)$$

The important property is that the $\hat{\psi}_j$ are independent and, in particular, the estimate $\hat{\psi}_j$ does not depend on the maximum degree of polynomials used.

To what degree should one take the polynomial?

Let S_j be the residual sum of squares which includes all degrees up to and including j. From the above results

$$F = \frac{S_{j-1} - S_j}{S_j}(N - j - 1)$$

is distributed as a Fisher–Snedecor $F(1, N - j - 1)$ variable if the jth degree is not justified.

From the tabulated value of the F distribution one can then give the prescription in Table 10.2.

Table 10.2. Maximum degree needed in polynomial approximation.

$N - j - 1$	2	3	4	6	8	12	20	60	120
Reject j^{th} order to 95% confidence level if F is smaller than	18.5	10.1	7.7	6	5.3	4.7	4.3	4	3.9

Similar methods can be designed for testing the variance of ε. In Table 10.3 we summarize tests of the mean and variance of a Normal distribution (in the one-dimensional case).

10.5.6. *Example of separate families of hypotheses*

Let us now turn from continuous families of hypotheses to separate parametric families. Take for example two families,

$$f(X, \phi) = \phi e^{-\phi X} \, ,$$

or the exponential distribution (4.37), and

$$g(X, \psi) = \psi^2 X e^{-\psi X} \, .$$

The parameters ϕ and ψ are unknown, but they can be estimated from the average $\overline{X}$ of N observations, using the maximum likelihood method:

$$\hat{\phi} = \frac{1}{\overline{X}}, \quad \hat{\psi} = \frac{2}{\overline{X}} \, .$$

Table 10.3. Normal law test: 1 dimension.

	Test of the mean — variance known	Test of the mean — variance unknown
asymp-totic	$-2 \ln \lambda = \dfrac{N}{\sigma^2}(\overline{X} - m_0)^2$ $\sim \chi^2(1)$ if m_0 is the true mean	$-2 \ln \lambda = N \ln \left[\dfrac{\dfrac{1}{N}\sum\limits_i (X_i - \overline{X})^2}{\dfrac{1}{N}\sum\limits_i (X_i - m_0)^2} \right]$ $= N \ln \left(\dfrac{S^2}{\hat{\sigma}^2} \right) \sim \chi^2(1)$ if m_0 is the true mean
finite sample	$-2 \ln \lambda \sim \chi^2(1)$ if m_0 is the true mean	$\lambda^{2/N} = \dfrac{\dfrac{1}{N}\sum\limits_i (X_i - m_0)^2}{\dfrac{1}{N}\sum\limits_i (X_i - \overline{X})^2} = \dfrac{\hat{\sigma}^2}{S^2}$ $= \dfrac{1}{1 + \frac{1}{N-1}t^2}$ $t = \dfrac{\sqrt{N}(\overline{X} - m_0)}{\dfrac{1}{N-1}\sum\limits_i (X_i - \overline{X})^2} \sim St(N{-}1)$ if m_0 is the true mean. $St(N{-}1)$ is Student's law of $N{-}1$ d.f.

	Test of the variance — mean known	Test of the variance — mean unknown
asymp-totic	$-2 \ln \lambda = \dfrac{N}{2}\left[\dfrac{\hat{\sigma}^2}{\sigma_0^2} + \ln \dfrac{\hat{\sigma}^2}{\sigma_0^2} - 1 \right] \sim \chi^2(1)$ if σ_0^2 is the true variance $\hat{\sigma}^2 = \dfrac{1}{N}\sum\limits_i (X_i = m_0)^2$	$-2 \ln \lambda = \dfrac{N}{2} \ln \dfrac{S^2}{\sigma_0^2} \sim \chi^2(1)$ if σ_0^2 is the true variance $S^2 = \dfrac{1}{N}\sum\limits_i (X_i - \overline{X})^2$
finite sample	$-2 \ln \lambda$ is a monotonic function of $\dfrac{\hat{\sigma}}{\sigma_0^2}$. $N\dfrac{\hat{\sigma}^2}{\sigma_0^2} \sim \chi^2(N)$ if σ_0^2 is the true variance	$-2 \ln \lambda$ is a monotonic function of $\dfrac{S^2}{\sigma_0^2}$ $(N-1)\dfrac{S^2}{\sigma_0^2} \sim \chi^2(N-1)$ if σ_0^2 is the true variance

Then the maximum likelihood ratio statistic gives us

$$-2\ln\lambda = 2\ln\frac{\prod\limits_{i=1}^{N} f(X_i,\hat{\phi})}{\prod\limits_{i=1}^{N} g(X_i,\hat{\psi})}$$

$$= -2\left[N\ln\hat{\phi} - \hat{\phi}N\overline{X} - 2N\ln\hat{\psi} - \sum_{i=1}^{N}\ln X_i + \hat{\psi}N\overline{X}\right] \quad (10.21)$$

or

$$-2N^{-1}\ln\lambda = -2\left[\ln\overline{X} - N^{-1}\sum_{i=1}^{N}\ln X_i + 1 - 2\ln 2\right]. \quad (10.22)$$

In this form we can now apply the central limit theorem to each sum, giving in the limit for large N,

$$-2N^{-1}\ln\lambda \to -2[\ln E(X) - E(\ln X) + 1 - 2\ln 2]. \quad (10.23)$$

The test statistic (10.22) is distributed around the value (10.23) according to a Normal law with variance

$$\sigma^2_{-2N^{-1}\ln\lambda}$$

$$= \frac{4}{N}\left\{\frac{\sigma_X^2}{E(X)^2} + E[(\ln^2(X)] - [E^2(\ln(X)] - 2\frac{E(X\ln X)}{E(X)} + 2E(\ln X)\right\}$$

$$(10.24)$$

where we have used the asymptotic expansion

$$\ln\overline{X} = \ln[E(X)] + \frac{1}{E(X)}[\overline{X} - E(X)] + \cdots.$$

Obviously the value of the expectations entering the above expressions depends on which hypothesis is true. All the expectations in the expression (10.24) can be evaluated analytically (we leave this as an exercise to the reader), with the results

$$\sigma^2_{-2N^{-1}\ln\lambda} = \begin{cases} \dfrac{4}{N}\left(\dfrac{\pi^2}{6} - 1\right) & \text{if } f \text{ is true} \\[3mm] \dfrac{4}{N}\left(\dfrac{\pi^2}{6} - \dfrac{1}{2}\right) & \text{if } g \text{ is true.} \end{cases} \quad (10.25)$$

10.5.7. *General methods for testing separate families*

(i) In the previous section we have seen that the distribution of the statistic (10.21) is in general not a χ^2 when f is true, but depends on N (and on the true value of the parameters [Cox]). This should not be taken to imply, however, that the likelihood ratio is worthless for testing separate families of hypotheses. If only its distribution is properly taken into account, it may well be used for constructing other test statistics. For instance [Cox], the test function

$$\frac{-2N^{-1}\ln\lambda - E(-2N^{-1}\ln\lambda)}{\sigma_{-2N^{-1}\ln\lambda}} \tag{10.26}$$

is asymptotically Normal if the expectations are taken with respect to the true hypothesis.

For the distributions f and g in the preceding section, the function (10.26) leads to a risk of first kind α, corresponding to the λ_α-point

$$\lambda_\alpha = \frac{1 + C - 3\ln 2}{\dfrac{1}{N}\left(\dfrac{\pi^2}{6} - 1\right)},$$

where $C \sim 0.577$ is the Euler constant, and a risk of second kind β, corresponding to the λ_β-point

$$\lambda_\beta = \frac{\dfrac{2\lambda_\alpha}{N}\left(\dfrac{\pi^2}{6} - 1\right) + 1 + \ln 2}{\dfrac{2}{N}\left(\dfrac{\pi^2}{6} - \dfrac{1}{2}\right)}.$$

When the moments of $-2\ln\lambda$ depend on the true value of the parameters, one may construct similar tests, replacing the unknown values by their estimates (and modifying the variance suitably).

(ii) A different method which avoids the necessity to compute these moments, is to construct *comprehensive* parametric hypotheses. For testing $f(\mathbf{X}, \boldsymbol{\phi})$ against $g(\mathbf{X}, \boldsymbol{\psi})$, a comprehensive family is

$$h(\mathbf{X}, \theta, \boldsymbol{\phi}, \boldsymbol{\psi}) = (1 - \theta)f(\mathbf{X}, \boldsymbol{\phi}) + \theta g(\mathbf{X}, \boldsymbol{\psi}).$$

The maximum likelihood ratio test can now be applied, testing the hypothesis

$$H_0: \quad \theta = 0, \quad \boldsymbol{\phi}, \boldsymbol{\psi} \quad \text{unspecified}$$

against the hypothesis

$$H_1 : \quad \theta \neq 0, \quad \phi, \psi \quad \text{unspecified}.$$

The test statistic is

$$-2 \ln \lambda = -2 \ln f(\mathbf{X}, \hat{\phi}') + 2 \ln[(1 - \hat{\theta}'')f(\mathbf{X}, \hat{\phi}'') + \hat{\theta}'' g(\mathbf{X}, \hat{\psi}'')], \quad (10.27)$$

where $\hat{\phi}'$ is the maximum likelihood estimate of ϕ, over the restricted space $\theta = 0$, while $\hat{\theta}''$, $\hat{\theta}''$, $\hat{\psi}''$ are the maximum likelihood estimates over the complete parameter space. It can now be shown that under the hypothesis H_0, $-2 \ln \lambda$ is distributed asymptotically as a $\chi^2(1)$ since one constraint, namely $\theta = 0$, has been imposed on the parameter space.

The power of the test can be evaluated from the fact that under the alternative hypothesis $-2 \ln \lambda$ behaves asymptotically as a non-central $\chi^2(1)$, and non-centrality parameter θ^2/I, where

$$I = E\left\{ \frac{[f(\mathbf{X}, \phi) - g(\mathbf{X}, \psi)]^2}{[(1 - \theta)f(\mathbf{X}, \phi) + \theta g(\mathbf{X}, \psi)]^2} \right\}.$$

In particular, for the alternative hypothesis $\theta = 1$,

$$I = E\left\{ \left[\frac{f(\mathbf{X}, \phi)}{g(\mathbf{X}, \psi)} - 1 \right]^2 \right\}.$$

Since this test compares $f(\mathbf{X}, \phi)$ with a mixture of families, it is not likely to be very powerful. That is, precision is lost in the formation of the comprehensive family.

(iii) Consider the direct likelihood ratio,

$$\ell = \frac{f(\mathbf{X}, \hat{\phi})}{g(\mathbf{X}, \hat{\psi})}, \quad (10.28)$$

where $\hat{\phi}, \hat{\psi}$ are the maximum likelihood estimates obtained by considering the two families separately. Now $g(\mathbf{X}, \hat{\psi})$ is the maximum of $h(\mathbf{X}, \theta, \phi, \psi)$ over the restricted region $\theta = 1$. It therefore follows that

$$g(\mathbf{X}, \hat{\psi}) \leq h(\mathbf{X}, \hat{\theta}'' \hat{\phi}'', \hat{\psi}'') \quad (10.29)$$

using the same notation as in Eq. (10.27). Combining Eqs. (10.28) and (10.29), we have

$$-2 \ln \ell = 2 \ln g(\mathbf{X}, \hat{\psi}) - 2 \ln f(\mathbf{X}, \hat{\phi})$$
$$\leq 2 \ln h(\mathbf{X}, \theta'', \phi'', \psi'') - 2 \ln f(\mathbf{X}, \hat{\phi}),$$

or
$$-2 \ln \ell \leq -2 \ln \lambda \,.$$

Then, if $-2 \ln \ell \geq k_\alpha$, it follows that $-2 \ln \lambda > k_\alpha$.

Consequently *a sufficiently large value of* $-2 \ln \ell$ *guarantees that the hypothesis* $f(\mathbf{X}, \phi)$ *would be rejected by the comprehensive test.* For example, at the 0.2% level, $k_\alpha = 9$. Then if $-\ln \ell \geq 4.5$, one can reasonably reject the hypothesis that the distribution was $f(\mathbf{X}, \phi)$. Such a criterion may be useful in practice, especially for large numbers of observations. In the example of the previous section this criterion is fulfilled for large N, and thus the test is consistent.

10.6. Tests and Decision Theory

We can now continue the discussion of Section 6.3.3 on what decision to take, given a test which distinguishes between the hypotheses H_0 and H_1. In Section 6.3.3 we concluded in the case of simple hypotheses that

  (i) the Neyman–Pearson test may be interpreted as a Bayesian decision rule;
 (ii) the choice of a significance level α is equivalent to giving *a priori* preference to the hypothesis H_0;
(iii) the classical test theory represents a particular formulation of a decision problem and is notably responsible for the asymmetry between H_0 and H_1;
(iv) the practical advantages of decision theory lie in its flexibility compared to classical theory and in the clear display of assumptions underlying the method.

In the next sections we shall demonstrate further the power of decision theory.

10.6.1. *Bayesian choice between families of distributions*

Suppose that one has to choose which of two families of distributions $f(\mathbf{X}|\phi)$ and $g(\mathbf{X}|\psi)$ best fits the data $\mathbf{X}$. Let us construct a new family of distributions, general enough to contain both f and g:
$$h(\mathbf{X}|\theta, \phi, \psi) = \theta f(\mathbf{X}|\phi) + (1 - \theta)g(\mathbf{X}, \psi) \,,$$

where $\theta = 0$ or 1 (but not intermediate). The decisions to be taken are: "Which family, if either, did the observations come from, and from which member of the family?"

Let us categorize the possible decisions as

d_0 : No choice is possible; results are ambiguous
d_1, ϕ^* : Family was $f(\mathbf{X}|\phi)$, with $\phi = \phi^*$
d_2, ψ^* : Family was $g(\mathbf{X}|\psi)$, with $\psi = \psi^*$.

Note that the last two decisions each represent a continuum of possible decisions.

In order to choose which decision to make, one needs a criterion. This may be taken to be minimum cost, with the cost function[*] $L(d, \theta, \phi, \psi)$ defined as in Table 10.4 below. This means that if (θ_1, ϕ) is the true state of nature, one loses β_1 by the decision "X is ambiguous". One loses $\alpha_1(\phi^* - \phi)^2$ if one chooses the correct family but the wrong parameter ϕ^*; etc.

Table 10.4. A cost function.

Decisions	True state of nature	
	$\theta = \theta_1 = 1, \phi$	$\theta = \theta_2 = 0, \psi$
d_0	β_1	β_2
d_1, ϕ^*	$\alpha_1(\phi^* - \phi)^2$	γ_1
d_2, ψ^*	γ_2	$\alpha_2(\psi^* - \psi)^2$

One must also specify one's degree of belief about the parameters θ, ϕ, ψ, in terms of a prior distribution. In this example we may assume that one expects family f with probability μ, and g with probability $1 - \mu$, while the prior distributions of ϕ and ψ are independent and given by $\pi(\phi)$ and $\pi(\psi)$.

The joint prior density for θ, ϕ, ψ can now be written

$$\pi(\theta, \phi, \psi) = \mu\pi(\phi)\pi(\psi)\delta(\theta) + (1 - \mu)\pi(\phi)\pi(\psi)\delta(1 - \theta) . \qquad (10.30)$$

The posterior density for θ, ϕ, and ψ, by Bayer theorem, can be written

$$\pi(\theta, \phi, \psi|\mathbf{X}) = k[\mu f(\mathbf{X}|\phi)\pi(\phi)\pi(\psi)\sigma(\theta) + (1 - \mu)g(\mathbf{X}|\psi)\pi(\phi)\pi(\psi)\sigma(1 - \theta)] . \qquad (10.31)$$

---

[*]Note that the *cost function* was called *loss function* in Chapter 6. We want to avoid double usage of the word *loss* in this chapter. Therefore *loss* is taken to have the meaning of error of first kind.

The normalizing constant k is given by

$$\frac{1}{k} = \mu F(\mathbf{X}) + (1 - \mu)G(\mathbf{X})$$

where

$$F(\mathbf{X}) = \int f(\mathbf{X}|\phi)\pi(\phi)d\phi \qquad (10.32)$$

and

$$G(\mathbf{X}) = \int g(\mathbf{X}|\psi)\pi(\psi)d\psi . \qquad (10.33)$$

One can now compute the posterior cost for each possible decision, averaging over the posterior distribution for (θ, ϕ, ψ). Given decision d_0, the posterior cost is

$$E[L(\theta, \phi, \psi, d_0|\mathbf{X})] = \int L(d_0, \theta, \phi, \psi)\pi(\theta, \phi, \psi|\mathbf{X})d\theta \, d\phi \, d\psi$$

$$= \frac{\beta_1\mu F(\mathbf{X}) + \beta_2(1 - \mu)G(\mathbf{X})}{\mu F(\mathbf{X}) + (1 - \mu)G(\mathbf{X})} . \qquad (10.34)$$

Similarly, the posterior costs, given decisions (d_1, ϕ^*) and (d_2, ψ^*), are

$$E[L(\theta, \phi^*, \psi, d_1|\mathbf{X}) = k\left[\int \alpha_1(\phi^* - \phi)^2\mu f(\mathbf{X}|\phi)\pi(\phi)d\phi + \gamma_1(1 - \mu)G(\mathbf{X})\right] \qquad (10.35)$$

$$E[L(\theta, \phi, \psi, ^* d_2|\mathbf{X})] = k\left[\gamma_2\mu F(\mathbf{X}) + \int \alpha_2(\psi^* - \psi)^2(1 - \mu)g(\mathbf{X}|\psi)\pi(\psi)d\psi\right] .$$

For each $\mathbf{X}$, one now chooses the decision which minimizes the posterior loss. For $E[L(\theta, \phi^*, \psi, d_1|\mathbf{X})]$ it is clear that the value ϕ^* to be considered is that which minimizes Eq. (10.35). This minimum is obtained when

$$\phi^* = \frac{\int \phi f(\mathbf{X}|\phi)\pi(\phi)d\phi}{\int f(\mathbf{X}|\phi)\pi(\phi)d\phi}$$

$$= E(\phi|\mathbf{X}), \qquad (10.36)$$

the mean of the posterior distribution for ϕ. Similarly

$$\psi^* = E(\psi|\mathbf{X}). \qquad (10.37)$$

(This result is a general one, when the cost is parabolic, proportional to the squared deviation.) At these decision values ϕ^*, ψ^*, the posterior costs become

$$E[L(\theta, \phi^*, \psi, d_1|\mathbf{X})] = \frac{\alpha_1 V(\phi|\mathbf{X})\mu F(\mathbf{X}) + \gamma_1(1-\mu)G(\mathbf{X})}{\mu F(\mathbf{X}) + (1-\mu)G(\mathbf{X})} \tag{10.38}$$

and

$$E[L(\theta, \phi, \psi^*, d_2|\mathbf{X})] = \frac{\gamma_2 \mu F(\mathbf{X}) + \alpha_2 V(\psi|\mathbf{X})(1-\mu)G(\mathbf{X})}{\mu F(\mathbf{X}) + (1-\mu)G(\mathbf{X})} \tag{10.39}$$

where $V(\psi|\mathbf{X})$ is the variance of the posterior distribution for ϕ.

The decision rule is now clear. Given observations $\mathbf{X}$, one chooses the minimum of Eqs. (10.34), (10.38) and (10.39). If a non-ambiguous decision is chosen the parameter value is given by Eq. (10.36), or Eq. (10.37). This decision rule is illustrated below.

Example: Let the two families of distributions be given by

$$f(X|\phi) = \phi e^{-\phi X}$$

$$g(X|\psi) = \psi^2 X e^{-\psi X},$$

and the prior distributions of ϕ and ψ by

$$\pi(\phi) = a^2 \phi e^{-a\phi}$$

$$\pi(\psi) = b^2 \psi e^{-b\psi}. \tag{10.40}$$

The choice (10.40) only implies that ϕ and ψ are known not to be zero, and of order of magnitude

$$E(\phi) = \frac{2}{a}, \quad E(\psi) = \frac{2}{b}.$$

Then it is easy to show that Eq. (10.32) becomes

$$F(\mathbf{X}) = a^2 \Gamma(N+2)(a+N\overline{X})^{-N-2},$$

where $\mathbf{X}$ is the average of N observations X_i. It follows that Eq. (10.36) becomes

$$\phi^* = (N+2)/(a+N\overline{X}).$$

Thus the variance of the posterior distribution for ϕ is

$$V(\phi|\mathbf{X}) = E[(\phi - \phi^*)^2] = (N+2)(a+N\overline{X})^{-2}.$$

Similarly, Eqs. (10.33) and (10.37),

$$G(\mathbf{X}) = b^2(b + N\overline{X})^{-2N-2} \prod_{i=1}^{N} X_i \Gamma(2N + 2)$$

$$\psi^* = (2N + 2)/(b + N\overline{X})$$

lead to the variance of the posterior distribution for ψ

$$V(\psi|\mathbf{X}) = E[(\psi - \psi^*)^2] = (2N + 2)(b + N\overline{X})^{-2}.$$

The posterior costs (10.34), (10.38) and (10.39) then become

$$E[L(d_0|\mathbf{X})] = \frac{1}{D}\left[\frac{\alpha_1\mu a^2\Gamma(N+2)}{(a+N\overline{X})^{N+2}} + \frac{\beta_2(1-\mu)b^2\Gamma(2N+2)}{(b+N\overline{X})^{2N+2}}\prod_{i=1}^{N}X_i\right], \quad (10.41)$$

$$E[L(d_1|\mathbf{X})] = \frac{1}{D}\left[\frac{\alpha_1\mu a^2\Gamma(N+3)}{(a+N\overline{X})^{N+4}} + \frac{\gamma_1(1-\mu)b^2\Gamma(2N+2)}{(b+N\overline{X})^{2N+2}}\prod_{i=1}^{N}X_i\right], \quad (10.42)$$

$$E[L(d_1|\mathbf{X})] = \frac{1}{D}\left[\frac{\gamma_1\mu a^2\Gamma(N+2)}{(a+N\overline{X})^{N+2}} + \frac{\gamma_1(1-\mu)b^2\Gamma(2N+3)}{(b+N\overline{X})^{2N+3}}\prod_{i=1}^{N}X_i\right], \quad (10.43)$$

with the common denominator

$$D = \frac{\mu a^2\Gamma(N+2)}{(a+N\overline{X})^{2N+2}} + \frac{(1-\mu)b^2\Gamma(2N+2)}{(b+N\overline{X})^{2N+2}}.$$

In order to gain some insight into the problem we assume that it is known that

$$a \approx 2b.$$

For large N we can use the Stirling formula for $\Gamma(N)$, and appropriate exponentials approximating the different powers of $(a + N\overline{X})$ and $(b + N\overline{X})$. Equations (10.41) to (10.43) then simplify to

$$E[L(d_0|\lambda)] \approx \frac{1}{D}\left[\beta_1\mu + \beta_2(1-\mu)e^{-\ln\lambda}\right] \quad (10.44)$$

$$E[L(d_1|\lambda)] \approx \frac{1}{D}\left[\frac{\alpha_1\mu}{N\overline{X}^2} + \gamma_1(1-\mu)e^{-\ln\lambda}\right], \quad (10.45)$$

$$E[L(d_2|\lambda)] \approx \frac{1}{D}\left[\gamma_2\mu + \alpha_2(1-\mu)\frac{2}{N\overline{X}^2}e^{-\ln\lambda}\right], \quad (10.46)$$

where

$$D \approx \mu + (1 - \mu)e^{-\ln \lambda}$$

and

$$\ln \lambda = N(\ln \overline{X} - N^{-1} \sum_{i=1}^{N} \ln X_i + 1 - 2\ln 2).$$

Thus it emerges that in the case $\alpha_1 = \alpha_2 = 0$ (no loss from estimation), this Bayesian test is a likelihood ratio test, since λ is the likelihood ratio

$$\lambda = f(\mathbf{X}, \hat{\phi})/g(\mathbf{X}, \hat{\psi}).$$

In Fig. 10.11 we plot the square brackets of Eqs. (10.44), (10.45), and (10.46) for this case. Not surprisingly, one would always prefer to take a decision (d_1, or d_2) when the cost incurred in taking no decision (d_0) is large enough.

The effect of α_1 or α_2 being non-zero is to increase the corresponding risks, because of the cost due to fluctuations in the estimate.

Note that ϕ^* and ψ^* approach the corresponding maximum likelihood estimates $\hat{\phi}$ and $\hat{\psi}$ as N goes to infinity. This is due to the decreasing importance of the prior knowledge for the choice of the estimate.

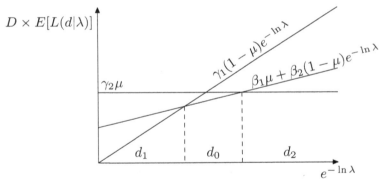

Fig. 10.11. Dependence of the posterior cost on the likelihood ratio for three decisions in the case described in the next.

10.6.2. *Sequential tests for optimum number of observations*

In constructing the cost function for a set of decision rules, it is natural to take into account also the cost of data acquisition. The optimal decision (in the Bayesian sense) then becomes a function of the number of observations. Classical decision rules with fixed number of observations) are not necessarily optimal, since they do not make use of the possibility to let the course of the experiment depend on the previous observations. Such procedures are called *sequential tests*.

Example

Suppose one has to make a decision to accept or reject a roll of film on the basis of how many good or bad pictures it contains.

The usual procedure is to decide on some proportion of good pictures at which the roll will be acceptable. Then a sample of N pictures is measured from the roll. On the basis of the proportion of good measurements, the roll is accepted or rejected.

The sampling plan is as follows:
Measure N pictures. Reject the roll if there are c or more bad pictures, otherwise accept the roll. This procedure is illustrated in Fig. 10.12.

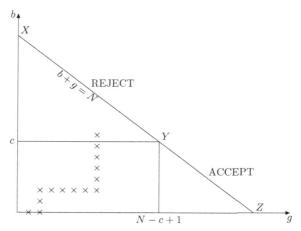

Fig. 10.12. Sampling plan for the usual sequential test in the film roll example. The number of good pictures $= g$, bad pictures $= b$.

The decision rule corresponds to dividing the line $XZ[b + g = N]$ into two regions, XY and YZ. If the observation (b, g) lies on XY, then reject the roll; if (b, g) is on YZ, accept the roll.

It is clear that an alternative procedure is as follows. Measure pictures on the roll one by one. Continue sampling until *either* c bad pictures are obtained in which case reject, *or* $N - c + 1$ good measurements are made, in which case accept the roll.

In the diagram, this corresponds to drawing the lines $b = c$ and $g = N - c + 1$. Clearly the number of pictures considered in the sequential plan will be fewer than in the first plan. In the case shown, the decisions reached would be identical, and obviously, the first procedure may in practice be carried out according to the second rule. It is therefore clear that the optimal decision has to be looked for in the class of sequential tests.

Let us consider a sequential test to decide between two simple hypotheses, $H_0 : \theta = \theta_0$, and $H_1 : \theta = \theta_1$. The following costs will contribute to the total loss function

$\ell_0 :$ cost due to error of first kind* (wrongly rejecting H_0)

$\ell_1 :$ cost due to error of second kind   (wrongly rejecting H_1)

$c :$ cost of acquistion per event.

The total loss function for N events is then given by Table 10.5 below.

Table 10.5. Loss function for two simple hypotheses.

Decisions	True state of nature	
	H_0	H_1
H_0 is true	Nc	$\ell_1 +$ Nc
H_1 is true	$\ell_0 +$ Nc	Nc

Let the probabilities of the errors of first and second kind be α and β, respectively, and let the expectation of the number of events necessary to take a decision be $E_1(\delta, N)$ when H_i is true. Taking the *a priori* probability of H_0 to be μ, the *posterior risk* for the decision rule δ becomes

$$r_\mu(\delta) = \mu[\alpha(\delta, N)\ell_0 + cE_0(\delta, N)] + (1 - \mu)[\beta(\delta, N)\ell_1 + cE_1(\delta, N)]. \quad (10.47)$$

The optimal decision is then at the minimum of Eq. (10.47) with respect to α, β or N.

Among non-sequential tests for N fixed, the Neyman–Pearson lemma, which is based on the likelihood ratio, is optimal (in the classical sense):

$$\lambda_N = \frac{L(\mathbf{X}|\theta_1)}{L(\mathbf{X}|\theta_0)} = \prod_{i=1}^{N} \frac{f(\mathbf{X}_i|\theta_1)}{f(\mathbf{X}_i|\theta_0)}. \quad (10.48)$$

The if $\lambda_N < c_\alpha$ one chooses H_0, and otherwise H_1, c_α being defined as in Section 10.3.1. A. Wald has suggested [Wald 1947] the following generalization:

choose two constants A and B; continue sampling while

$$B < \lambda_N < A \tag{10.49}$$

and stop when the inequalities are broken. If $\lambda_N < B$, decide for $H_0 : \theta = \theta_0$, and if $\lambda_N > A$, decide for $H_1 : \theta = \theta_1$.

The following fundamental theorems then hold [Lehmann, p. 98–115]:

(i) Decision theory approach

The sequential probability ratio test minimizes the posterior risk (10.47) with

$$A = \frac{\mu}{1-\mu} \cdot \frac{1-\mu_1}{\mu_1}$$

$$B = \frac{\mu}{1-\mu} \cdot \frac{1-\mu_0}{\mu_0}$$

where μ_i are the *a priori* probabilities corresponding to the case where the optimal decision would be to reject H_i without any observation. (They always exist and $\mu_0 \leq \mu_1$.)

(ii) Classical theory approach

Among all tests (sequential or not) for which

$$P(\text{rejecting } H_0|H_0) \quad \leq \quad \alpha \quad \text{(loss)}$$
$$P(\text{accepting } H_0|H_1) \quad \leq \quad \beta \quad \text{(contamination)}$$

and for which $E_0(N)$ and $E_1(N)$ are finite, the sequential probability ratio test minimizes at the same time $E_0(N)$ and $E_1(N)$.

Moreover, one has the approximate equalities

$$A \simeq \frac{1-\beta}{\alpha} \quad \text{and} \quad B \simeq \frac{\beta}{1-\alpha}. \tag{10.50}$$

The comparison between the two approaches may be made [Lehmann] in a way similar to Section 6.3.3.

When the hypotheses are not simple, the optimal properties of the Neyman–Pearson test do not extend to the corresponding sequential test. This is especially unfortunate since the most interesting application is to continuous parameter families of hypotheses. The usual procedure is then to reformulate the problem so that the "simple hypothesis" approach may be used although it is no longer necessarily optimal. For example, a common situation is that of deciding between

$$H_0 : \theta \geq \theta_0 \quad \text{and} \quad H_1 : \theta < \theta_0 .$$

This can be reformulated as

$$H_0 : \theta > \theta_0 \quad \text{and} \quad H_1 : \theta < \theta_1 \,, \quad \theta_1 < \theta_0 \,.$$

Example

Let us return to the film example using this trick. θ is then the proportion of bad pictures. The roll of film has to be accepted or rejected with significance level α, and power $1 - \beta$. Equation (10.48) becomes

$$\lambda_N = \frac{\binom{N}{b} \theta_1^b (1 - \theta_1)^{N-b}}{\binom{N}{b} \theta_0^b (1 - \theta_0)^{N-b}} \,.$$

The sequential probability ratio test is given by Eq. (10.49).

$$B < \left(\frac{\theta_1}{\theta_0}\right) \left(\frac{1 - \theta_1}{1 - \theta_0}\right)^{N-b} < A \,.$$

Then

$$\ln B < b \ln \frac{\theta_1}{\theta_0} + (N - b) \ln \frac{1 - \theta_1}{1 - \theta_0} < \ln A$$

where B and A are given by Eqs. (10.50). The acceptance and rejection regions of this test are shown in Fig. 10.13.

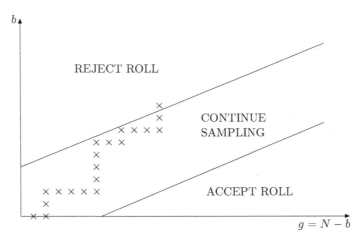

Fig. 10.13. Sample plan for the film problem, reformulated as a Neyman–Pearson test of simple hypotheses.

10.6.3. *Sequential probability ratio test for a continuous family of hypotheses*

Suppose that a sequential probability ratio test for two values, θ_0 and θ_1 of the parameters has been designed. The power function (then also called operating characteristic) for the test for discriminating $f(X|\theta_0)$ against $f(X|\theta)$ may be shown [Lehmann] to be

$$\beta(\theta) \simeq \frac{A^{h(\theta)} - 1}{A^{h(\theta)} - B^{h(\theta)}}$$

where $h(\theta)$ is the non-zero solution of

$$\int \left[\frac{f(X|\theta_1)}{f(X|\theta_0)}\right]^{h(\theta)} f(X|\theta)dX = 1.$$

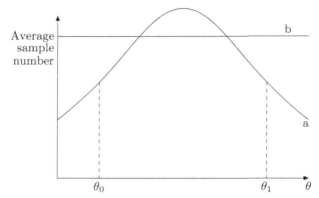

Fig. 10.14. Example of (a) sequential probability ratio test, and (b) Neyman–Pearson test discussed in Section 10.6.3.

The average sample number as a function of θ is given by

$$E(N|\theta) \simeq \frac{\beta(\theta)\ln B - [1 - \beta(\theta)]\ln A}{E\left\{\log\left[\frac{f(X|\theta_1)}{f(X|\theta_0)}\right]\bigg|\theta\right\}}$$

provided the denominator is not zero.

The sequential probability ratio test is optimal at the two fixed points θ_0 and θ_1, but often the sample size in between lies above the equivalent single sample size (see Fig. 10.14). This shows that the sequential probability ratio test is likely not to be the best sequential test in the case of composite hypotheses.

10.7. Summary of Optimal Tests

In Table 10.6 below we summarize our advice about what test to choose, depending on whether the hypotheses are simple or composite.

A case not dealt with in Table 10.6 is the situation when one wants to test an hypothesis against all other possible hypotheses. Such *goodness-of-fit* tests are dealt with in Chap 11.

Table 10.6. Existence of optimal tests.

Simple hypothesis against simple hypothesis	– One optimal test exists: Neyman–Pearson test (Sec. 10.3.1). – The expected number of observations may be decreased by using sequential probability ratio test (Sec. 10.6.2).
Composite hypotheses	– In general, no uniformly most powerful test. One exception: one-sided test for the exponential family (Sec. 10.4.2). Therefore one should demand reasonable properties: unbiasedness (Sec. 10.2.3), most powerful test in region of interest (Sec. 10.2.4), maximum local power (Sec. 10.4.3). – Continuous family of hypotheses (constraints on the parameters, test on significance of extra parameters). A general method: likelihood ratio test (Sec. 10.5). Asymptotically the encountered distributions are simple (χ^2) (Sec. 10.5.3). For finite samples the distributions are classical in the Normal law case (Sec. 10.5.5, Table 10.2). – Separate families of hypotheses (no relation between the hypotheses). General method: [Sec. 10.5.7, (i)]. Comprehensive model and use of maximum likelihood ratio test [Sec. 10.5.7, (ii)]. Significance of a likelihood ratio in the general case [Sec. 10.5.7, (iii)].

Chapter 11

GOODNESS-OF-FIT TESTS

As in the previous chapter, we are again concerned with the test of a null hypothesis H_0 with a test statistic T, in a critical region w_α, at a significance level α. Unlike the previous situations, however, the alternative hypothesis, H_1 is now the set of all possible alternatives to H_0. Thus H_1 cannot be formulated, the risk of second kind, β, is unknown, and the power of the test is undefined. Since it is in general impossible to know whether one test is more powerful than another, the theoretical basis for goodness-of-fit (GOF) testing is much less satisfactory than the basis for classical hypothesis testing. However, the problem is so important and the methods are sufficiently good, that GOF testing is quantitatively the most successful area of statistics. In particular, Pearson's venerable Chi-square test [see Section 11.2] is undoubtedly the most heavily used method in all of statistics.

In Section 11.1, we present the general procedure for making a GOF test. In Sections 11.2 and 11.3 we shall discuss tests for histograms, in Section 11.4 tests free of binning, and Section 11.5 will be concerned with the application of these results to some often-met problems. In Section 11.6 we shall see how to combine two or more (independent) tests.

11.1. GOF Testing: From the Test Statistic to the P-value

Goodness-of-fit tests compare the experimental data with their p.d.f. under the null hypothesis H_0, leading to the statement: if H_0 were true and the experiment were repeated many times, one would obtain data as far away (or further) from H_0 as the observed data with probability P. The quantity P is then called the *P-value* of the test for this data set and hypothesis. A small value of P is taken as evidence against H_0, which the physicist calls a *bad fit*.

It is clear from the above that in order to construct a GOF test we need:

(1) A *test statistic*, that is a function of the data and of H_0, which is a measure of the "distance" between the data and the hypothesis, and

(2) A way to calculate the probability of exceeding the observed value of the test statistic for H_0. That is, a function to map the value of the test statistic into a *P-value*.

If the data X are discrete and our test statistic is $t = t(X)$ which takes on the value $t_0 = t(X_0)$ for the data X_0, the P-value would be given by:

$$P_X = \sum_{X:t \geq t_0} P(X|H_0), \qquad (11.1)$$

where the sum is taken over all values of X for which $t(X) \geq t_0$.

Let us take a simple example of discrete counting data. We wish to test a theory which predicts the decay rate of a radioactive sample to be 17.3 decays per hour. We leave the counter active for one hour and measure 12 counts. Then we want to know if the observation N=12 is compatible with the expectation $\mu = 17.3$. The obvious test statistic is the absolute difference $|N - \mu|$, and assuming that the probability of n decays is given by the Poisson distribution, we can calculate the P-value by taking the sum in Eq. (11.1):

$$P_{12} = \sum_{n:|n-\mu| \geq 5.3} \frac{e^{-\mu}\mu^n}{n!} = \sum_{n=1}^{12} \frac{e^{-17.3}17.3^n}{n!} + \sum_{n=23}^{\infty} \frac{e^{-17.3}17.3^n}{n!}.$$

Using the relationships between the cumulative Poisson distribution and the cumulative χ^2 distribution given in Section 4.1.3, one can evaluate the above P-value, which is $P_{12} = 0.229$. The interpretation is that the observation is not significantly different from the expected value, since one should observe a number of counts at least as far from the expected value about 23% of the time.

When the data are continuous, the sum in Eq. (11.1) becomes an integral:

$$P_X = \int_{X:t>t_0} P(X|H_0), \qquad (11.2)$$

and this now becomes so complicated to compute that one tries to avoid using this form. Instead, one looks for a test statistic such that the distribution of

t is known independently of H_0. Such a test is called a *distribution-free test*. In the rest of this chapter we consider only distribution-free tests, such that the P-value does not depend on the details of the hypothesis H_0, but only on the value of t, and possibly one or two integers such as the number of events, the number of bins in a histogram, or the number of constraints in a fit. Then the mapping from t to P-value can be calculated once for all and published in tables, of which the well-known χ^2 tables are an example.

We have already pointed out that by combining events into *histogram bins* (called *data classes* in the statistical literature), some information is lost: the position of each event inside the bin. The loss of information may be negligible if the bin width is small compared with the experimental resolution, but in general one must expect tests on binned data to be inferior to tests on individual events. Unfortunately, the requirement of distribution-free tests restricts the choice of tests for unbinned data, and we present only those based on the *order statistics* (or *empirical distribution function*) in Section 11.4. Moreover, this class is limited to data depending on only one random variable, and to hypotheses H_0 which do not depend on parameters $\boldsymbol{\theta}$ to be estimated from the data.

When data are combined into histograms, more tests are available, but it is very important that the number of events per bin is sufficiently great to justify the approximations made in deriving the methods: the good properties of many tests (e.g. independence of distribution) are true only asymptotically. Such considerations may seriously limit the use of goodness-of-fit tests in many dimensions.

11.2. Pearson's Chi-square Test for Histograms

Karl Pearson made use of the asymptotic Normality of a multinomial p.d.f. in order to find the (asymptotic) distribution of another statistic, of the form (8.17)

$$(\mathbf{n} - N\mathbf{p})^T \, \underset{\sim}{V}^{-1} \, (\mathbf{n} - N\mathbf{p})$$

where $\underset{\sim}{V}$ is the covariance matrix of the observations $\mathbf{n}$. Knowing that this matrix is of rank $(k-1)$ because of the relation $\Sigma_i^k n_i = N$, we will take $(k-1)$ terms. For definiteness, let us take the first $(k-1)$ terms (we shall restore the symmetry in the k terms later). Let $\underset{\sim}{W}$ be the $(k-1) \times (k-1)$ matrix obtained from $\underset{\sim}{V}$ by dropping the kth row and column. One can show easily that

$$N(\underset{\sim}{W}^{-1})_{ij} = \frac{1}{p_i}\delta_{ij} + \frac{1}{p_k} \, .$$

Now consider the statistic

$$T = (\mathbf{n} - N\mathbf{p})^T \, \underset{\sim}{W}^{-1} \, (\mathbf{n} - N\mathbf{p}) \tag{11.3}$$

where only the first $(k-1)$ components in $\mathbf{n}$ and $\mathbf{p}$ are considered. The random variable $(\mathbf{n} - N\mathbf{p})$ is, asymptotically, distributed multinormally with covariance matrix $\underset{\sim}{W}$. Thus T has, in the same limit, a $\chi^2(k-1)$ distribution.

The choice of the $(k-1)$ terms of Eq. (11.3) being arbitrary, let us find an equivalent expression for T which treats all bins symmetrically:

$$T = \frac{1}{N} \left\{ \sum_{i=1}^{k-1} \frac{(n_i - Np_i)^2}{p_i} + \frac{1}{p_k} \sum_{i=1}^{k-1} \sum_{j=1}^{k-1} (n_i - Np_i)(n_j - Np_j) \right\}$$

$$= \frac{1}{N} \left\{ \sum_{i=1}^{k-1} \frac{(n_i - Np_i)^2}{p_i} + \frac{1}{p_k} \left[\sum_{i=1}^{k-1} (n_i - Np_i) \right]^2 \right\}$$

or finally

$$T = \frac{1}{N} \sum_{i=1}^{k} \frac{(n_i - Np_i)^2}{p_i} = \frac{1}{N} \sum_{i=1}^{k} \frac{n_i^2}{p_i} - N \,. \tag{11.4}$$

This is the usual χ^2 goodness-of-fit test for histograms. The distribution of (11.4) is generally accepted as close enough to $\chi^2(k-1)$ when all the expected numbers of events per bin (Np_i) are greater than 5. Cochran relaxes this restriction, claiming the approximation to be good if not more than 20% of the bins have expectations between 1 and 5 [Cochran].

The method used above to get around the singularity of the covariance matrix $\underset{\sim}{V}$ may not seem very convincing, but the result (11.4) can be proved rigorously [Fourgeaud, p.298–303].

11.2.1. *Moments of the Pearson statistic*

Let us call H_T the true hypothesis, with probability content p_i per bin. Then

$$E(T|H_T) = \sum_i \frac{q_i(1-q_i)}{p_i} + N \left(\sum_i \frac{q_i^2}{p_i} - 1 \right) \tag{11.5}$$

and when $q_i = p_i$

$$E(T|H_0) = k - 1 \,.$$

One can show that $E(T)$ is minimal when $q_i = p_i$ by minimizing $E(T|H_T)$ with respect to q_i, subject to $\Sigma_{i=1}^k q_i = 1$. Hence

$$E(T|H_0) = k - 1$$

$$E(T|H_T) \geq k - 1$$

whatever N (finite statistics).

If one chooses the binning in such a way that all bins have equal probability content $p_i = 1/k$ under H_0, one can calculate easily the following moments [Kendall, II, p. 434]:

$$V(T|H_0) = 2(k-1)$$

$$V(T|H_T) = 4(N-1)k^2 \left\{ \sum_i q_i^3 - \left(\sum_i q_i^2 \right)^2 \right\} + 2k^2 \left\{ \sum_i q_i^2 - \left(\sum_i q_i^2 \right)^2 \right\}$$

$$E(T|H_T) = (k-1) + (N-1) \left\{ k \sum_i q_i^2 - 1 \right\} .$$

11.2.2. *Chi-square test with estimation of parameters*

If the parent distribution depends on a parameter $\boldsymbol{\theta}$, to be estimated from the data, one does not expect the T statistic to behave as a $\chi^2(k-1)$. Consider two methods of estimating $\boldsymbol{\theta}$: maximum likelihood estimates based either on binned or on individual observations.

In the first case (called "multinomial" maximum likelihood estimation in Section 8.4.5) one maximizes

$$L(\mathbf{n}|\boldsymbol{\theta}) = N! \prod_{i=1}^k \frac{[p_i(\boldsymbol{\theta})]^{n_i}}{n_1!}$$

or its logarithm with respect to θ_j, $j = 1, \ldots, r$, when $\boldsymbol{\theta}$ has r dimensions. Then one can show that the T statistic computed using these estimates behaves asymptotically as a $\chi^2(k-r-1)$ distribution. Intuitively, one has lost r degrees of freedom in estimating $\boldsymbol{\theta}$.

In the second case, one uses all available information to estimate $\boldsymbol{\theta}$ (which should be a better method), and binned observations to form T. One can show [Kendall, II, p. 430] that the cumulative distribution of T is intermediary to a $\chi^2(k-1)$ (which holds when θ is fixed) and a $\chi^2(k-r-1)$ (which holds for the first estimation method). The test is no longer distribution-free, but when k is large and r small, the two boundaries $\chi^2(k-1)$ and $\chi^2(k-r-1)$ become close enough to make the test practically distribution-free (see Fig. 9.2 for cumulative χ^2 distributions).

11.2.3. *Choosing optimal bin size*

In practice, one must decide how to bin the observations. Too few bins carry too little information, but too many bins may lead to too few events per bin. Most of the results we have given are only true asymptotically, as for instance the Normal limit of a multinomial law.

A rule of thumb for the bin size is that, given the number of bins k, the bins should be chosen to have equal probability contents under H_0. The T test is consistent (hence asymptotically unbiased) whatever the binning. However, for finite statistics this test is not, in general, unbiased. One of the reasons for using equiprobable bins is that, in such a case, one can prove that the test is locally unbiased. We shall not given the detailed proof [Kendall II, p. 435] but only sketch it briefly.

Let us define $\varepsilon_i = p_i - (1/k)$ and let us study the power of the test as a function of ε. The equiprobable case corresponds to $\varepsilon = 0$, and local unbiasedness means that $P(\varepsilon) \geq P(0)$ for ε small. Thus one expands $P(\varepsilon)$ in a Taylor series and the result is

$$P(\varepsilon) = P(0) + A \sum_i \varepsilon_i^2 + \cdots$$

where one can show that $A > 0$.

The distribution of T has been described in the previous sections on the assumption that the bins are independent of the observations. However, it turns out that even if the bin positions are estimated from the data, the asymptotic distribution of T is not affected.

In particular:

— estimated parameters can be used in designing a binning scheme with equal probabilities
— one may start from a large number of equal size bins, then group together adjacent bins until the required probability contents per bin are obtained
— evidently, one should not try different binnings, then choose that one which gives the lowest T. That test statistic would not be distributed as a χ^2!

If one devises a binning scheme based on the expected distribution under H_0, one can use tables of the distribution if it is classical enough. In most practical cases, however, one must find the cumulative distribution $F(X)$ by integration, then divide the divide the interval of variation of $F(X)$, $[0,1]$ into k equal parts,

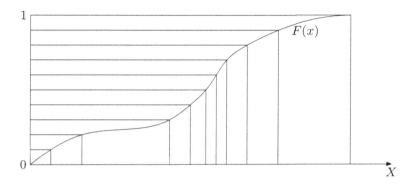

Fig. 11.1. A procedure to find equiprobable bins.

to find the corresponding bins in the variable X (Fig. 11.1). As long as X is one-dimensional there is no arbitrariness in this procedure.

When $\mathbf{X} = X_1, \ldots, X_n$ is n-dimensional, one could compute the projected distribution $F(X_1)$ and define bins in the same way, but then one loses all information about the other components. To recover it, one may compute $F(X_2|X_1)$ for each bin in X_1, then apply the same scheme and finally obtain k^2 bins where no information about $X_3, \ldots, X_n$ is taken into account. This procedure can be iterated n times to yield

$$Z_1 = F_1(X_1)$$
$$Z_2 = F_2(X_2|X_1) \tag{11.6}$$
$$Z_i = F_i(X_i|X_1, \ldots, X_{i-1}), \quad i = 1, n.$$

Thus, one has a transformation $\mathbf{Z} = R(\mathbf{X})$ and, by construction, Z is uniformly distributed inside the n-dimensional hypercube, of extension $[0,1]$ in every dimension. One can now divide this hypercube into equal cells

$$c_{j_1 j_2 \cdots j_n} = \left\{ \mathbf{Z}; \frac{j_i - 1}{m_i} \le Z_i \le \frac{j_i}{m_i}, \quad i = 1, \ldots, n \right\} \tag{11.7}$$

with $j_i = 1, \ldots, m_i$ if one chooses a partition into m_i parts in each dimension. (To be correct, one should in fact use a slightly more detailed notation since the limits of the cells now appear in both of two adjacent cells!) Then the T statistic becomes

$$T = \frac{k}{N} \sum_{i=1}^{k} n_i^2 - N$$

with

$$k = \prod_{i=1}^{n} m_i \, .$$

The total number of cells cannot, in this way, be adjusted to a preselected value (for example one could not use a prime number). However, it is never very important to adjust precisely k. We shall see how this number can be optimized.

We have kept open the possibility of different partitions, keeping the m_i not necessarily equal. In fact, one may have physical reasons to use different partitions. For example, one component may not be very sensitive to the hypothesis tested whereas another may be, and one may want the total number of cells to be fixed. Or, when one does not fold the resolution function into H_0, one should not use bins smaller than the resolution; thus the partition scheme (11.7), although reasonably symmetrical with respect to all components, may not be the best one for a specific problem. For example, one may divide the n-dimensional space of $\mathbf{X}$ into hypervolumes of equal probability contents which may have some physical meaning, relevant to the problem studied. Very often this amounts to a change of coordinates where some of the new coordinates carry no information and can be integrated out.

Let us now optimize the number of bins k. We first search for the limiting power as N increases. Since the test is consistent, the power of the test against any fixed hypothesis becomes 1 asymptotically. Therefore, in order to find the limiting power function one considers a sequence of alternative hypotheses H_T where $q_i = p_i[1 + (\Delta_i/\sqrt{N})]$ which converges towards H_0 as N increases (Δ_i is fixed). The distribution of T under H_T is then asymptotically a non-central χ^2 with non-centrality parameter $K = \Sigma_i p_i \Delta_i^2$ and $(k-r-1)$ degrees of freedom (r parameters being estimated by maximum likelihood on binned observations).

We now choose, in the class of tests with equiprobable bins, the optimal test. Since there is no UMP within this class, something else must be imposed. Choose for instance, a power level p_0 (that is, contamination $1 - p_0$) which can be considered acceptable. The power function $p(k, K)$ along $p = p_0$ defines a relation $K_0(k)$ (see Fig. 11.2), which we shall minimize with respect to k.

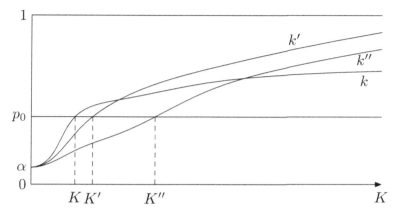

Fig. 11.2. Example of power functions for different number of histogram bins (k).

Using a Normal approximation to the non-central χ^2 distribution for H_T it is possible to write $p(k, K)$

$$p = \phi\left[\sqrt{2}(N-1)\left(\frac{b}{k}\right)^{5/2} - \lambda_\alpha\right].$$

Solving for k we get the optimum

$$k = b\left[\frac{\sqrt{2}(N-1)}{\lambda_\alpha + \lambda_{1-p_0}}\right]^{2/5} \tag{11.8}$$

where λ_α is the α-point of the standard Normal distribution.

In the case of a simple hypothesis (no free parameter) and $p_0 = 0.5$ a value $b = 4$ has been obtained [Mann]. In general, a figure for b between 2 and 4 should be used. From Eq. (11.8) we see that k varies as $N^{2/5}$. Moreover, the expected number of events per bin $N/k \sim N^{3/5}$ should not be too small, say ≥ 5. We give a few practical numbers in Table 11.1.

Let us now summarize the recommendations:

(i) Determine the number of bins using Eq. (11.8) with $b \sim 2$ to 4.
(ii) If then N/k turns out to be too small, decrease k to satisfy $N/k \geq 5$.
(iii) Define the k bins such as to have equal probability contents, either from the theoretical distribution under H_0, or from the data. Note that if the random variable observed is multidimensional, one still has different binnings corresponding to equal probability contents in k bins.

(iv) If parameters have to be estimated, use maximum likelihood estimation based on individual observations, but remember that the test statistic is not fully distribution-free (see Section 11.2.2).

Table 11.1. Values of $k(N(/k)$ from Eq. (11.8), using $b = 4$, $\lambda_{0.5} = 0$, $\lambda_{0.2} = 0.84$, $\lambda_{0.05} = 1.64$, $\lambda_{0.01} = 2.33$. The probability to accept H_0 when false is $1 - p_0$.

N	α	p_0	
		0.5	0.8
200	0.01	27 (7.4)	24 (8.3)
	0.05	31 (6.5)	27 (7.4)
500	0.01	39 (13)	35 (14)
	0.05	45 (11)	39 (13)

11.3. Other Tests on Binned Data

11.3.1. *Runs test*

In addition to the necessity of binning the observations, a drawback of the T statistic is that the signs of the deviations $(n_i - Np_i)$ are lost. The *runs test* is based on the sign of such deviations. The main interest of this test is that, *for simple hypotheses*, it is independent of the χ^2 test on the same bins and thus brings in new information.

Under hypothesis H_0, all patterns of signs are equally probable. This simple fact allows us to write the following results [Wilks, p. 154]. Let M be the number of positive deviations, N the number of negative deviations, and R the total number of runs, where a *run* is a sequence of deviations of the same sign, preceded and followed by a deviation of opposite sign (unless at the end of the range of the variable studied). Then

$$P(R = 2s) = \frac{2\binom{M-1}{s-1}\binom{N-1}{s-1}}{\binom{M+N}{M}}$$

$$P(R = 2s - 1) = \frac{\binom{M-1}{s-2}\binom{N-1}{s-1} + \binom{M-1}{s-1}\binom{N-1}{s-2}}{\binom{M+N}{M}}.$$

The critical region is defined as improbably low values of $R : R \leq R_{min}$. Given the probability of R, one can compute R_{min} corresponding to the significance level required. The expectation and variance of R are

$$E(R) = 1 + \frac{2MN}{M+N}$$

$$V(R) = \frac{2MN(2MN - M - N)}{(M+N)^2(M+N-1)}.$$

Although the runs test is usually not as powerful as Pearson's χ^2 test, it is (asymptotically) independent of it and hence the two can be combined to produce an especially important test as discussed in Section 11.6.

11.3.2. *Empty cell test, order statistics*

The null hypothesis is that the observations $\mathbf{X}$ have a distribution function $F_0(\mathbf{X})$. Let us divide the X-space into M cells of equal probability content, as in Section 11.2.3, and let the observations be distributed over the cells in such a way that S_0 cells contain no observations. Then S_0 is the *number of empty cells*. The probability function of S_0, under the null hypothesis, is

$$P(S_0 = s) = \frac{1}{M^N} \binom{M}{s} \sum_{i=0}^{M-s} \binom{M-s}{i} (-1)^i (M - s - i)^N.$$

The possible values of s are $(K, K+1, \ldots, M-1)$, where $K = \max(0, M - N)$.

Tables of this distribution have been calculated. The expectation and variance of S_0 are

$$E(S_0) = M\left(1 - \frac{1}{M}\right)^N$$

$$V(S_0) = M(M-1)\left(1 - \frac{2}{M}\right)^N + M\left(1 - \frac{1}{M}\right)^N - M^2\left(1 - \frac{1}{M}\right)^{2N}.$$

When M and N approach infinity in such a way that $\rho = N/M > 0$, S_0 is asymptotically Normally distributed as

$$N\{Me^{-\rho}, M[e^{-\rho} - e^{-2\rho}(1+\rho)]\}$$

from which approximate tests can easily be formed. It is recommended in the literature to use $\rho = 1.255$, since the distribution then tends fastest to Normality.

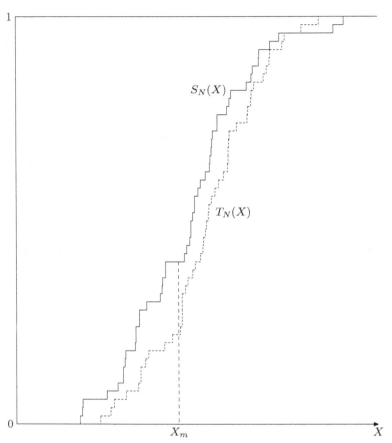

Fig. 11.3. Example of two cumulative distributions, $S_N(X)$ and $T_N(X)$, of the type (11.9). For these two data sets, the maximum distance $S_N - T_N$ occurs at $X = X_m$.

Let us at this stage introduce the *order statistics* $X_{(i)}$. Given N independent observations $X_1, \ldots, X_N$ of the random variable X, let us reorder the observations in ascending order, so that $X_{(i)} \le X_{(2)} \le, \ldots, \le X_{(N)}$ (this is always permissible since the observations are independent). The ordered observations $X_{(i)}$ are called the order statistics. Their cumulative distribution (Fig. 11.3) is called the *empirical distribution function* or EDF:

$$S_N(X) = \begin{cases} 0 & X < X_{(1)} \\ i/n & \text{for} \quad X_{(i)} \le X < X_{(i+1)}, \quad i = 1, \ldots, \quad N - 1. \\ 1 & X_{(N)} \le X \end{cases} \tag{11.9}$$

Note that $S_N(X)$ always increases in steps of equal height, N^{-1}.

By definition of the cumulative distribution function $F(X)$, Eq. (2.16), and from the law of large numbers (Section 3.3)

$$\lim_{N \to \infty} \{P[S_N(X) = F(X)]\} = 1 .$$

In Section 11.4 we shall consider different norms on the difference $S_N(X) - F(X)$ as test statistics.

Let us now design a two-sample empty cell test for the comparison of two experimental distributions. Let the first set consist of N observations X_i with order statistics $X_{(r)}$, $r = 1, \ldots, N$. Define $N + 1$ bins I, which coincide with the order statistics, bin I_1 extending from $-\infty$ to $X_{(1)}$ and bin I_{N+1} extending from $X_{(N)}$ to ∞. From the second set of M observations Y_j, find the number r_i of observations Y falling in the bin I_i.

Let S_0' be the number of Y-empty bins I_i with $r_i = 0$. The probability function of S_0' is

$$P(S_0 = s) = \frac{\binom{N+1}{s}\binom{M-1}{N-s}}{\binom{N+M}{N}} .$$

The possible values of s are $(K, K + 1, \ldots, N)$, and $K = \max[0, N + 1 - M]$. Asymptotically, the distribution of S_0' becomes Normal, with mean $(N + 1)/(1 + \ell)$ and variance $(N + 1)\ell^2/(1 + \ell)^3$, where $\ell = M/N$.

11.3.3. *Neyman–Barton smooth test*

In order to design tests to be independent of the distribution function specified by the null hypothesis H_0 it is necessary to transform the data into some standard variable. Neyman and Barton used the transformation (11.6). They considered the one-dimensional case, transforming from observations

$$X_i, \quad i = 1, \ldots, N, \qquad \text{to} \qquad Y_i = F_0(X_i) .$$

Under H_0, the Y_i are uniformly distributed between 0 and 1. The tests are then constructed for specified departures from uniformity. The alternative hypothesis H_s is given as

$H_s : Y_i$ are distributed with density function $f(Y|H_s)$
where

$$f(Y|H_s) = c(\theta_1 \cdots \theta_s) \exp \left(1 + \sum_{r=1}^{s} \theta_r \ell_r(Y) \right) .$$

The functions $\ell_r(y)$ are the Legendre polynomials of order r, orthonormal on $(0,1)$. One then tests

$$H_0 : \theta_1 = \theta_2 = \cdots = \theta_s = 0$$

or

$$H_0 : \sum_{r=1}^{s} \theta_r^2 = 0 \tag{11.10}$$

against any other combination.

Writing down the likelihood function for the Y_i, one has

$$L(\mathbf{Y}|\boldsymbol{\theta}) = [c(\boldsymbol{\theta})]^N \exp\left[\sum_{r=1}^{s} \theta_r \sum_{i=1}^{N} \ell_r(Y_i)\right].$$

Therefore

$$t_r = \sum_{i=1}^{N} \ell_r(Y_i)$$

is a sufficient statistic for θ_r. Since H_0 has been written in the form (11.10), it seems reasonable to use as a test statistic some function of $t_r{}^2$, for instance

$$p_s^2 = \frac{1}{N} \sum_{r=1}^{s} t_r^2$$

$$= \frac{1}{N} \sum_{r=1}^{s} \left[\sum_{i=1}^{N} \ell_r(Y_i)\right]^2.$$

Under the hypothesis H_S, it turns out that p_S^2 is asymptotically distributed as a non-central $\chi^2(s, K)$ with non-centrality parameter

$$K = N \sum_{r=1}^{s} \theta_r^2.$$

The significance levels for the null hypothesis are therefore obtained from the central $\chi^2(s)$ distribution. It has been shown that $p_1{}^2$ and $p_2{}^2$ are adequately approximated by $\chi^2(1)$ and $\chi^2(2)$ if $N \geq 20$.

Suppose now that the observations X_i, $i = 1, \ldots, N$ have been binned into k bins, with probabilities

$$p_i = p_{0i} \text{ under } H_0$$

and

$$p_i = p_{si} \text{ under the alternative } H_s .$$

The transformation to a uniform distribution is then given by

$$Y_i' = \sum_{j=1}^{i-1} p_{0j} + \frac{1}{2} p_{0i} .$$

We now require a set of polynomials $P_r(Y')$, which are orthonormal with respect to summation over the p_{0i},

$$\sum_{i=1}^{k} p_{0i} P_r(Y_i') P_t(Y_i') = \delta_{rt} . \tag{11.11}$$

The test statistic becomes

$$p_{k-1}^2 = \sum_{r=1}^{k-1} u_r^2 = \frac{1}{N} \sum_{r=1}^{k-1} \left[\sum_{i=1}^{k} n_i P_r(Y_i') \right]^2$$

where n_i is the number of observations in the ith bin. This reduces, by virtue of Eq. (11.11) exactly to Eq. (11.4):

$$p_{k-1}^2 = \sum_{i=1}^{k} \frac{n_i^2}{N p_{0i}} - N = T .$$

Thus p_{k-1}^2 is identical with the χ^2 test of Section 11.1.2. The statistics p_r^2 of order $r < k - 1$ can be defined as components, or partitions of the T statistic.

The beauty of the statistics p_r^2, when the data are binned, is that they select the appropriate functions of the Y_i for the test in hand to have maximum power; that is, they isolate the important components of T.

The results for the χ^2 test under composite hypotheses (where parameters are estimated from the data) carry over to the binned p_k^2 tests, the degrees of freedom reducing by one for each parameter estimated by the multinomial maximum likelihood method.

11.4. Tests Free of Binning

As already noted, tests on unbinned data should, in principle, be better than tests on binned data. The likelihood of the data would appear to be a good candidate at first sight. Unfortunately, this carries little information as a test statistic, as we shall see in Section 11.4.4.

The more successful tests are based on comparing the cumulative distribution function $F(X)$ under H_0 with the equivalent distribution of the data [Eq. (11.9)]. The test statistic is some measure of the "distance" between the experimental and hypothetical distribution functions. For the Smirnov–Cramér–Von Mises test, Section 11.4.1, this is the average square difference between two functions, and the Kolmogorov test, Section 11.4.2, takes the maximum or minimum difference.

11.4.1. *Smirnov–Cramér–von Mises test*

Consider the statistic

$$W^2 = \int_{-\infty}^{\infty} [S_N(X) - F(X)]^2 f(X) dX \,, \tag{11.12}$$

where $f(X)$ is the p.d.f. corresponding to the hypothesis H_0, $F(X)$ is the cumulative distribution, and $S_N(X)$ is defined by Eqs. (11.9). Inserting Eqs. (11.9) into Eq. (11.12), one can obtain

$$W^2 = \int_{-\infty}^{X_1} F^2(X) dF(X) + \sum_{i=1}^{N-1} \int_{X_i}^{X_{i+1}} \left[\frac{i}{N} - F(X) \right]^2 dF(X)$$

$$+ \int_{X_N}^{\infty} [1 - F(X)]^2 dF(X)$$

$$= \frac{1}{N} \left\{ \frac{1}{12N} + \sum_{i=1}^{N} \left[F(X_i) - \frac{2i-1}{2N} \right]^2 \right\} , \tag{11.13}$$

using the properties $F(-\infty) \equiv 0$, $F(+\infty) \equiv 1$.

For a fixed value of X, $S_N(X)$ has a binomial distribution, with moments

$$E[S_N(X)] = F(X)$$

$$E\{[S_N(X) - F(X)]^2\} = \frac{1}{N} F(X)[1 - F(X)] \,.$$

Thus the statistic (11.13) has mean and variance

$$E(W^2) = \frac{1}{N} \int_0^1 F(1 - F) dF = \frac{1}{6N}$$

$$V(W^2) = E(W^4) - [E(W^2)]^2 = \frac{4N-3}{180N^3} \,.$$

The distribution of W^2 is completely independent of the distribution of X, even for finite N. This can be seen easily with the change of variable $Y = F(X)$. The definition of W^2 becomes

$$W^2 = \int_0^1 [S_N(Y) - Y]^2 dY$$

which does not depend any more on $f(X)$.

The asymptotic characteristic function of NW^2 is [Smirnov]:

$$\lim_{N \to \infty} E(e^{it \, NW^2}) = \sqrt{\frac{\sqrt{2it}}{\sin \sqrt{2it}}}.$$ (11.14)

By inversion of Eq. (11.14), one can compute the critical values in Table 11.2 [Anderson]:

Table 11.2. Critical values of Smirnov statistic.

Test size α	Critical value of NW^2
0.10	0.347
0.05	0.461
0.01	0.743
0.001	1.168

It has been shown [Marshall] that, to the accuracy of this table, the asymptotic limit is reached when $N \geq 3$.

When H_0 is composite, W^2 is not in general distribution-free. When X is many-dimensional, the test also fails, unless the components are independent. However, one can form a test to compare two distributions, $F(X)$ and $G(X)$. Let the number of observations be N and M, respectively, and let the hypothesis be H_0: $F(X) = G(X)$. Then the test statistic is

$$W^2 = \int_{-\infty}^{\infty} [S_N(X) - S_M(X)]^2 d \left[\frac{NF(X) + MG(X)}{N + M} \right].$$ (11.15)

The asymptotic characteristic function (11.14) then holds for the quantity

$$\frac{MN}{M + N} W^2,$$

where W^2 is given by Eq. (11.15).

11.4.2. *Kolmogorov test*

The test statistic is now the maximum deviation of the observed distribution $S_N(X)$, Eqs. (11.9), from the distribution $F(X)$ expected under H_0. This is defined either as

$$D_N = \max |S_N(X) - F(X)| \qquad \text{for all } X$$

or as

$$D_N^\pm = \max \{\pm[S_N(X) - F(X)]\} \quad \text{for all } X,$$

when one is considering only one-sided tests. It can be shown that the limiting distribution of $\sqrt{N}D_N$ is

$$\lim_{N\to\infty} P(\sqrt{N}D_N > z) = 2\sum_{r=1}^{\infty}(-1)^{r-1}\exp(-2r^2z^2) \qquad (11.16)$$

and that of $\sqrt{N}D_N^\pm$ is

$$\lim_{N\to\infty} P(\sqrt{N}D_N^\pm > z) = \exp(-2z^2). \qquad (11.17)$$

Alternatively, the probability statement in Eq. (11.17) can be restated as

$$\lim_{N\to\infty} P[2N(D_N^\pm)^2 \le 2z] = 1 - e^{-2z^2}. \qquad (11.18)$$

Thus $4N(D_N^\pm)^2$ have a $\chi^2(2)$ distribution. The limiting distributions (11.16), (11.17), and (11.18) are considered valid for $N \gtrsim 80$. In Table 11.3 we give some critical values of $\sqrt{N}D_N$ for the distribution (11.16).

Table 11.3. Critical values of Kolmogorov statistic.

Test size α	Critical value of $\sqrt{N}D_N$
0.01	1.63
0.05	1.36
0.10	1.22
0.20	1.07

The equivalent statistic for comparing two distributions $S_N(X)$ and $S_M(X)$, Fig. 11.3, is

$$D_{MN} = \max|S_N(X) - S_M(X)| \qquad \text{for all } X$$

or, for one-sided tests

$$D^{\pm}_{MN} = \max \{\pm[S_N(X) - S_M(X)]\} \qquad \text{for all } X.$$

Then $\sqrt{MN/(M+N)}D_{MN}$ has the limiting distribution (11.16) and $\sqrt{MN/(M+N)}D^{\pm}_{MN}$ have the limiting distribution (11.17).

Finally, one may invert the probability statement about D_N to set up a *confidence belt* for $F(X)$. The statement

$$P\{D_N = \max |S_N(X) - F(X)| > d_\alpha\} = \alpha$$

defines d_α as the α-point of D_N. If follows that

$$P\{S_N(X) - d_\alpha \le F(X) \le S_N(X) + d_\alpha\} = 1 - \alpha.$$

Therefore, setting up a belt $\pm d_\alpha$ about $(S_N(X)$, the probability that $F(X)$ is entirely within the belt is $1 - \alpha$ (similarly d_α can be used to set up one-sided bounds). One can thus compute the number of observations necessary to obtain $F(X)$ to any accuracy. Suppose for example that one wants $F(X)$ to precision 0.05 with probability 0.99, then one needs $N = (1.628/0.05)^2 \sim 1000$ observations.

11.4.3. *More refined tests based on the EDF*

Users of the Kolmogorov test will probably notice that the maximum difference D_N or D_{MN} almost always occurs near the middle of the range of X. This has led Anderson and Darling to propose an improved test which gives more weight to the ends of the range. The *Anderson–Darling test* [Anderson] is an example of a definite improvement on the Kolmogorov test, but one which comes at the expense of losing the *distribution-free* property. This means that the P-value must be computed differently depending on whether one is testing for Normally distributed data or uniformly distributed data, for example. Tests of this kind are outside the scope of this book; the reader is referred to the literature [D'Agostino].

11.4.4. *Use of the likelihood function*

Suppose that the N observations $\mathbf{X}$ have p.d.f. $f(\mathbf{X})$ and log-likelihood function

$$\lambda = \sum_{i=1}^{N} \ln f(X_i).$$

If no parameter needs to be estimated from the data, one can, in principle, compute the expectation and the variance of λ,

$$E_{\mathbf{X}}(\lambda) = N \int \ln f(\mathbf{X}) \cdot f(\mathbf{X}) d\mathbf{X},$$

$$V_{\mathbf{X}}(\lambda) = N \int [\ln f(\mathbf{X}) - N^{-1} E_{\mathbf{X}}(\lambda)]^2 f(\mathbf{X}) d\mathbf{X},$$

and even higher moments, if one feels that the Normality assumption is not good enough.

When parameters have to be estimated by the maximum likelihood method, the calculations become much more complicated, but the expected distribution of λ can always be found approximately by Monte Carlo methods. It is then possible to calculate a P-value given by the probability of λ being at least as far from the expected value as the observed (maximum) log-likelihood.

Unfortunately, the value of the likelihood does not make a good GOF test statistic. This can be seen in different ways, but the first clue should come when we judge whether the likelihood is a measure of the "distance" between the data and the hypothesis. At first glance, we might expect it to be a good measure, since we know the maximum of the likelihood gives the best fit to the data. But in m.l. estimation, we are using the likelihood for fixed data as a function of the parameters in the hypothesis, whereas in GOF testing we use the likelihood for a fixed hypothesis as a function of the data, which is very different.

Suppose for example that the hypothesis under test H_0 is the uniform distribution (which can be arranged by a simple coordinate transformation). For this hypothesis, it is easily seen that the likelihood has no power at all as a GOF test statistic, since all data sets (with the same number of events) have exactly the same value of the likelihood function, no matter how well they fit the hypothesis of uniformity.

More extensive studies show a variety of examples where the value of the likelihood function has no power as a GOF statistic and no examples where it can be recommended [Heinrich].

11.5. Applications

11.5.1. *Observation of a fine structure*

The situation often arises in experimental physics that some important feature (a resonance, a spectral line) manifests itself as a relatively narrow signal (a peak or a dip) superimposed on a smooth background, see Fig. 11.4. The first question is then: do the observations suggest fine structure in the region AB? If yes, the next problem will be to estimate parameters, such as size and position of the signal.

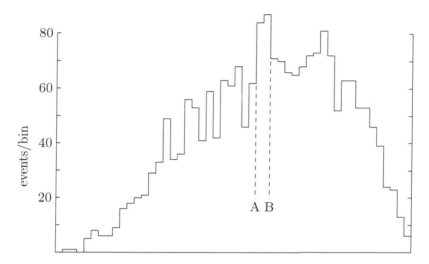

Fig. 11.4. Experimental histogram with a fine structure in the region AB.

Alternatively, one may wish to determine the probability of observing a statistical fluctuation of a given size in a given region AB, or in an arbitrary region of the same width. Clearly then, the null hypothesis (H_0: no true physical effect in the region AB) requires some knowledge of the shape of the background or noise. While sometimes this may be known with great accuracy, often the only reasonable assumption is that of "smoothness", which usually amounts to representing the background by a low-order polynomial.

Note that if the full range of the random variable is much greater than AB, the test should not be applied to the full range because that may easily lead to accepting H_0 although it is false (error of second kind). This is so because the larger the region under consideration, the greater is the probability that one statistical fluctuation of width AB occurs somewhere. Thus the only knowledge required of the background is a prediction of its size and variance in the region AB.

Let us describe the background by a function $b(\mathbf{X}, \boldsymbol{\theta})$ of the observations $\mathbf{X}$ and unknown parameters $\boldsymbol{\theta}$. The estimates $\widehat{\boldsymbol{\theta}}$ and the covariance matrix $\underset{\sim}{V}$

can be obtained by the methods of Chapters 7 and 8 (excluding the region AB), to give

$$\hat{b}_{AB} = \int_A^B b(\mathbf{X}, \widehat{\boldsymbol{\theta}}) d\mathbf{X} \, .$$

Since $\hat{b}_{AB}$ is a function of $\widehat{\boldsymbol{\theta}}$, one can compute its variance by the usual methods of change of variable (Section 2.3.2),

$$\hat{\sigma}_{AB}^2 = \mathbf{D}^T \, \underset{\sim}{V} \, \mathbf{D} \, ,$$

where $\mathbf{D}$ is the vector of derivatives

$$D_i = \left. \frac{\partial \hat{b}_{AB}}{\partial \theta_i} \right|_{\hat{\theta}_i} = \int_A^B \frac{\partial}{\partial \hat{\theta}_i} b(\mathbf{X}, \widehat{\boldsymbol{\theta}}) d\mathbf{X} \, .$$

Let the number of observations in AB be n_{AB}. The natural test statistic for determining whether n_{AB} is significantly different from $\hat{b}_{AB}$ is

$$T = \frac{(n_{AB} - \hat{b}_{AB})^2}{V(n_{AB} - \hat{b}_{AB})} \, ,$$

where $V(n_{AB} - \hat{b}_{AB})$ is the variance. Under the hypothesis H_0 (see Section 4.1.2)

$$E(n_{AB}) = b_{AB} \, ,$$

$$V(n_{AB}) = b_{AB} \, ,$$

and b_{AB} is estimated by $\hat{b}_{AB}$. Thus

$$V(n_{AB} - \hat{b}_{AB}) \approx \hat{b}_{AB} + \hat{\sigma}_{AB}^2 - 2 \, \text{cov} \, (n_{AB}, \hat{b}_{AB}) \, .$$

Since we excluded the region AB when estimating $\boldsymbol{\theta}$, n_{AB} and $\hat{b}_{AB}$ are uncorrelated:

$$V(n_{AB} - \hat{b}_{AB}) \approx \hat{b}_{AB} + \hat{\sigma}_{AB}^2 \tag{11.19}$$

and

$$T \approx \frac{(n_{AB} - \hat{b}_{AB})^2}{\hat{b}_{AB} + \hat{\sigma}_{AB}^2} \, . \tag{11.20}$$

when n_{AB} is large enough, it is Normally distributed around $\hat{b}^{AB}$, and T behaves as a $\chi^2(1)$. Equivalently, one often expresses the statistic (11.20) in terms of *standard deviations*

$$d = \sqrt{T} \tag{11.21}$$

One then speaks of the *odds* against exceeding a given number of standard deviations, see Table 11.4. (Warning — since the approximation that n_{AB} is Normal becomes worse as d increases, the odds in this table for large d may be misleading.)

Table 11.4. Odds against exceeding d standard deviations.

d	Odds	
1	2.15	: 1
2	21.0	: 1
3	370	: 1
4	16,000	: 1
5	1.7×10^6	: 1

Until now, we have implicitly assumed the region AB to be *selected independently of the observations*. For example, the fine structure may already have been observed in previous experiments and its existence and shape may be well established. However, if the choice of region AB is based on the data, the numbers in Table 11.4 are no longer appropriate, since they do not account for the probability of the occurrence of a signal in any arbitrary place of the full region.

To illustrate this, consider signals which are only one bin wide. Let p be the probability of exceeding d standard deviations in a given bin. When no bin is specified in advance, the probability of exceeding d standard deviations in at least one bin out of k bins is obviously

$$q = 1 - (1 - p)^k \approx kp\,.$$

For instance, in a histogram of 40 bins, a 3 standard deviation effect in a given bin has the same significance as a 4 standard deviation effect in any one (unspecified) bin. In general, if for a specified bin j the probability of d_j exceeding the α-point λ_α is

$$P(d_j > \lambda_\alpha) = \alpha\,,$$

then

$$P(\text{exactly } \ell \text{ bins have } d_j > \lambda_\alpha) = \binom{k}{\ell}\alpha^\ell(1 - \alpha)^{k-\ell}$$

and

$$P(\text{at least } \ell \text{ bins have } d_j > \lambda_\alpha) = \sum_{m \geq \ell}^{k} \binom{k}{m} \alpha^m (1 - \alpha)^{k-m} .$$

What happens then if one considers signals of width two bins, three bins, etc.? If the decision what signal width to consider is taken prior to the experiment, one may as well choose the bins accordingly, so as to obtain only signals one bin wide. If, on the contrary, the decision is taken on the basis of the data, the problem is much more complicated. In fact, it is the problem of the resolution function, encountered earlier (e.g. Section 4.3.4).

If the width of a fine structure is expected to be small compared to the width of the experimental resolution function, its signal will be difficult to observe: a narrow peak above background is seen lower and wider, and a narrow dip is seen shallower and wider. It is therefore important to include the resolution function in evaluating the expected effect.

Suppose that an observed signal has been decided significant, on the basis of the statistics (11.20) or (11.21). That implies that one has rejected the null hypothesis

$H_0 :$ no true physical effect in the region AB .

Next one wants to proceed to measure the size of the signal. One can still estimate it by $s = (n_{AB} - \hat{b}_{AB})$, but the variance of the estimate is no longer given by Eq. (11.19). The problem is not a goodness-of-fit test any more, but a test of H_1 (assumed true) against H_0, where

H_1: true physical effect of size s and background of size b_{AB} in region AB. The variance (11.19) is then replaced by

$$V(n_{AB} - \hat{b}_{AB}) \approx n_{AB} + \hat{\sigma}_{AB}^2 , \tag{11.22}$$

which still lacks correlation terms for the same reason as before.

Let the risk of second kind be

$$\beta = p(d \leq \lambda_\alpha | H_1) ,$$

where d is defined by Eq. (11.21). Under H_1 the expectation and variance of n_{AB} are

$$E(n_{AB}) = V(n_{AB}) = b_{AB} + s ,$$

and s is approximately distributed as $N(\mu, \sigma^2)$, with

$$\mu = \frac{s}{\sqrt{\hat{b}_{AB} + \hat{\sigma}_{AB}^2}} , \qquad \sigma^2 = \frac{\hat{b}_{AB} + s + \hat{\sigma}_{AB}^2}{\hat{b}_{AB} + \hat{\sigma}_{AB}^2} .$$

It follows that
$$\beta = \phi \left(\frac{\lambda_\alpha \sqrt{\hat{b}_{AB} + \hat{\sigma}^2_{AB}} - s}{\sqrt{\hat{b}_{AB} + s + \hat{\sigma}^2_{AB}}} \right).$$
(11.23)

11.5.2. *Combining independent estimates*

Suppose that one wants to get a better estimate of a set of r parameters $\boldsymbol{\theta}$ by combining independent estimates $\widehat{\boldsymbol{\theta}}$ from N different experiments, and suppose that the covariance matrix of the estimates in experiment J is $\underset{\sim}{V}_J$. The correct procedure would be to summarize the information about $\boldsymbol{\theta}$ by the likelihood of the experiment, or possibly by a sufficient statistic, but neither one is usually known (and the latter rarely exists).

Using only the estimates $\widehat{\boldsymbol{\theta}}_J$ and the covariance matrix $\underset{\sim}{V}_J$ implies Normally distributed estimates, which is a good approximation only when the number of observations is large enough. Unfortunately, one may often be forced to make even stronger approximations. This happens when (as is usual) the experimenter quotes an estimate $\widehat{\boldsymbol{\theta}}_i$ and a 68.3% confidence interval $\pm \varepsilon_i$ for each parameter i, as if they were uncorrelated (although he may have properly used the full covariance matrix $\underset{\sim}{V}$ in determining ε). Then one can only approximate $\underset{\sim}{V}$ by the diagonal matrix with elements ε_i^2.

We shall assume Normality of the estimates, and covariance matrices known. Suppose that one decides to discard $N-M$ experiments which carry little information (small precision) compared with the experiment with most information. If the Normality assumption indeed is justified for the discarded experiments, little information is then lost. If it is not justified, those experiments might have distorted the over-all result badly.

Obviously one has the problem of loss (of Normally distributed estimates) and contamination (of estimates erroneously assumed Normal), to which no general solution exists.

Suppose that one were to believe blindly in the remaining M experimental results. Using the Normality assumption, one could then easily construct a sufficient statistic summarizing all results in a new estimate. However, one may envisage that some estimates are biased or that their variances are underestimated or overestimated. One could therefore first make a *compatibility check* of those results, and then maybe reject some of them.

Let the null hypothesis be

H_0: all experiments represent observations from distributions with the same mean and with correctly specified distributions.

To find a test statistic we turn to the least squares method. Let $\widehat{\widehat{\boldsymbol{\theta}}}$ be the estimate for which the *residual sum of squares*

$$Q^2(\boldsymbol{\theta}) = \sum_{J=1}^{M} (\boldsymbol{\theta} - \widehat{\boldsymbol{\theta}}_j)^T \, \underline{V}_J^{-1} \, (\boldsymbol{\theta} - \widehat{\boldsymbol{\theta}})$$

has its minimum $Q_{\min}^2 = Q^2(\widehat{\widehat{\boldsymbol{\theta}}})$. Solving for

$$\left. \frac{\partial Q^2(\boldsymbol{\theta})}{\partial \boldsymbol{\theta}} \right|_{\boldsymbol{\theta}=\widehat{\widehat{\boldsymbol{\theta}}}} = 0$$

given

$$\widehat{\widehat{\boldsymbol{\theta}}} = \left(\sum_{J=1}^{M} \underline{V}_J^{-1} \right)^{-1} \cdot \left(\sum_{J=1}^{M} \underline{V}_J^{-1} \, \widehat{\boldsymbol{\theta}}_J \right). \tag{11.24}$$

The variance of the estimate (11.24) is

$$\underline{V} = \left(\sum_{J=1}^{M} \underline{V}_J^{-1} \right)^{-1}.$$

Note that although the different experiments were assumed to be uncorrelated, one can generalize the method to the case when they are correlated, as for example, when all experiments make use of the same constant of nature, known with non-negligible precision.

The test uses the statistic $Q_{\min}^2$, known (from the Normality assumption) to behave as a $\chi^2[r(M-1)]$, if M experiments are used and $\boldsymbol{\theta}$ has r dimensions. The experiments are compatible at confidence level $(1 - \alpha)$ if

$$Q_{\min}^2 \le \lambda_\alpha \tag{11.25}$$

where λ_α is the α-point of $\chi^2[r(M-1)]$. If $r = 1$, the above procedure simplifies to the well-known *weighted means* procedure:

$$\hat{\theta} = \frac{\displaystyle\sum_{J=1}^{M} \sigma_j^{-2} \hat{\theta}_J}{\displaystyle\sum_{J=1}^{M} \sigma_J^{-2}} \tag{11.26}$$

where σ_J^2 is the variance $V(\hat{\theta}_J)$ of the estimate of θ in experiment J.

When the whole set of M experiments has passed the test (11.25), they are compatible, and there are no more problems concerning the best estimate

(11.24) or (11.26). Let us therefore turn to the case when the experiments, by the test (11.25), are incompatible. How does one separate the poor experiments, responsible for the failure of the test, from the good ones?

In the first place, one should turn to physics arguments instead of statistical tests. Because, even when it is obvious that the incompatibility is due to one particular experiment I, statistics cannot tell whether experiment I is wrong and the remaining $M - 1$ experiments right, or vice versa. Thus, experiments can safely be discarded only on physical arguments (if there are any).

Let us discuss two specific ways of dealing statistically with incompatibility.

(ii) Isolation of one experiment, I

One can test each experiment using as a test statistic the *full* quantities. For this one uses the difference between $\hat{\hat{\theta}}$ and $\hat{\theta}_I$:

$$T_I = (\hat{\hat{\theta}} - \hat{\theta}_I)^T \underset{\sim}{W}_I^{-1} (\hat{\hat{\theta}} - \hat{\theta}_I),$$

where $\underset{\sim}{W}_I$ is the variance of this difference. Let us calculate $\underset{\sim}{W}_I$ in the following way:

$$\underset{\sim}{W}_I = E\left[(\hat{\hat{\theta}} - \hat{\theta}_I)(\hat{\hat{\theta}} - \hat{\theta}_I)^T\right]$$

$$= E\left\{ [(\hat{\hat{\theta}} - \theta) - (\hat{\theta}_I - \theta)] [(\hat{\hat{\theta}} - \theta) - (\hat{\theta}_I - \theta]^T \right\}$$

$$= E\left[(\hat{\hat{\theta}} - \theta)(\hat{\hat{\theta}} - \theta)^T\right] - 2E\left[(\hat{\hat{\theta}} - \theta)(\hat{\theta}_I - \theta)^T\right] + E\left[(\hat{\theta}_I - \theta)(\hat{\theta}_I - \theta)^T\right]$$

$$= \underset{\sim}{V} - 2E\left[(\hat{\hat{\theta}} - \theta)(\hat{\theta}_I - \theta)^T\right] + \underset{\sim}{V}_I .$$

The cross term can be found using the expression (11.24) of $\hat{\hat{\theta}}$,

$$E\left[(\hat{\hat{\theta}} - \theta)(\hat{\theta}_I - \theta)^T\right] = \underset{\sim}{V} \sum_J \underset{\sim}{V}_J^{-1} E\left[(\hat{\theta}_J - \theta)(\hat{\theta}_I - \theta)^T\right].$$

Since we have assumed no correlations between the experiments, only the terms with $J = I$ remain, with final result $\underset{\sim}{V}$. Thus one has

$$\underset{\sim}{W}_I = \underset{\sim}{V}_I - \underset{\sim}{V},$$

or

$$T_I = (\hat{\hat{\theta}} - \hat{\theta}_I)^T (\underset{\sim}{V}_I - \underset{\sim}{V})^{-1} (\hat{\hat{\theta}} - \hat{\theta}_I). \tag{11.27}$$

If the experiments are correlated, the above formulation still holds, and off-diagonal terms also contribute. If $\boldsymbol{\theta}$ is one-dimensional, Eq. (11.27) simplifies to

$$T_I = \frac{(\hat{\hat{\theta}} - \hat{\theta}_I)^2}{\sigma_I^2 - \sigma^2}, \tag{11.28}$$

and the *pull* of experiment I on $\hat{\hat{\theta}}$ is

$$\text{pull}_I = \frac{\hat{\hat{\theta}} - \hat{\theta}_I}{\sqrt{\sigma_I^2 - \sigma^2}}. \tag{11.29}$$

Although Eq. (11.29) is widely known, the negative sign in the radical still puzzles many users. It obviously comes from the strong correlation $(\rho = \sigma/\sigma_I)$ between $\hat{\hat{\theta}}$ and $\hat{\theta}_I$.

The statistic (11.29) has the standard Normal distribution $N(0,1)$: Normal by assumption and standard by construction.

(ii) Penalizing all M experiments by a scale factor

When one admits the possibility that the measured variances σ_j^2 of a single parameter θ are known only relatively, to within a common factor k^2, some of the information present in the data must be used to estimate the *scale factor k*. Clearly then, any tests for inconsistency among the data will be less sensitive.

To test whether a single observation $\hat{\theta}_I$ is consistent with the rest, one may use the test statistic

$$T_I' = \frac{(\hat{\hat{\theta}} - \hat{\theta}_I)^2}{\hat{k}^2 (\sigma_I^2 - \sigma^2)}. \tag{11.30}$$

From Eq. (8.24), the scale factor is estimated by

$$\hat{k}\sqrt{\frac{Q_{\min}^2}{M-1}}.$$

The statistic

$$\tau_1 = \sqrt{T_I'} = \frac{\hat{\hat{\theta}} - \hat{\theta}_I}{\hat{k}\sqrt{\sigma_I^2 - \sigma^2}}$$

has a distribution with p.d.f. given by[*]

$$p(\tau_1) = \frac{1}{\sqrt{\pi(M-1)}} \frac{\Gamma\left(\dfrac{M-1}{2}\right)}{\Gamma\left(\dfrac{M-2}{2}\right)} \left(1 - \frac{\tau_1^2}{M-1}\right)^{(M-4)/2} \; ; M \geq 3 \,.$$

Note that τ_1 has a finite range:

$$-\sqrt{M-1} \leq \tau_1 \leq \sqrt{M-1} \,,$$

and that for $M = 4$, its p.d.f. is constant. For large M, it approaches the standard Normal distribution (see Table 11.5).

Table 11.5

Number of experiments M	Two-sided significance levels for τ_1				
	10%	5%	1%	0.1%	End-pt.
4	1.56	1.64	1.72	1.73	1.732
6	1.63	1.81	2.05	2.18	2.236
8	1.64	1.87	2.21	2.45	2.646
10	1.65	1.90	2.29	2.62	3.000
∞ (Normal)	1.64	1.96	2.58	3.29	∞

An alternative and simpler procedure is to calculate $\hat{\bar{\theta}}$, $\hat{k}$, and σ, ignoring experiment I. The test statistic then becomes

$$\tau_2 = \frac{\hat{\bar{\theta}} - \hat{\theta}_I}{\hat{k}\sqrt{\sigma^2 + \sigma_I^2}} \,.$$

Then τ_2 has a student's t-distribution with $M-2$ degrees of freedom.

11.5.3. *Comparing distributions*

Just as the Kolmogorov statistic permitted one to compare two one-dimensional distributions of unbinned data, one would also like to have a method for comparing several histograms. For this purpose we shall return to the χ^2 statistic (Section 11.2).

---

[*]We are grateful to Dr. W. J. Leigh for this information.

Let the number of histograms be r, each with k bins. Let the ith bin of the jth histogram contain n_{ij} observations, so that

$$\sum_{i=1}^{k} n_{ij} = N_j, \quad j = 1, r,$$

$$\sum_{j=1}^{r} n_{ij} = m_i, \quad i = 1, k.$$

The null hypothesis will be specified by a parent p.d.f. of probability content p_i in the ith bin, common to all r histograms. The p_i's have to be estimated before the test can be formulated. Let μ_j be the expectation of N_j, then the likelihood of the observations is

$$L(\mathbf{n}|\mathbf{p}) = \prod_{j=1}^{r} \left\{ \left(\frac{\mu_j^{N_j} e^{-\mu_j}}{N_j!} \right) \cdot (N_j!) \left(\prod_{i=1}^{k} \frac{p_i^{n_{ij}}}{n_{ij}!} \right) \right\}$$

$$= \left[\left(\sum_i m_i \right)! \prod_{i=1}^{k} \left(\frac{p_i^{m_i}}{m_i!} \right) \right]$$

$$\times \left[\frac{\prod_{j=1}^{r} N_j!}{\left(\sum_{j=1}^{r} N_j \right)!} \right] \cdot \left[\prod_{i=1}^{k} \frac{m_i!}{\sum_{j=1}^{r} (n_{ij}!)} \right] \cdot \left(\prod_{j=1}^{r} \frac{\mu_j^{N_j} e^{-\mu_j}}{N_j!} \right). \quad (11.31)$$

In forming $\partial \ln L / \partial p_i$ we note that only the first bracket in Eq. (11.31) depends on p_i. Thus the derivation of $\hat{p}_i$ is identical to the derivation of $\hat{q}_i$ in Section 11.1, and we find:

$$\hat{p}_i = \frac{m_i}{\sum_{i=1}^{k} m_i}.$$

Let us take as test statistic

$$T = \sum_{i=1}^{k} \sum_{j=1}^{r} \frac{[n_{ij} - E(n_{ij})]^2}{E(n_{ij})}, \quad (11.32)$$

where $E(n_{ij}) = p_i E(N_j)$. Replacing p_i and $E(N_j)$ by their estimates $\hat{p}_i$ and N_j, respectively, it follows that

$$E(n_{ij}) = \frac{m_i}{\sum_i m_i} N_j = \frac{\sum_{\ell=1}^{r} n_{i\ell}}{\sum_{\ell=1}^{r} \sum_{q=1}^{k} n_{q\ell}} N_j \,.$$

T behaves as a χ^2 in the asymptotic limit, when the number of events is large enough for n_{ij} to be Normal, $N\{E(n_{ij}), E(n_{ij})\}$. The degrees of freedom can be computed from the number of observations (rk) and the number of estimated parameters: r values of N_j and $(k-1)$ values of p_i (since they sum up to $1 : \Sigma_i p_i = 1$). Thus we have for T a $\chi^2[(k-1) \cdot (r-1)]$ distribution.

The statistic (11.32) can be written in different ways:

$$T = \left(\sum_{j=1}^{r} N_j \right) \cdot \left[\sum_{j=1}^{r} \left(\frac{1}{N_j} \sum_{i=1}^{k} \frac{n_{ij}^2}{m_i} \right) - 1 \right]$$

$$= \frac{\left(\sum_{j=1}^{r} N_j \right)}{\left(\prod_{j=1}^{r} N_j \right)} \cdot \left\{ \sum_{i=1}^{k} \left[\frac{\sum_{i=1}^{k} \left(n_{ij}^2 \prod_{\ell \neq j}^{r} N_\ell \right)}{m_i} \right] - \prod_{j=1}^{r} N_j \right\} \qquad (11.33)$$

$$= \left(\sum_{j=1}^{r} N_j \right) \cdot \left\{ (r-1) - \sum_{j=1}^{r} \sum_{\ell > j}^{r} \left[\frac{N_j + N_\ell}{N_j N_\ell} \sum_{i=1}^{k} \left(\frac{n_{ij} N_{i\ell}}{m_i} \right) \right] \right\} \,.$$

For the simple case $r = 2$, Eq. (11.33) may be rewritten

$$T = \left(\frac{N_1 + N_2}{N_1 N_2} \right) \cdot \left[(N_1 + N_2) \sum_{i=1}^{k} \left(\frac{n_{i1}^2}{m_i} \right) - N_1^2 \right]$$

$$= \left(\frac{N_1 + N_2}{N_1 N_2} \right) \cdot \left[(N_1 + N_2) \sum_{i=1}^{k} \left(\frac{n_{i2}^2}{m_i} \right) - N_2^2 \right] \,.$$

If, without reference to the observations, it is possible to divide the bins into two or more subsets, within which we want to test the similarity of the distributions, then it is possible to *partition* the total χ^2 statistic into components, each of which has, independently, a χ^2 distribution. Such a division can often improve the precision of the χ^2-test. This of course, depends on, *a priori*, being able to divide the bins into sets within which, or between which, one expects deviations to occur.

We illustrate the procedure by an example. Suppose one can divide the total of k bins into two sets, A and B, within which one expects differences to lie. The specification of the sets A and B is made *a priori*, without reference to the observations. Then the total statistic T can be *partitioned* into three statistics

$$T = T_A + T_B + T_{A|B},$$

each with a χ^2 distribution. These statistics are defined by

$$T_A = \frac{(N_1 + N_2)^2}{N_1 N_2} \left[\sum_{i \in A} \left(\frac{n_{i1}^2}{m_i} \right) - \left(\frac{n_{A1}^2}{m_A} \right) \right]$$

$$T_B = \frac{(N_1 + N_2)^2}{N_1 N_2} \left[\sum_{i \in B} \left(\frac{n_{i1}^2}{m_i} \right) - \left(\frac{n_{B1}^2}{m_B} \right) \right]$$

$$T_{A|B} = \frac{(N_1 + N_2)^2}{N_1 N_2} \left[\frac{n_{A1}^2}{m_A} + \frac{n_{B1}^2}{m_B} - \frac{N_1^2}{N_1 + N_2} \right]$$

where we have used $n_{Aj} = \Sigma_{i \in A} n_{ij}$, $m_A = \Sigma_{i \in A} m_i$ and similar expressions for B. By construction, these statistics have the distributions $\chi^2(k_A - 1)$, $\chi^2(k_B - 1)$ and $\chi^2(1)$, respectively, where k_A, k_B are the numbers of bins in the sets A and B.

These results generalize in a straightforward way to partition into more than two sets. It must always be borne in mind, however, that any such division must be made *without reference to the observations*.

11.6. Combining Independent Tests

It may happen that two or more different tests may be applied to the same data, or the same test applied to different set of data, in such a way that although no one test is significant by itself, the combination of tests becomes significant. When using a *combined test*, one must of course know the properties of the individual tests involved, and in addition two new problems arise:

(a) establishing that the individual tests are *independent*
(b) finding the *significance level* of the combined test.

11.6.1. *Independence of tests*

Each of the tests to be combined is defined by its test statistic. The tests are said to be *independent* if the test statistics are independent random variables, as defined in Section 2.2.2. This means that the expression for the probability (given the null hypothesis) of obtaining given values of the test statistics t_i factorizes into separate parts, each depending on only one statistic:

$$P(t_1, t_2 | H_0) = P(t_1 | H_0) \cdot P(t_2 | H_0). \qquad (11.34)$$

In practice, it may be difficult actually to write down the probability (11.34), although it can often be seen, simply by considering the way the different tests are constructed, whether such a factorization would, in principle, be possible.

Of the tests described in this chapter, only the Pearson χ^2 test and the runs test are (asymptotically) independent [Kendall, II, p. 442]. Intuitively this is clear, since Pearson's test does not depend on the ordering of the bins or on the signs of the deviations in each bin, while this is the only information used in the runs test. In fact, Pearson's test, although probably the most generally used test of fit, has been criticized for its lack of power precisely because it does not take account of this information. However, if the two tests are combined, this criticism is no longer valid.

When H_0 is not a truly simple hypothesis (i.e., when one or more parameters have been estimated from the data), Pearson's test is still asymptotically independent of the runs test, but in this case the distribution of Pearson's statistic is only approximately known (Section 11.2.2), and the distribution of runs is, in general, not known at all, so that the combination is of no practical value.

11.6.2. *Significance level of the combined test*

Suppose that two independent tests have been applied to a given set of data, yielding individual significance levels $p_1 = \alpha_1$ and $p_2 = \alpha_2$. It might be supposed then that the over-all significance level α (i.e., the probability that p_1 and p_2 are such that $p_1 p_2 < \alpha_1 \alpha_2$) would be just $\alpha = \alpha_1 \alpha_2$. The reasoning would be: supposing H_0 true, the probability of finding $p_1 \leq \alpha_1$ and simultaneously $p_2 \leq \alpha_2$ is certainly $\alpha_1 \alpha_2$. Like so many statements about probability, this one *is* true, but misleading and gives the *wrong* over-all significance level α [Wallis]. The difficulty is that ($p_1 \leq \alpha_1$ and $p_2 \leq \alpha_2$) is a sufficient condition for ($p_1 p_2 \leq \alpha_1 \alpha_2$), but it is not a necessary one, so that the probability of the product is not the product of the probabilities.

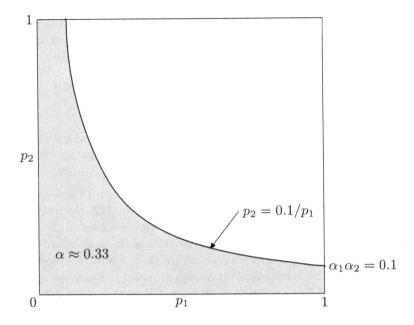

Fig. 11.5. Combining two continuous tests.

In fact, when the variables involved are continuous so that the distributions of individual confidence levels p_1 and p_2 are uniform (under H_0) between zero and one, the problem is straightforward. The probability that $p_1 p_2 \leq \alpha_1 \alpha_2$ is obtained by integrating under the curve in Fig. 11.5, yielding

$$\alpha = \alpha_1 \alpha_2 [1 - \ln(\alpha_1 \alpha_2)] \,, \tag{11.35}$$

which is larger than $\alpha_1 \alpha_2$.

For more than two tests, the analytic expressions become complicated, but it turns out that by making a simple transformation, one obtains a χ^2 distribution. Let there be N independent tests with (continuous) confidence levels α_i, and take

$$\alpha' = -2 \ln \prod_{i=1}^{N} \alpha_i \,.$$

Then α' is distributed (under H_0) as a $\chi^2(2N)$ [Fisher].

When one or more of the test statistics is discontinuous (for example, can take on only integer values) the preceding analysis is no longer valid [Wallis] although it may be a good approximation if the discontinuous statistic is "almost" continuous (can take on many values). To see how the calculation must be modified, let us consider the combining of one continuous test and the discontinuous test (this is the case of interest for combining the Pearson χ^2 test with the runs test). In this case, the set of possible values of α_1 and α_2 does not populate uniformly the unit square, but instead is limited to certain values corresponding, for example, to the vertical lines in Fig. 11.6. As before, we draw the curve $p_2 = \alpha_1\alpha_2/p_1$ so that all points beneath the curve have products less than $\alpha_1\alpha_2$, and it is desired to evaluate α, the total probability of allowed points under this curve. This is now given not by the total area under the curve, but instead by the area shaded in Fig. 11.6. Since this is always less than the total area under the curve, it follows that the true confidence level is *smaller* than that given by the "continuous" approximation, Eq. (11.35).

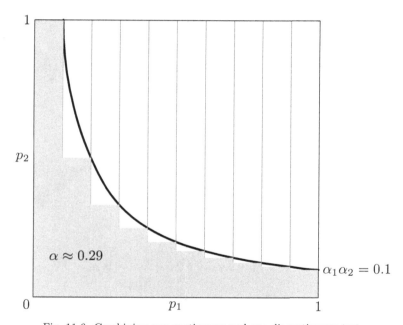

Fig. 11.6. Combining one continuous and one discontinuous test.

REFERENCES

Abramovitz: M. Abramowitz and I. Stegun, Handbook of Mathematical Functions, Dover Publications Inc., New York, reprinted from: National Bureau of Standards, Applied Mathematica Series, Washington, 1964.

Adair: R. Adair and H. Kasha, Analysis of some results of quark searches, Phys. Rev. Letters **23**, 1355 (1969).

Anderson: T. W. Anderson and D. A. Darling, Asymptotic theory of certain goodness-of-fit criteria based on stochastic processes, Ann. Math. Statist. **23**, 193 (1952).

Bayes: T. Bayes, An essay towards solving a problem in the doctrine of changes, Biometrika **45**, 293 (1958). Reprint of 1763 original.

Cochran 1934: W. G. Cochran, The distribution of quadratic forms in a normal system with applications to the analysis of covariance, Proc. Camb. Phil. Soc. **30**, 178 (1934).

Cochran 1952: W. G. Cochran, The chi-squared test of goodness-of-fit, Ann. Math. Statist. **23**, 315 (1952).

Cochran 1954: W. G. Cochran, Some methods for strengthening the common chi-squared tests, Biometrics **10**, 417 (1954).

Cousins 1994: Robert D. Cousins, A method which eliminates the discreteness in Poisson confidence limits and lessens the effect of moving cuts specifically to eliminate candidate events, Nucl. Instr. Meth. Phys. Res. **A337**, 557 (1994).

Cousins 1995: Robert D. Cousins, Why isn't every physicist a Bayesian?, Am. J. Physics **63**, 398–410 (1995).

Cox: D. R. Cox, Tests of separate families of hypotheses, Proc. 4th Berkeley

Symposium (University of California Press, 1961) Vol. 1, p. 105.

Cramér: H. Cramér, Mathematical Methods of Statistics, Princeton University Press, 1946.

Crow 1959: E. L. Crow and R. S. Gardner, Biometrika **46**, 441 (1959).

Crow 1967: E. L. Crow and M. M. Siddiqui, Robust estimation of location, J. Amer. Stat. Ass. **62**, 353 (1967).

D'Agostino: Ralph B. D'Agostino and Michael A. Stephens, eds., Goodness-of-Fit Techniques (Marcel Dekker, New York and Basel, 1986).

Evans: D. A. Evans and W. H. Barkas, Exact treatment of search statistics, Nucl. Instr. Meth. **56**, 289 (1967).

Feldman: Gary J. Feldman and Robert D. Cousins, Phys. Rev. **D57**, 3873 (1998).

Feller: W. Feller, An Introduction to Probability Theory and its Applications, John Wiley and Sons, New York, 1966 and 1968 (Two vols.).

Finetti: Bruno de Finetti, Theory of Probability: A Critical Introductory Treatment (2 vols.) John Wiley, New York (1974).

Fisher: R. A. Fisher, Statistical methods for research workers, Oliver and Boyd, Edinburgh and London, 1958, re-issued in Statistical Methods, Experimental Design, and Scientific Inference, J. H. Bennett, ed., Oxford Univ. Press (1990).

Fourgeaud: C. Fourgeaud and A. Fuchs, Statistique, Dunod, Paris, 1967.

Gill: Philip E. Gill, Walter Murray and Margaret H. Wright, Practical Optimization, Academic Press (1981).

Hahn: G. J. Hahn and S. S. Shapiro, Statistical Models in Engineering, John Wiley and Sons, New York, 1967.

Haldane: J. B. S. Haldane and S. M. Smith, The sampling distribution of a maximum likelihood estimate, Biometrika **43**, 96 (1956).

Heinrich: Joel Heinrich, Pitfalls of Goodness-of-fit from likelihood, in the Proceedings of PHYSTAT2003, SLAC, Stanford, California.
http://www.slac.stanford.edu/econf/C030908/papers/MOCT001.pdf

Hogg: R. V. Hogg and A. T. Craig, Sufficient statistics in elementary distribution theory, Sankhya **17**, 209 (1956).

Huber: P. J. Huber, Robust estimation of a location parameter, Ann. Math. Statist. **35**, 73–101 (1964).

James 1975: F. James and M. Roos, Comput. Phys. Commun. **10**, 343 (1975). Since this publication, the program Minuit has gone through considerable evolution and has been translated into several programming languages and incoporated into many other programs. It is currently available in Fortran 77 and C++ to research institutes and universities from the CERN Program Library.

James 1980: F. James, Monte Carlo theory and practice, Rep. Prog. Phys. **43**, 1145–1189 (1980).

James 2000: F. James, L. Lyons and Y. Perrin (eds.), Workshop on Confidence Limits, CERN Yellow Report 2000–005, Geneva, 17–18 January 2000. Also available at
http://ph-dep.web.cern.ch/ph-dep/Events/CLW/papers.html
The recommendation to publish the likelihood function occurs in the Panel Discussion, on pp. 279–280 of the printed version.

Jeffreys: Harold Jeffreys, Theory of Probability, 3rd edition, Clarendon, Oxford, 1983.

Kass: Robert E. Kass and Larry Wasserman, Formal rules for selecting prior distributions: A review and annotated bibliography, J. Am. Statist. Assoc. **91**, 1343–70 (1996).

Kendall: M. G. Kendall and A. Stuart, The Advanced Theory of Statistics, Charles Griffin and Co. Ltd., London. Vol. 1: 1963, Vol. 2: 1967, Vol. 3: 1968. Later editions of this series of volumes have a modified author list. For the most recent edition at the time of writing, see [Stuart].

Knop: R. E. Knop, Error in estimation of total events, Rev. Sci. Instrum. **41**, 1518 (1970).

Kotz: Samuel Kotz and Norman L. Johnson, eds., Breakthroughs in Statistics (2 Vols.), Springer, New York (1993).

Kullback: S. Kullback, Information theory and statistics (Dover Publications Inc., New York, 1968).

Lawley: D. N. Lawley, A general method for approximating to the distributionof likelihood ratio criteria, Biometrika **43**, 295 (1956).

Lehmann: E. L. Lehmann, Testing statistical hypotheses (John Wiley and Sons, New York, 1959).

Lindley 1965: D. V. Lindley, Introduction to probability and statistics (Cambridge University Press, London, 1965), 2 Vols.

Lindley 1961: D. V. Lindley, The use of prior probability in statistical inference and decisions, Proc. 4th Berkeley Symp. on Mathematical Statistics and Probability (University of California Press, Berkeley and Los Angeles, 1961), Vol. 1, p. 453.

Mann: H. B. Mann and A. Wald, On the choice of the number of intervals in the application of the chi-squared test, Ann. Math. Statist. **18**, 50 (1942).

Marshall: A. W. Marshall, The small sample distribution of nW_n^2, Ann. Math. Statist. **29**, 307 (1958).

McCusker: C. McCusker and I. Cairns, Evidence of quarks in air-shower cores, Phys. Rev. Letters **23**, 658 (1969).

NBS: NBS, Probability tables for the analysis of extreme value data (National Bureau of Standards, Applied Mathematics Series, 22, Washington, 1953).

Neyman: J. Neyman, Philosoph. Trans. R. Soc. London A236 (1937) 333. Reprinted in A Selection of Early Statistical Papers on J. Neyman, Univ. of California Press, Berkeley, 1967, pp. 250–289.

O'Hagan: Anthony O'Hagan, Bayesian Inference, Vol. 2B of Kendall's Advanced Theory of Statistics, Arnold Publishers, London, 1994.

Pearson: K. Pearson, On the criterion that a given system of deviations from the probable in the case of a correlated system of variables is such that it can reasonably be supposed to have arisen from random sampling, Philos. Mag. Ser. **5**, 157–175 (1900). Reprinted with comments in [Kotz, vol. 2].

Quenouille: M. H. Quenouille, Notes on bias in estimation, Biometrika **43**, 353 (1956).

Rao: C. R. Rao, Efficient estimates and optimum inference procedures in large samples, J. Roy. Statist. Soc. **B24**, 46 (1962).

Rice: J. R. Rice and J. S. White, Norms for smoothing and estimation, SIAM Review **6**, 243 (1964).

Shannon 1948: C. E. Shannon, A mathematical theory of communication, Bell System Technical Journal **27**, 379–423 and 623–656 (1948).

Shannon 1949: C. E. Shannon and W. Weaver, The mathematical theory of communication (University of Illinois Press, Urbana, 1949).

Smirnov: N. V. Smirnov, Sur la distribution de W^2, C. R. Acad. Sci. Paris **202**, 449 (1936).

Solmitz: F. Solmitz, Analysis of experiments in particle physics, Annual Reviews of Nuclear Science **14** (1964).

Stevens: W. L. Stevens, Biometrika **37**, 117 (1950).

Stuart: Alan Stuart, Keith Ord and Steven Arnold, Kendall's Advanced Theory of Statistics, Sixth Edition, Arnold Publishers, Vol. 2A, Classical Inference and the Linear Model, 1999. Vol. 2B, on Bayesian Inference, is by [O'Hagan].

Tortrat: A. Tortrat, Principes de statistique mathématique (Dunod, Paris, 1961).

Tukey: J. W. Tukey, The future of data analysis, Ann. Math. Statist. **33**, 1–67 (1962).

Wald 1947: A. Wald, Theory and application of sequential probability ratio tests (John Wiley and Sons, New York, 1947).

Wald 1950: A. Wald, Statistical decision function (John Wiley and Sons, New York, 1950).

Wallis: W. A. Wallis, Compounding probabilities from independent significance tests, Econometrica **10**, 229 (1942).

Wiener: N. Wiener, Cybernetics (John Wiley and Sons, New York, 1948).

Wilks: S. S. Wilks, Mathematical Statistics (John Wiley and Sons, New York, 1962).

SUBJECT INDEX